Phelps Dodg[e]

WORLDWIDE OPERATIONS AS OF JUNE 1999

- Phelps Dodge Mining Company
- Phelps Dodge Industries

VISION & ENTERPRISE

⟨⟩

Where there is no vision, the people perish.
—Proverbs 19:18

Vision is the art of seeing things invisible.
—Jonathan Swift

Nos. 19 & 21 Cliff Street,
Between John and Fulton Streets.

Carlos A. Schwantes

VISION &

THE UNIVERSITY OF ARIZONA PRESS & PHELPS DODGE CORPORATION

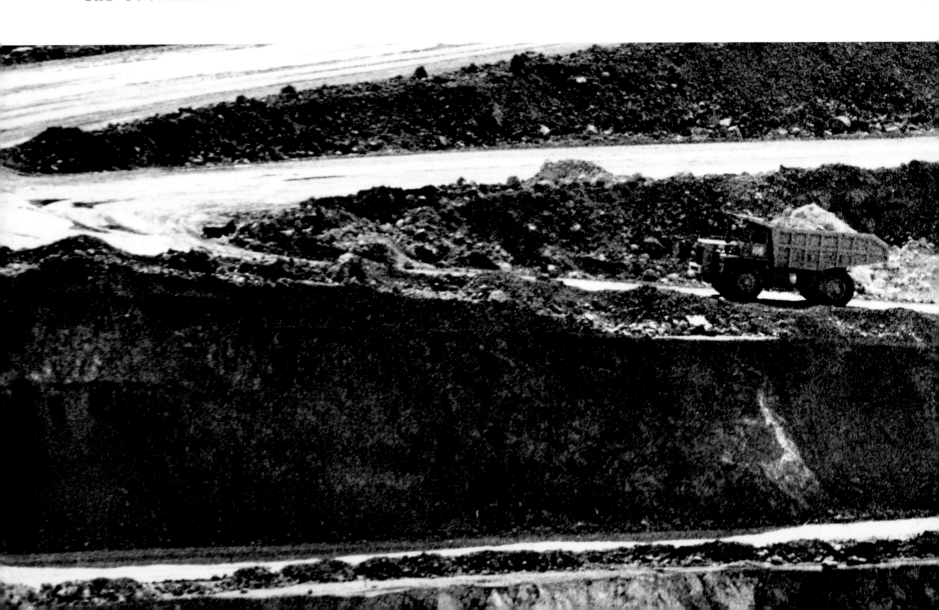

ENTERPRISE

Exploring the History of Phelps Dodge Corporation

The research, writing, and publication of this book were funded and supported by Phelps Dodge Corporation.

First printing
The University of Arizona Press
© 2000 The Arizona Board of Regents
This book is printed on acid-free, archival-quality paper.
Printed in China
05   04   03   02   01   00   6   5   4   3   2   1

Library of Congress Cataloging-in-Publication Data
Schwantes, Carlos A., 1945–
    Vision and enterprise : exploring the history of Phelps
Dodge Corporation / Carlos A. Schwantes.
       p.   cm.
    Includes bibliographical references and index.
ISBN 0-8165-1943-9 (acid-free paper)
    1. Phelps Dodge Corporation—History.   2. Metal
trade—United States—History.   I. Title.
HD9506.U64 P387 2000
338.7'622343'0973—dc21
                           99-6758
                            CIP

British Library Cataloguing-in-Publication Data
A catalog record for this book is available from the
British Library.

OPPOSITE: Two
UNDERGROUND MINERS *pose
with their pneumatic drill in this
1920s photograph from the
Moctezuma Copper Company, a
once active Phelps Dodge property
in northern Mexico.*

CONTENTS

PREFACE

Personal Perspectives on Phelps Dodge

WHAT'S IN A NAME? Once when Douglas Yearley played golf on a course near his corporation's Phoenix headquarters, the stranger paired with him innocently asked what he did for a living. The man who was at the time chairman, president, and chief executive officer of a $4 billion enterprise employing more than sixteen thousand people in twenty-seven countries responded modestly, "I work for Phelps Dodge." For a couple of holes the stranger silently pondered the unfamiliar name. Then he politely asked, "How's the car business?" One thing Phelps Dodge never manufactured or sold was automobiles.[1]

Back in 1914 when the Dodge brothers, two Detroit machinists, built their first automobiles, Phelps, Dodge & Company of New York was already eighty years old. The firm had previously achieved distinction as one of America's foremost mercantile and metals businesses; and today as an industrial giant it ranks near the top of the list of

the oldest large-scale enterprises in the United States. Based now in Arizona, the modern Phelps Dodge Corporation mines or manufactures a variety of basic materials, most notably copper and carbon black, that help automobiles and tires run better and last longer. Indeed, a common thread running through its lengthy history is specialization in the basic substances of civilization. What that meant in terms of both its day-to-day business operations as well as its participation in the economic transformation of the American Southwest may surprise people unfamiliar with the name Phelps Dodge.

The Copper People and More

ABOVE ALL ELSE, Phelps Dodge Corporation during the twentieth century was synonymous with copper. At one point in the late 1990s, its mines yielded a mind-boggling 1.75 billion pounds of copper a year. It produced between one-quarter and one-third of all red metal mined in the United States, a statistic that ranked Phelps Dodge as the largest nonfuel-mining company in North America. Around the globe only one other enterprise, the state-owned mining company of Chile (CODELCO), accounted for more copper than did Phelps Dodge.

Perhaps even more incredible is how it extracted so much metal from ore deposits that averaged less than six-tenths of 1 percent copper and that before the application of modern technologies were scorned as worthless. The feat of recovery might be compared to the challenge of removing a single raisin from a bowl of bran flakes, but only after the raisin had first been put through a blender and chopped so fine as to be virtually invisible.

You can't do that at the breakfast table, but at its open-pit mines Phelps Dodge does something similar by working around the clock to recover finely disseminated copper. Furthermore, its scientists and engineers search for environmentally friendly methods that the company will use to recover copper treasures even more efficiently in the new millennium. In research no less than in worldwide exploration for additional sources of minerals, Phelps Dodge people possessing widely different educational backgrounds employ a formidable mix of personal knowledge and cutting-edge technologies to peer at least ten to twenty years into the future to ensure their corporation's prosperity. Seekers have in common an angle of vision that permits them to see, like the legendary Superman with his X-ray vision, beyond the obvious into processes and places hidden from

individuals who lack the specialized training.

Each year a torrent of red metal pours downstream from its mines, smelters, and refinery to reach fabricating plants owned by Phelps Dodge and other firms that make industrial goods. Manufacturers in the United States will use the largest amounts of copper to produce building wire, fixtures and conduits for plumbing and heating, and the unseen but always vital components of automotive electrical and air conditioning systems. In all, Americans typically will consume no less than seventeen hundred pounds of copper throughout their lives.

Here's another Phelps Dodge fact worth considering: its copper may also end up in products ranging from a new generation of computer chips to special alloyed strings for classical, acoustical, and electric guitars. Phelps Dodge does not manufacture the computer chips themselves, and certainly not the guitars, but it ranks as a significant maker of the brass overwrap wire used for high-quality guitar strings. These come from the South Carolina plant where it also makes specialty conductors that energize Boeing model 777 jetliners! Once the Phelps Dodge slogan was "From Mine to Market." Today it would be accurate to speak of Phelps Dodge copper traveling from mine to concert stage—or even into deep space, because when the Hubble telescope scans distant galaxies and beams dramatic images back to earth it does so with the aid of specialty conductors made by Phelps Dodge.

In the United States, the corporation's landscape of copper production encompasses several giant mine and mill complexes located in Arizona and New Mexico, a refinery and rod mill in Texas, another rod mill in Connecticut, and

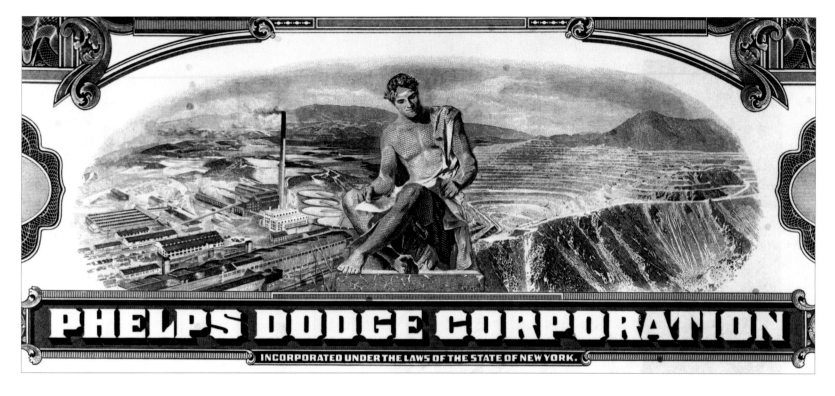

PHELPS DODGE CORPORATION

INCORPORATED UNDER THE LAWS OF THE STATE OF NEW YORK.

ILLUSTRATING HOW *complex landscapes of production define Phelps Dodge Corporation is the engraving that appears on current shares of its common stock, which showcases the historic mine, mill, and smelter complex in Morenci, Arizona.*

manufactories in states as distant as Indiana, New Jersey, and South Carolina. There is a logic to it all. From tons of low-grade ore, Phelps Dodge creates something of value as it extracts metallic copper by two basic but very dissimilar processes. The newer one uses various solutions—a combination of highly selective reagents and ordinary water—that yield cathodes of virtually pure copper. In contrast to solution extraction–electrowinning (SX/EW), the traditional method concentrates copper, then purifies it by fire at Phelps Dodge's two large smelters in New Mexico, and finally refines it into cathodes in El Paso, Texas. But the stream of copper does not stop there.

The company's two big continuous-cast rod mills melt cathodes produced either by fire or solution extraction and transform the shape of

the metal from thin sheets into a seemingly endless rod slightly larger in diameter than an ordinary pencil. This form of copper serves as the essential feedstock from which all types of wire and cable are made. Phelps Dodge sells coils of copper rod as well as various specialty shapes to outside customers, but many of the heavy coils will continue by truck or train from the rod mills of Phelps Dodge Mining Company to the several fabricating plants of Phelps Dodge Industries before the copper finally leaves the capacious hands of the corporation that first mined it.

For example, near the banks of the Rio Grande a few miles south of the El Paso refinery and rod mill complex is a modern fabricating plant operated by Phelps Dodge Magnet Wire Company. In this and other state-of-the-art facilities located in the United States, Mexico, and Austria, Phelps Dodge Magnet Wire makes an array of surprisingly complex and hardworking conductors that are as essential to electric motors and related equipment as blood is to the human body. Other specialty types of wire and cable come from the plants of Phelps Dodge High Performance Conductors located in New Jersey, Georgia, and South Carolina. Machinery runs around the clock to draw and spool wires and cables, some of which are considerably finer than a strand of human hair. From a single pound of copper Phelps Dodge can draw more than three million feet of fine-gauge wire—enough to extend from Baltimore to Boston.[2]

Apart from magnet wire and high-performance conductors, Phelps Dodge plants in the United States once manufactured telephone wires and cables, high-voltage electrical cables, and building wire for home and commercial uses, as well as an extraordinary array of copper, brass, and bronze plumbing fixtures and pipes. In the early and mid-1980s, during a period of restructuring required to survive the near collapse of world copper prices, it jettisoned all subsidiary operations except for Phelps Dodge Magnet Wire and Phelps Dodge International. The radical surgery together with adoption of solution extraction technology that greatly lowered the cost of producing copper repositioned Phelps

⌒ LOOKING SOMEWHAT LIKE *an adult version of the children's toy known as the Slinky are these still-warm coils of continuous-cast copper rod photographed in El Paso in 1982. Mills in Texas and Connecticut make Phelps Dodge Mining Company the world's largest supplier of rod, the stock from which electrical and communications wire and cable are manufactured.*

Dodge to take full advantage of a later upswing in the price of red metal.

By the end of the 1980s the rejuvenated enterprise added three new subsidiaries, the Columbian Chemicals Company, Accuride Corporation (sold in 1997), and family-owned Hudson International Conductors, the latter business leading to creation of Phelps Dodge High Performance Conductors in the mid-1990s. The various components of Phelps Dodge Industries in turn encompassed individual plants and additional subsidiary operations, many of them located in Europe, South America, and Asia. As once was true of the British Empire, the sun never sets on the many mining and manufacturing activities of Phelps Dodge.

Like the mining properties it acquired over the years, each new manufacturing subsidiary of Phelps Dodge came with a lengthy history of its own. And whether old or newly acquired, each business retained a distinctive personality, though all had much in common as producers of industrial goods.

DESPITE ITS SIZE and longevity, Phelps Dodge Corporation remains a stranger to many people because it neither advertises nor retails its industrial goods to a mass market. The sole exception currently is Triangle brand copper sulfate, a product the El Paso refinery sells in big brown-and-white bags in agricultural supply stores. From the time of its founding in 1834 until the present, Phelps Dodge manufactured or supplied basic products that other enterprises utilized to make consumer goods they sold under their own names or brand labels, such as General Motors, Whirlpool, or Zenith.

In the stream of industrial production, Phelps Dodge traditionally occupied the headwaters, the place of beginnings; and there it has stood for nearly 170 years as a pioneer in the creation of value. The result: in everything from washing machines to communication satellites, Phelps Dodge's copper, carbon black, and specialty wire and cable products will turn up just about everywhere. Phelps Dodge may not be a familiar or household name, yet its products are vital to every modern household.

I type these words and sentences on a personal computer protected from overheating by a tiny fan, the copper heart of which is composed of magnet wire probably made by Phelps Dodge Industries. Carbon black from Columbian Chemicals imparts clarity and impact to the output of my laser printer, and it is indispensable to my automobile. If its tires lacked the strength that carbon black imparts to rubber, they would grow old and worn out after less than five thousand miles of use, far short of the forty thousand miles that manufacturers typically guarantee today. About 20 percent of the weight of any tire is carbon black. It also forms a vital component of the exterior paint, door seals, motor mounts, timing belts, and windshield wiper blades found on every modern automobile.

Further, each time I start my car, switch on a light, or talk on the telephone I'm likely to encounter copper produced by Phelps Dodge. The average automobile contains more than fifty pounds of copper and copper alloys, and some luxury models may use as many as 150 electric motors. Every one of them is dead without magnet wire.

My home is typical of many in America in

that it contains more than thirty electric motors to power everything from garage-door openers and garbage disposals to clocks and video recorders. Any images on my television screen or computer monitor would shrink to a tiny point of light were it not for a collar of magnet wire around the neck of the picture tube that spreads the beam of electrons to create lifelike color images. I know for certain that household wiring once made by Phelps Dodge transmits electric power to all my appliances. The basic component of the heavy-duty copper cables used to connect my stereo speakers and amplifier was more than likely manufactured by Phelps Dodge High Performance Conductors.

Outside the home, notable structures that contain Phelps Dodge wires and pipes include the James Madison Building of the Library of Congress, the twin towers of the World Trade Center, Rockefeller Center, Kennedy International Airport, and the Verrazano Narrows Suspension Bridge. The copper trail continues seemingly without end. Back in the 1950s and 1960s the company made countless miles of special shipboard cable to power and control the USS *Nautilus* and her sister atomic submarines, and it extruded copper for massive atom smashers and for heat exchangers in electric power plants. Although subsidiaries in the United States no longer manufacture many electrical or plumbing products closely identified with the Phelps Dodge name in years past, they still make magnet wire and high-performance conductors for so many different applications that Phelps Dodge's handiwork will continue to be found where least expected.

Beginning in the mid-1950s, the corporation

PHELPS DODGE COPPER PRODUCTS CORPORATION

Featured in this circular are some of the products manufactured by Phelps Dodge Copper Products Corporation at plants located at Yonkers, New York; Bayway, New Jersey; and Los Angeles, California.

KEY PARTS in world's largest atom smasher

COPPER EXTRUSIONS

Within the heart of Brookhaven National Laboratory's ring-shaped Synchrotron, world's mightiest atom-smashing machine now under construction, will be more than 400 tons of the purest copper—extruded into a unique shape. These extrusions form an essential part of gigantic electromagnets which help keep the proton "on the track" at speeds as great as 180,000 miles per second.

ABOVE: ADVERTISEMENTS SHOWED HOW *copper extrusions from Phelps Dodge formed part of the world's largest atom smasher at Brookhaven National Laboratory.*

OPPOSITE: THIS NOVEMBER 1963 *photograph of the converter aisle at the now dismantled smelter in Douglas, Arizona, shows why heavy industry was once synonymous with the muscle of America.*

established a series of joint ventures abroad that are today major manufacturers of wire and cable. Some of those enterprises resulted in wry and perhaps unintended consequences, such as happened in the late 1990s when Phelps Dodge, the copper company, easily ranked among Latin America's largest producers of *aluminum* wires and cables after the corporation formed joint ventures in Brazil and Venezuela. Ironies, no less than surprise connections, are what make the Phelps Dodge story somewhat like mining itself—one rich vein of history or mineral-rich ore unexpectedly leads to another. The story of Phelps Dodge Corporation is a tale of romance, luck, and hard work combined in improbable ways that would read like fiction were it not all well grounded in historical fact.

Phelps Dodge in American History

FOR NEARLY SEVENTEEN decades—since 1834 when Andrew Jackson occupied the White House—Phelps Dodge participated in American history in ways most people can't imagine. Take D day, for example. Photographs of troops storming ashore on June 6, 1944, in the largest amphibious invasion in history are familiar enough, yet I wonder how many people realize that the Allied war machine would have been stuck on its hard-won Normandy beachhead or would have sputtered and stalled during its final drive against the Nazis had not Phelps Dodge secretly manufactured special metallic pipelines that stretched beneath the English Channel and partway across France to supply much-needed

petroleum. When it came to furnishing copper wire to extend America's transcontinental telegraph line to the West Coast in 1861, Phelps Dodge was there too, as it was when Levi Strauss first reinforced a pair of his heavyweight denim trousers with copper rivets. Over the years more than sixty million pounds of copper have ended up as rivets in more than a billion pairs of jeans manufactured by Levi Strauss & Company.

As important as copper is to Phelps Dodge, let us talk first about cotton. When American history speaks of King Cotton and his enormous economic influence in the years before the Civil War, it is really describing the activities of merchants like Anson G. Phelps and his sons-in-law, William E. Dodge and Daniel James, who arranged to transport cotton from warehouses in Savannah, New Orleans, and other southern ports to mills and markets in England and return with shiploads of sheet iron, copper, and tinplate, the universal packaging material of its day. When John D. Rockefeller first sold cans of Standard Oil's kerosene, the tinplate came from Phelps Dodge or one of its competitors in the import-export business.

Consider also Phelps Dodge's role in the one American revolution that historians have yet to popularize. Let's call it the revolution in household systems. Back in the 1850s and 1860s, the years when Abraham Lincoln rose to prominence, American homes lacked every basic system we take for granted today to provide us with clean water, sewage disposal, electricity at the flick of a switch, telephone connections, and central heating. Before Lincoln moved to the

White House he lived in a Springfield, Illinois, home that included a three-hole outhouse in the backyard, now carefully preserved by the National Park Service.

By 1900 so many new amenities had become available to urban Americans that they could boast of more comfortable homes than the Sun King, Louis XIV, amidst the splendor of his Palace of Versailles could ever imagine. How much would the French monarch have paid for several flush toilets to alleviate the odor created by almost three hundred portable privies scattered throughout his sprawling palace? Rural electrification during the 1930s extended the revolution in household systems from city to farm and significantly reduced the isolation and hardship once synonymous with country life. It was during this underappreciated revolution in everyday life that Phelps Dodge emerged as a major supplier of copper and brass for the wires, pipes, and fixtures that redefined personal comfort and convenience.

Also during the depression decade of the 1930s, when producers struggled to balance an oversupply of copper with a weak market, Phelps Dodge pursued an imaginative solution by teaming up with the legendary industrial designer R. Buckminster Fuller to design a modular all-copper bathroom. Though definitely not a commercial success, one Dymaxion Bathroom was publicly displayed at New York's Museum of Modern Art and thus became probably the only bathroom ever so honored.

In the American Southwest, where copper for the Dymaxion Bathroom originated, Phelps Dodge became a builder of railroads and towns.

To design Tyrone, New Mexico (intended to be the most beautiful company town in America and quite possibly the world), it hired one of the nation's top-ranked architects, Bertram Grosvenor Goodhue, who drew up plans for the model community that Phelps Dodge constructed atop the Continental Divide in 1915. In addition, Phelps Dodge at one time owned no fewer than twenty-five hundred houses (none with Dymaxion Bathrooms), about a million square feet of commercial space, several Phelps Dodge Mercantile stores, half a dozen hospital buildings or clinics, water and electric utilities, and even several newspapers. It has sold most of those properties, but Phelps Dodge remains among the top corporate landholders in Arizona and New Mexico, its mining company division controlling deeded or leased lands that total about a million acres.

The corporate landscape at Phelps Dodge is no mere figure of speech. Its landholdings cover an area nearly twice the size of Rhode Island. In fact, if all its land were somehow extruded to form a corridor thirty feet wide, the width of a typical two-lane highway, it would form a ribbon encircling the earth at the equator eleven times. (The dizzying statistic doesn't include at least two thousand miles of underground passageways beneath the old mining towns of Bisbee and Jerome, Arizona.)

There is a reason for Phelps Dodge's landed empire. Primarily it resulted from the corporation's quest for water needed by its mill and smelter complexes in the arid Southwest. The largest single block of land is the Hidalgo property in the arid boot heel of southwestern

THE PHELPS DODGE *landscape included attractive company towns like Tyrone. The New Mexico mountain community was so charming that after mining ended there in the early 1920s a portion of the town still functioned as a popular resort.*

THE INSET SHOWS *a portion of the original Phelps Dodge Mercantile building in Tyrone.*

New Mexico, where groundwater from some half a million acres is utilized by one of Phelps Dodge's two state-of-the-art copper smelters. Phelps Dodge Mining Company has a few smaller holdings in Nevada and west Texas.

Especially in early twentieth-century Arizona and New Mexico, the name Phelps Dodge meant far more than single-minded business enterprise. In those two states the company was both a pioneer developer and an abiding presence. Employment at Phelps Dodge was more than a job—in many places it was a way of life handed down from one generation to the next. Because of its size and perceived power in the sparsely populated Southwest, the company sometimes became a convenient target for critics. It is only a myth that the state budget of Arizona once had to be approved initially by Phelps Dodge, but questions of power, real and imagined, are nonetheless part of the Phelps Dodge story.

Hard-won Success

THE PHELPS DODGE story is instructive not just for what it reveals about the evolution of one Fortune 500 corporation but also for what it says about success and failure in the evolving world of business since Phelps and Dodge formed their partnership back in 1834. During the course of the next 170 years the landscape of American enterprise would become littered with the remains of tens of thousands of partnerships and corporations that failed, some of them older and far larger than Phelps Dodge. Where is the partnership of Russell, Majors, and Waddell today, the powerful trio who helped win the West

with a fleet of freight wagons that lumbered across the high plains in the days of Kit Carson and Wild Bill Hickok, or Abbot-Downing, known for the finely crafted Concord stagecoaches that came to symbolize frontier America? What about the Studebaker brothers, Pennsylvania Railroad, or Eastern Airlines? There once was a time when more than five hundred companies made automobiles in the United States, but now only a few well-known names remain. The list of examples from American business is endless.

How was it that Phelps Dodge successfully evolved from a genteel, New York–based cotton- and metal-trading firm into a western mining giant, and how did Phelps Dodge surmount the problems of an industry notable for boom-and-bust cycles to become a robust multinational enterprise? In Arizona alone, more than four hundred thousand mining claims were recorded over the years, and more than four thousand mining companies formed. How was it that Phelps Dodge succeeded where so many others failed? What was so special about its ability to combine and sustain vision and enterprise?

This is not to suggest that Phelps Dodge avoided adversity. Quite the contrary. As you will learn, the original Phelps Dodge partnership was born out of family tragedy. Over the years the company experienced its full share of disappointing setbacks and near disasters, but always it emerged a survivor. Copper, it should be noted here, is the ultimate traded commodity. Its price is determined daily on terminal markets in London and New York, and Phelps Dodge has no control over that. In the late 1990s if the annual average price of copper fell by as little as a penny per pound, it lowered the pretax earnings of Phelps Dodge by some $18 million. When the price of copper crashed during the early 1980s, it nearly dragged down Phelps Dodge. In the company's long history no era was more grim.

"We've been tempered by the fire of near bankruptcy," reflected the chairman and chief executive officer Douglas C. Yearley in early 1990. "When you're in a cyclical business and you've been through the storm we came through, I think you tend to be conservative." Conservative but also progressive and efficient, that is Phelps Dodge's modern angle of vision. Unlike another mining giant, Anaconda, its onetime competitor, which in the early 1980s was still reeling from earlier expropriation of its huge mine holdings in Chile, Phelps Dodge managed to step carefully back from the brink, make significant changes to its business

portfolio, dramatically lower its production costs, and then march boldly forward along a slightly different route to claim its place among the world's largest copper companies.[3]

In fact, at the dawn of a new millennium Phelps Dodge is perhaps more vigorous and responsive than at any time in its long history. Today its employees work hard as a team not just to ensure their company's survival in a rapidly changing business climate but also to help it thrive. Like all successful enterprises, modern Phelps Dodge Corporation strives to adapt to an economic world in which seemingly the only constant is change.

Why Vision and Enterprise? *Origins and Strategies*

IN THE HISTORY of any business enterprise, complex technological or financial matters are often best understood in terms of real people and everyday life. Unfortunately, American history in the past has often presented the rise of big business in the late nineteenth century as if large-scale enterprises resembled disembodied spirits that arose unexpectedly from the post–Civil War landscape to frighten citizens who worried a great deal in print about trusts, monopolies, robber barons, and other assorted bogeymen (real or imagined). It became easy to forget that real people, most of whom bore no resemblance to the often caricatured robber barons, made decisions that were responsible for the rise of big business. Sometimes they made mistakes, but that is a risk inherent in the human condition.

Among the real people who became synonymous with American enterprise is James Douglas, the man who skillfully led Phelps Dodge into the brave new world of Arizona copper at a time when the Southwest remained a remote and raw frontier territory and its copper mines represented the ultimate business risk. Douglas, the first president of Phelps Dodge after it evolved from a partnership into a corporation late in 1908, was a self-taught metallurgist who had originally trained as a theologian and then as a medical doctor. With his neatly trimmed white beard and his professorial inclinations that included a fondness for writing learned essays on a variety of subjects, notably history, he was the antithesis of the popular stereotype of a business baron. A man of truly encompassing vision, he guided Phelps Dodge not only into Arizona Territory but also into the complex and rapidly evolving world of metal mining and processing—and he did so successfully. Douglas quipped that he prepared himself for this sometimes topsy-turvy enterprise by first running a mental hospital in his Canadian homeland.

That's hardly the whole story. In the desert Southwest where Phelps Dodge mined its copper, Douglas envisioned and constructed a far-reaching system of production that included mines, mills, smelters, and railroads as well as modern towns, waterworks, hospitals, schools, and homes that still define significant sections of the landscape between Phoenix and El Paso. Because of its wide-ranging activities—critics would label them examples of corporate paternalism—Phelps Dodge Corporation was never just another narrowly focused business enterprise.

It was because of encounters with extraordinary and far-seeing people like Dr. Douglas that I chose the title *Vision and Enterprise* for this book. Prophets of old knew that where there was no vision the people perished. The same is true for corporations today. That is the main reason I wrote *Vision and Enterprise* in terms of corporate biography, seeking to weave together the story of several generations of Phelps Dodge leaders who successfully adjusted the corporation's angle of vision in response to the changing world of business. Continual adjustment to change is a primary secret of Phelps Dodge's longevity. At the same time, amazingly, corporate leadership managed to hold fast to many core values that dated back to 1834—and that is another secret of Phelps Dodge's longevity.

More than a century after Dr. Douglas first guided Phelps Dodge into Arizona copper, one of his successors as chairman (and an executive who like Douglas appreciated the value of history) encouraged research, writing, and publication of *Vision and Enterprise*, for as Douglas Yearley advised *Enterprise* magazine in 1992, "we never want to forget the lessons of the past" as they apply to the management of Phelps Dodge. That kind of history-minded perspective in business leaders is something I applaud and support, and it explains why the challenge of researching and writing *Vision and Enterprise* proved so pleasant.[4]

No part of my work was more enjoyable than interviewing various people associated with Phelps Dodge during the past several decades. At the studios of CNBC television I met L. William Seidman, who because of successive careers as economics adviser to President Gerald Ford, vice-chairman of Phelps Dodge Corporation, business school dean, and head of the Federal Deposit Insurance Corporation was known to media colleagues as the Iron Man. A man of youthful vigor even as he neared the age of eighty, Seidman commuted to work in busy downtown Washington, D.C., by bicycle, a feat that added to his legendary status.

When Richard Pendleton, Jr., a retired senior vice-president, reminisced about his many years at Phelps Dodge it was over lunch at the New York Yacht Club. Afterwards he gave me a brief tour of the nautical hideaway just off Fifth Avenue where Arthur Curtiss James, a Phelps Dodge heir and once reportedly the nation's largest individual investor in railroad securities, held sway as commodore, having succeeded the legendary Cornelius Vanderbilt. A model of James's yacht is among the many dozens on display. Here too a magisterial portrait of the banker J. P. Morgan keeps watch to ensure that members maintain proper decorum in the time-honored tradition. Missing from its accustomed place, alas, noted a rueful Pendleton, himself an avid yachtsman, was the America's Cup trophy, which represented the pinnacle of all sailing races.

On the opposite coast, I vividly recall the warm spring afternoon I spent with Warren and Eleanor Fenzi in Santa Barbara, California. The day provided a brief moment in paradise when flowers perfumed the air and the blue Pacific Ocean shimmered in the distance. We talked about Fenzi's father, Cammillo, the gifted landscape architect whom Phelps Dodge hired to beautify its railroad stations and company towns, and about Warren himself, who began with Phelps Dodge as a junior engineer in 1937 and rose through the ranks to serve as the corporation's president from 1975 to 1980. Fenzi's own angle of vision encompassed almost three-quarters of a century of company history.

Several months earlier I spent a January afternoon in Safford, Arizona, where I enjoyed conversing at length with Pedro Gomez and his daughter and son-in-law, Velia and Simon Peru. Our time together was highlighted by a feast of Velia's Mexican treats. Gomez, an immigrant who at the age of two walked north with his family from Jalisco, Mexico, some ninety years ago, moved from El Paso to the Clifton-Morenci mining district in Arizona Territory in 1910, where he lived until 1996. The family of his wife, Marina, arrived in the Morenci area even earlier,

in the 1880s. Gomez recalled that many people who moved north from Mexico preferred agricultural work, but their country also had a long history of mining. He himself quit school after the sixth grade to help support his family. That was back in 1914 when the price of copper dipped to less than sixteen cents per pound and deprived his father of steady work in the mines and mills.

When I interviewed Pete Gomez in 1997 he was the patriarch of three generations of Phelps Dodge employees. He began working for the company in 1922 as a boilermaker's helper. He quickly learned the trade and became a specialist in steam locomotive repair on the railroad that linked Morenci and Clifton. Gomez retired in 1971, but he lived to see something "I never dreamed" possible: his grandson, Ramiro G. Peru, became a senior vice-president of Phelps Dodge Corporation early in 1997 (and also chief financial officer two years later). Theirs is a great American success story. The corporation's

history is replete with such stories, starting with that of Anson Greene Phelps himself.[5]

It is not possible to describe all the pleasant encounters with people who helped me better understand and explore personally the many landscapes of Phelps Dodge, both the contemporary as well as the half-forgotten historical ones. I've provided a lengthy list of their names in Sources and Suggestions for Further Reading.

THE ROOTS OF *Vision and Enterprise* extend back to a museum catalog I edited in the early 1990s called *Bisbee: Urban Outpost on the Frontier,* which was funded in part by the National Endowment for the Humanities and published by the University of Arizona Press. The study of a Phelps Dodge mining community came to the attention of corporation executives late in 1995 who urged me to write a modern history of Phelps Dodge. The challenge was intriguing, but it entailed professional risks as well. Very likely I would be accused of wearing a "copper collar," a pejorative term used in Arizona to belittle legislators, journalists, or anyone else who is perceived as supportive of the state's all-important copper industry.

So let me emphasize that, while I sought and generously received advice and information from Phelps Dodge employees on many complicated matters of technology and finance in order to explain their intricacies better to nontechnical readers, no one at Phelps Dodge Corporation told me how or what to write. Though Phelps Dodge generously provided both travel and research assistance, I remained fully independent in thought and action. Publication of *Vision and*

⤳ Romance and Revolution ⤳

PEOPLE OFTEN SPEAK *of the romance of mining, and many books have been written on the subject, but people seldom consider the romance of manufacturing. Phelps Dodge Industries and its predecessor enterprises have their share of dramatic tales too, as I'll relate later in the account of how George Jacobs cooked up the first varnish for magnet wire in his wife's kitchen—and lived to tell the tale. That was not because she became so angry but because the smelly mixture he baked on wires in her oven was so volatile. Still, it made them multimillionaires.*

Each of the hundreds of people I've interviewed added fascinating details to the modern romance of Phelps Dodge, in both mining and manufacturing, but perhaps no one told a more entertaining story than Carlos Quiroz, a native of Chile where the following drama unfolded. We visited together in May 1998 in the Santiago offices of COCESA (an acronym for Cobre Cerrillos, S.A.), one of Phelps Dodge International's several wire and cable manufactories in South America.

Quiroz is a handsome, silver-haired man who speaks fluent English. He smiles easily, and his eyes fairly sparkled as he recounted his several decades at COCESA—which included the tragicomic events of September 11, 1973. When that day began he had little to smile about. Some two months earlier the Marxist government of Salvador Allende had forcibly nationalized COCESA and placed day-to-day operations in the hands of its own specially selected "interventor." The new man, after first expressing surprise that he was not overseeing an automobile-manufacturing operation, grandly informed Quiroz that he was to be terminated as plant manager as soon as the government found an engineer of its own choosing (political leaning) with comparable skills. Yet on several occasions when Quiroz tried to resign, his new boss refused to let him go.

Enterprise by a major university press and the manuscript's review by a panel of scholars of the press's own choosing underscore that commitment. Having already written or edited more than a dozen books, I honored anew the obligation that a trained historian has to relate past events as they actually happened.

For example, with regard to the controversial Bisbee deportation of 1917, the corporation sent me six large boxes of old records hitherto inaccessible to historians. Phelps Dodge attached no strings to the archival materials and asked only that I treat the perplexing event in its full historical context. The same advice guided my discussion of the 1983 copper strike, an emotionally charged event that significantly altered the course of labor relations not just at Phelps Dodge but also across the United States during the late twentieth century.

My aim was to tell the corporation's story in terms of biography as well as by personal examination of the highly visible landscapes that Phelps Dodge had created or acquired over the years. Three decades of teaching United States history at the college level and lecturing aboard cruise ships and trains have convinced me that

Perhaps it didn't matter. For two or more hours each day production ground to a halt as the interventor encouraged all workers to gather around a newly erected in-plant stage to listen to political lectures by Cuban revolutionaries or to be uplifted by visiting musicians. Spreading economic chaos, notably the growing lack of bread and other basic necessities of life, contributed to Chile's national malaise that culminated on September 11 in a rocket attack on the imposing presidential palace, La Moneda, and the military overthrow of Allende, who used an AK-47 given him by Fidel Castro to commit suicide six hours after the fighting erupted. Even before Quiroz saw the Hawker Hunter jets circling downtown Santiago and black smoke billowing on the horizon, he sensed both the rapidly rising tension and imminent revolution.

He recalled too how concerned everyone at COCESA was when a large canvas-covered truck unexpectedly turned off the main highway and sped to the back corner of the plant's parking lot. There its driver jumped out and fled on foot, never to be seen again. The back of the vehicle seemed to bristle with gun barrels aimed in all directions. No less concerned was a military contingent stationed outside the main gate. A gun battle seemed imminent.

Finally, after what seemed like hours, not a sound coming from the strange truck, cautious managers crept outside to investigate what turned out to be a cargo of frozen lambs. Their stiff legs protruded menacingly from between the slats. Each employee was delighted to take a lamb home for what turned out to be a two-day curfew. Some had little else in their houses to eat.

After two days the new government, in one of its first acts of privatization, returned the COCESA plant to Phelps Dodge. Quiroz headed back to work to find that Allende's appointed interventor had hastily fled across the nearby Andes to Argentina, and accompanying him was a plant secretary! Quiroz, who never learned what became of the romantic couple, concluded his story by flashing a broad smile that spoke volumes about his and COCESA's ultimate triumph over adversity.

people respond best to the visual aspects of history, such as when they personally tour a battlefield or travel along the Columbia River in the wake of the early nineteenth-century explorers Lewis and Clark. Firsthand exploration of Phelps Dodge landscapes only confirmed my belief in the value of field history.

In Arizona, the location of Phelps Dodge's headquarters since 1987, the Grand Canyon without doubt forms the state's most dramatic natural landscape. There a skilled geologist can explain many chapters of the earth's history to curious onlookers. In southeastern Arizona, about two hundred miles from the Grand Canyon, is one of the world's largest open-pit copper mines. It forms the centerpiece of a Phelps Dodge landscape that sprawls across several square miles of rugged mountains and high deserts. No single photograph, not even a large panoramic one, can possibly encompass it all. Yet at Morenci an observer can note how complex questions of technology, finance, and labor relations become observable and thus intelligible in a built environment that includes hulking industrial structures along with a modern shopping mall and workers' tidy homes. With the aid of historical photographs, maps, and personal interviews with longtime employees, I sought to understand everything this vast and visually arresting landscape revealed about the changing price of copper and its impact on individual and corporate aspirations, about matters of technology and the environment, and about many other things I'll discuss on later pages.

Here in summary is my strategy: to combine the elements of biography, cultural landscape study, and history to tell a story of vision and enterprise that is solidly grounded in fact but accessible and perhaps even interesting to general readers. My narrative tends to form what people in the mining industry might describe as a simplified flow sheet. The typical one provides a quick overview of the basic technologies and processes required to produce a metal like copper. The simplified flow sheet reveals only the most important parts of production, however, and here it should be noted that the same thing is true for a book of this length. In truth, many volumes would need to be written to capture the full history of Phelps Dodge. Over the years the corporation and its subsidiary firms have acquired or sold numerous businesses, both large and small, and with each transaction Phelps Dodge enlarged its own corporate history.

LEFT: The entrance to the Mine Division of Phelps Dodge Morenci in 1996 as defined by the overhead conveyor system that takes copper ore to the concentrator out of sight to the left. Viewing the vast landscape of production from this portal barely hints at the hive of activity just out of sight.

OPPOSITE: On the move in one small corner of the Morenci mine: these now historic "muck trucks" were superseded during the 1990s by gigantic vehicles capable of carrying 240- and 320-ton loads.

Unfortunately, to provide detailed histories of any but the most significant of all those businesses would result in a massive tome not unlike a metropolitan telephone directory—a book frequently consulted for tidbits of information but never really read.

Mining companies in particular can easily generate many big numbers, abundant abbreviations, and technical language that overwhelm general readers and make any history slow going. Where appropriate I have rounded off numbers and avoided most abbreviations, even the familiar initials PD. For quick reference, readers will find a glossary of technical terms at the back of *Vision and Enterprise.*

Because written history involves numerous choices about what to include or leave out, I shoulder full responsibility for what appears on the pages that follow, including any errors of interpretation or fact. To my knowledge, moreover, no relative of mine has ever worked for Phelps Dodge. My personal introduction to corporate America began at the bottom, literally, when I earned college dollars by cleaning restrooms, waiting tables, and shipping pharmaceuticals for Eli Lilly and Company. I spent two additional summers in Indianapolis packing boxes of textbooks for the publisher Ginn and Company (the people who introduced America's beginning readers to the legendary Dick and Jane characters). While working at Ginn I joined the Teamsters union.

Still later, in Michigan, I donned asbestos gloves and endured the acrid smell of burning rubber to cook heavy-duty drain hoses that the Ball Corporation supplied to washing machine manufacturers. As a graduate student in history at the University of Michigan I enjoyed frequent tours of Ford's sprawling River Rouge plant where slabs of red-hot steel rumbled in one door and in due course a shiny new Mustang rolled out another. I mention this background only because I have no intention of telling the story of Phelps Dodge entirely from the top down. Before I put on a white collar, I wore the blue. Such is my personal angle of vision.

As RECORDED ON *the pages that follow, the history of Phelps Dodge Corporation spans nearly 170 years, from its founding in 1834 until mid-1999. The narrative cannot record all the changes that occurred after that closing date, when the book entered production, but many key events are summarized in Chapter 19, which begins on page 433.*

HEAVY MACHINERY *works the open-pit mine at Santa Rita, New Mexico, in the 1930s. Now part of the Chino Mines Company in which Phelps Dodge Corporation holds a two-thirds interest, the site is one of the pioneer open-pit mines in the world.*

VISION & ENTERPRISE

*My greatest pleasure has been writing history, because it involves reading
history, not in a perfunctory manner but as a study.*
—James Douglas, first president of Phelps Dodge after incorporation in 1908

Our history and name have value that should be preserved.
—Douglas C. Yearley, chairman of Phelps Dodge Corporation, speaking in 1992

The past isn't dead; it isn't even past yet.—William Faulkner

Chapter One

THE VIEW *from*
DR. DOUGLAS'S DESK

JOIN ME AT the large rolltop desk where Dr. James Douglas worked during his frequent visits to Bisbee, Arizona. We are in the former headquarters building of the Copper Queen Consolidated Mining Company, which since 1897 has been a silent witness to decisions made in these rooms and plans carried out to mine copper ore in the surrounding hills. Off-duty miners once congregated on the shaded grounds outside to argue politics and exchange gossip. In their place gather teenage boys with their clattering skateboards.

The two-story red brick structure is now the Bisbee Mining and Historical Museum, a repository for some twenty-six thousand photographs as well as numerous artifacts that document a century of changes in the land that surrounds us. The building, located at the corner of Main Street and Brewery Gulch, was originally strategically positioned near Phelps Dodge's El Paso & Southwestern Railroad station, the O.K. Livery Stable, and the several mining operations that energized the area. They're all just memories, and rooms that served as offices for mine executives are no longer astir with clerks and bookkeepers, geologists and engineers.

The general office building *of the Copper Queen Consolidated Mining Company was located at the busy heart of turn-of-the-century Bisbee, Arizona. Why the transfer wagons were lined up for their collective portrait in 1903 is not known. Early-day photographers commonly failed to label their images and thus left succeeding generations to guess at the facts.*

What became the Copper Queen Branch of Phelps Dodge Corporation moved out of the building in 1961. Where I sit to type these words was once the main reception area, and the big desk with its many cubbyholes was probably located then in an adjoining room, the general manager's office. The original oak paneling still covers the walls and ceiling to project an aura of authority. All in all, this is a most evocative site from which to launch a history of Phelps Dodge Corporation from 1834 to the present.

Still visible high on the flanks of Copper Queen Hill across the street is the original *glory hole*. The large opening was the main entrance to the Copper Queen Mine and the massive deposit of oxidized copper, iron, and manganese that made the district famous and Phelps Dodge an Arizona institution. Nearly two thousand miles of passageways called *adits* and *drifts* and vertical openings called *shafts* and *raises* create a mazelike underground landscape of production. In miners' lingo, areas where copper was extracted were called *stopes*. During nearly a century of active mining here, some thirty-three shafts reached down into the earth, and tall, gaunt headframes still mark the entrances to several of them.

In the early years, valuable ore was extracted from the glory hole, crushed in a small breaker, and dropped down a wooden chute into a furnace shed, where it was smelted to remove copper. Mountain springs supplied water needed to run a small steam engine and cool the smelter's metal jacket. Workmen fashioned posts and beams for the original furnace shed from undressed tree trunks. Despite these crude beginnings, the Copper Queen Consolidated Mining Company prospered and contributed mightily to Phelps

Dodge's profits year after year. In total, it rewarded stockholders with approximately $100 million in dividends. Before all active mining ceased in Bisbee in 1975, its mines yielded fortunes in lead, zinc, manganese, gold, and silver—more silver and gold than any other place in Arizona—but above all else it was copper, some eight billion pounds of it, that determined the fate of Bisbee.

As early as 1890, and little more than a decade after its founding, Bisbee emerged as the greatest of all southwestern mining camps. That year the Copper Queen alone produced some 752,000 pounds of copper per month, an amount equivalent to the combined output of all copper mines in the United States only a few decades earlier, and the total would continue to rise for many years to come. During the first half of the twentieth century the Copper Queen often ranked among the two or three richest copper mines on earth and was the main reason Bisbee became known as the Queen of the Copper Camps and its high school issued diplomas printed on sheets of copper.

In addition, the mines of Bisbee yielded some two hundred different minerals, including blue azurite, red cuprite, and bright green malachite. Astonishingly beautiful specimens are on display at the Smithsonian's National Museum of Natural History in Washington, D.C., where exhibits of all kinds attract more than six million visitors each year. Many of the Smithsonian's Bisbee gems were donated to it by an ardent collector, James Douglas, one of the most multi-talented and influential men in Phelps Dodge history. Above me here in Bisbee hangs his portrait. Because Douglas was fascinated by both

technology and the challenge of writing history, I'd like to imagine his personal delight in the idea that someone used a modern laptop computer at his antique wooden desk to tell the story of Phelps Dodge Corporation. The desk provides a superb vantage point from which to survey the grand corporate landscape before exploring it in intimate detail in succeeding chapters.

The Hand of History

D r. Douglas never lived in Bisbee. During regular visits from New York he often spent time inspecting mine and smelter operations with local managers. He wrote many letters and sometimes even brought along a "magic lantern," an early slide projector he used to give illustrated travel lectures at his beloved Copper Queen Library. For three decades and more, Douglas wielded extraordinary power, often from this desk; yet in neither private nor public life did he fit the stereotype of a copper baron or a captain of industry, to use terms popularized by the press of his day. By any standard he was a gifted business leader who presided over the mining operations of Phelps Dodge, and eventually the whole corporation. Leadership gave Douglas power to do more than merely read or write history. He made it with gusto—and not just here in Bisbee, but across broad sections of the Southwest where he fashioned many of the urban and industrial landscapes defined by Phelps Dodge's mines, smelters, and railroads.

From Douglas's desk our imaginations transport us back through time to the modest countinghouse of Phelps, Dodge & Company located in the waterfront mercantile district of New York's lower East Side and forward to the Fortune 500 corporation that Phelps Dodge is today. Here in Bisbee, time and place and personality intersect, for without James Douglas and the Copper Queen Consolidated Mining Company it would be impossible to understand why Phelps, Dodge & Company abruptly turned west in 1881 or how radically the firm changed during the years that followed. It finally quit the import-export business altogether in 1906.

About every fifty years since its founding in 1834, Phelps Dodge has reinvented itself. Perhaps the regularity of such dramatic transformations is only coincidental, and perhaps reinventing is too strong a word to describe the way Phelps Dodge has periodically refocused its angle of vision and redeployed its financial and human resources. Yet in mid-1881, within the space of a single month, the mercantile partnership that had so long been a fixture of New York's polite society suddenly laced up a pair of miner's heavy boots and planted one foot firmly in Arizona's Clifton-Morenci area and the other in Bisbee.

It was a giant first step into the sometimes rough-and-tumble world of copper mining, and it can truthfully be said that Phelps Dodge never looked back with regret. In fact, a century later Phelps Dodge moved its corporate head-quarters from Park Avenue in New York to Central Avenue in Phoenix. Back in 1881, newly incorporated Phoenix could boast of almost two thousand residents, but that was nothing compared to New York, the nation's largest city, which recorded a population of slightly more than one million. Even infamous Tombstone, with its makeshift population of some seven thousand miners and the usual swarm of hangers-on lured by a mineral bonanza, ranked ahead of Phoenix, while Bisbee claimed three hundred residents. That same year, 1881, when Dr. Douglas first walked the tough streets of Tombstone and nearby Bisbee, the Earp brothers and Doc Holliday settled old grudges with the Clantons and McLowrys in a blazing gunfight near Tombstone's O.K. Corral.

Out in Arizona, Phelps Dodge never hesitated to follow where its first veins of copper might lead, both literally and figuratively. It pursued copper south across the border in Nacozari, Mexico, and later in Ajo and Jerome, Arizona, and east in Tyrone and Santa Rita, New Mexico. As it perfected its skills in finding and processing elusive red metal, Phelps Dodge in still more recent years developed the fabulous Candelaria mine in Chile and pursued other ore bodies in Arizona, among them four largely undeveloped deposits north of Safford.

About fifty years after Douglas first steered Phelps Dodge toward Arizona Territory, one of his successors, Louis Shattuck Cates, boldly defied the Great Depression of the 1930s and led Phelps Dodge into the business of refining copper and manufacturing a variety of copper-related products. The enterprise extended from mine to market, as it was commonly phrased at the time. Cates also committed the company to purchasing new mines and to launching development of the great open-pit complex at Morenci that now produces more copper than any other mine in North America. Far from being intimidated by the Great Depression, the hard-driving Cates elbowed the gloom aside and seized one new business opportunity after another for Phelps Dodge, much as Douglas had done in the 1880s when he pursued every chance to explore for uncharted mineral riches in Arizona Territory. Cates and Douglas possessed vastly different personalities and styles, but the "professor," as Douglas liked to be addressed, and the no-nonsense mining engineer shared a willingness to take calculated risks that paid rich rewards for Phelps Dodge.

Fast-forward another fifty years, to the early 1980s, when Chairman George B. Munroe wrestled with the challenges created as the price of copper in constant dollars plummeted even lower than during the depression decade of the 1930s and the costs of production skyrocketed. Some vocal skeptics doubted that Phelps Dodge

DR. JAMES DOUGLAS *(1837–1918) was renowned for his even temper. He had "a very gentle nature," reminisced Arthur H. Churchill, who began fifty years of service in Phelps Dodge's New York offices in 1881, the same year the company first dispatched Douglas to Arizona.*

could survive the deadly combination, but Munroe—a Rhodes scholar, Harvard-educated lawyer, and contemplative executive in the manner of Dr. Douglas—together with his lieutenants steered Phelps Dodge past one shoal after another. That was no easy feat of navigation. When a lengthy labor dispute erupted into violence at Clifton and Morenci, it only added to the grimness of those troubled times.

Once again Phelps Dodge altered its angle of vision. The corporation embraced a sophisticated new technology that yielded nearly pure copper at far less cost than ever before, and at the same time it lessened its dependence on copper by replacing several industries that fabricated copper products with others that used no copper at all. By the late 1980s Phelps Dodge Corporation was rejuvenated and fully ready to implement new strategies of growth and development. One measure of its remarkable recovery was the price of a share of Phelps Dodge common stock, which investors valued at slightly more than $6 in 1984 (adjusted for a later two-for-one split) but at almost $90 by early 1997 before it settled back because of the world's oversupply of red metal.

For Phelps Dodge Corporation, some of the most important historical events that determined its angle of vision occurred along the winding streets of Bisbee. In fact, the whole town

⤳ IN EARLY-TWENTIETH-CENTURY BISBEE, *the general office building of the Copper Queen Consolidated Mining Company was located just in front of the Mediterranean-style Copper Queen Hotel—to the left in this image. Both buildings still stand and are centers of community life.*

～ Portraits Keep Watch ～

ONE OF GEORGE MUNROE'S *lieutenants during the dark days of the early 1980s was Douglas C. Yearley, who served as the corporation's chairman during the 1990s. Yearley welcomed the prosperity of that decade's early and middle years, yet he could never forget, even if he wanted to, the harsh lessons Phelps Dodge learned when it fought for its life. Every time he entered or left his office in the corporation's twin-towered headquarters in Phoenix, he could not escape the hand of history on his shoulder—or being measured against his predecessors of the past sixteen decades. Along the broad hallways hang lifelike oil paintings of James Douglas and Louis Cates. Here too are portraits of George Munroe and his successor G. Robert ("Bull") Durham, as well as Robert Page, Walter Douglas, and others who headed Phelps Dodge. Every member of the corporate family tree keeps silent watch.*

To reach his office, Yearley must also tacitly acknowledge the abiding presence of Mr. Phelps, Mr. Dodge, and Mr. James. Phelps, gray-haired and jaws clenched, gazes sternly from beneath arched eyebrows as if to scrutinize with a measure of disbelief the panoramic painting of the great open-pit mine at Morenci, which hangs on the opposite wall of the corporation's boardroom. William E. Dodge, with the fashionably long sideburns of his day, looks at Phelps, whose portrait is flanked by the ruddy and cherubic likeness of a second son-in-law, Daniel James. These are the founders.

We can thank Dr. Douglas for adding portraits of the next generation of leaders. It was he who suggested to board members in 1908 that paintings of D. Willis James and William E. Dodge, Jr., be ordered at company expense and hung in the boardroom, then located at 99 John Street in New York. He wanted to recognize the importance of those two cousins in redirecting Phelps Dodge into the metal mines of Arizona. Cleveland H. Dodge moved to amend the resolution to commission a portrait of Douglas himself and add it to the gallery, though he was never related to any of the founding families by blood or marriage. The painting remains a remarkable tribute to Douglas's importance to Phelps Dodge.

Overhead in the corporation's modern boardroom and portrait gallery hangs a reflective ceiling of brightly polished copper panels. No one can miss the symbolism. Nor can anyone doubt from the portraits that the name Phelps Dodge is synonymous with history. "Anytime I take people around the office," observed Yearley, "I walk them by the various portraits and talk about history." Everything about the display exudes tradition.

(Based on an interview with Douglas C. Yearley, Scottsdale, Arizona, February 17, 1997.)

is a delight to anyone who seeks to evoke the past. Whether you sit at the desk of Dr. Douglas or stroll along twisting and uneven streets so typical of early mining camps, you will sense ghosts of the past near at hand. They reside in grand buildings like the Pythian Castle and Copper Queen Hotel no less than in the former general offices of the Copper Queen Consolidated Mining Company. Few bars now line Brewery Gulch, once the boozy heart of the town's commercial district, but Bisbee itself is certainly no ghost town. In fact, in its "Best of Arizona" tribute for 1997, the Phoenix newspaper, the *Arizona Republic*, ranked Bisbee's turn-of-the-century landscape as the best downtown in the entire state.

Bisbee is physically isolated in the canyons of the Mule Mountains of southeastern Arizona, about six miles north of the Mexican border. At an elevation of 5,300 feet and surrounded by hills that top 7,300 feet, Bisbee is an oasis that receives more moisture and less summer heat than the surrounding drylands. Although tourists find Bisbee's mountainous location appealing and even picturesque, the sole reason for its founding was mining.

"That depends on the price of copper" was a refrain once commonly heard on Bisbee's meandering streets, and it is still true. The price of copper always equaled prosperity or adversity. A rise in price in London or the opening of a new mine in Chile invariably affected daily life in old Bisbee—as it does in other Phelps Dodge towns today. The vernacular of mining was the common language of all residents regardless of their economic status or ethnic background. Even now, the low hills surrounding the oldest portion of

Bisbee reveal a red hue caused by iron in oxidized porphyry copper deposits.

Before Phelps Dodge first invested in Bisbee mines in 1881, the camp could be described as equidistant from nowhere in particular and nowhere at all. How the New York mercantile firm of Phelps, Dodge & Company became inextricably linked to a remote mining camp like Bisbee can be summed up in two words: James Douglas. It does not detract at all from the importance of the two cousins, Dodge and James, who dispatched Douglas to Arizona Territory to suggest that his angle of vision quickly became that of the Phelps Dodge partners. In Bisbee it is still impossible to separate the names Phelps Dodge and James Douglas. To speak of one is to infer the other.

More than any other individual, Douglas forged the link between two distinct periods of Phelps Dodge history: the mercantile era before 1881 and the mining era that followed, when the firm risked its first dollars digging for riches beneath the desert earth of Arizona Territory. This in turn would lead to the mine-to-market era envisioned by Louis Cates fifty years later. For all this, the Douglas name was never added to the corporate title, but because he was a modest and unassuming man, that omission probably did not trouble him at all.

How Professor Douglas Transformed Phelps Dodge

THE STORY USUALLY told to explain how Douglas and Phelps Dodge hitched their respective fortunes to the same mineral bonanza in the frontier West goes like this. On a wintry morning early in January 1881, a solitary figure hurried down the streets lined with busy warehouses and shops that fronted the East River piers. Here numerous sailing ships from around the world exchanged their cargoes.

The stranger, a tall and handsome man named William Church, abruptly turned into the red brick building at No. 11 Cliff Street. Here Phelps, Dodge & Company conducted business on two floors, retail on the lower and wholesale on the upper. Church brushed past salesmen and shipping clerks and boldly climbed the steps to the second floor. He was a man with a mission, but he must certainly have paused when he reached the head of the stairs. With the only light provided by tiny windows and flickering gaslights, the scene must have appeared dark and foreboding, perhaps even cramped and intimidating, to a man accustomed to the bright skies and open spaces of Arizona Territory. Once his eyes adjusted to the dimness he slowly surveyed a large room filled with desks where bookkeepers and stock clerks busily jotted notes and records. This was a typical countinghouse, one of many that formed the commercial heart of the nation's largest city but like nothing Church knew out West.

The Phelps Dodge offices that William Church entered in New York had no women employees and no mechanical typewriters, and that too was typical for businesses at the time. A single telephone was at the head of the stairs, but the newfangled technology was so scratchy sounding in operation that none of the partners would use it. They delegated the task of jotting down incoming messages to a junior clerk. At the

back wall was located a small private office with a sign on the door that read William Earl Dodge. The firm's most senior partner still came to his office occasionally. If Church glanced around he might have noticed that the only other enclosed space was a small washroom used by clerks and partners alike.

William E. Dodge, Jr., worked at one of the large rolltop desks near the center of the large room. Out of his hearing the clerks often referred to him as the Earl. Perhaps that was a play on his middle name intended to distinguish him from his elderly father, or perhaps it was because he had the refined manners of English nobility. At another of the largest desks worked D. Willis James, forceful and at times even irascible. The firm's many clerks regarded him with a mixture of awe and fear. One longtime employee, Arthur H. Churchill, who worked for Phelps, Dodge & Company from 1881 until the early 1930s, recalled that if he made an error Mr. James would bawl him out before the whole office staff as a "stupid ass." The younger Dodge bore a grudge longer; James would flare up but the storm was soon over.

Church approached the desk of William Earl Dodge, Jr. Along with James he managed most of the day-to-day activities of the now venerable mercantile firm that engaged in the import-export business and in the manufacture of kettles, copper wire, and buttons for uniforms. Closely allied with Phelps, Dodge & Company at the time were the Ansonia Brass and Copper Company and the Ansonia Clock Company, which operated immense factories in Ansonia, Connecticut (a town named for Anson Phelps), and across the East River in Brooklyn, although

NEW YORK'S SOUTH STREET AREA *as seen from Maiden Lane in the late 1820s. In this bustling neighborhood fronting the East River the partnerships of Phelps & Peck and later Phelps, Dodge & Company located their offices and warehouses. Merchandise was typically stored on the upper floors of these buildings; the front portion of their first floors was often a showroom filled with newly arrived goods set out for customers to inspect and sample.*

fire had recently destroyed that plant. The stranger from Arizona presented no business card or letter of introduction—he was not interested in the impressive manufacturing enterprises that then formed the Phelps Dodge landscape—but he boldly requested a loan of $30,000 (perhaps as much as $50,000 according to some accounts) to develop several claims his Detroit Copper Mining Company held near Clifton, a camp located about 125 miles northeast of Bisbee.

At the time neither Bisbee nor Clifton, nor even the entire territory of Arizona, meant a thing to the partners, and for an upstart like Church to approach a New York firm as distinguished and conservative as Phelps, Dodge & Company without a letter or card of introduction was an egregious breach of business etiquette. Even worse, Church brought no engineer's report or samples of ore from the properties he proposed to develop. Who did Church think he was?

More important, what did the Phelps Dodge partners think of this impertinent stranger? They had absolutely no idea who Church was, and in the past they had seldom extended credit to anyone, and certainly not to a total stranger. Had Church approached the desk of the imperious Mr. James first, he might quickly have been shown the door and for good measure blasted as a "stupid ass." But something about Church's boldness and determination intrigued the courtly Mr. Dodge, who encouraged Church to tell him more about his Arizona mine ventures. He must have spun a fascinating tale, for although Church left Phelps Dodge that day empty-handed, he made a lasting and favorable impression on his listeners. Church's

obvious sincerity and self-confidence so impressed the partners that they decided to at least evaluate his proposal. One reason they did so was that their Ansonia, Connecticut, brass works had only recently begun treating copper from an Arizona mine called the Copper Queen, and another reason was that recent changes in their business, notably the departure of the Stokeses (one of the founding families of the original partnership), had left the remaining partners with an unusual amount of money on hand to invest in promising new ventures.

A short time later—perhaps even the same day—James Douglas climbed the wooden stairs leading up to the offices of Phelps, Dodge & Company. It was his first visit, and therein lay yet another remarkable coincidence. Douglas entered the tightly knit family enterprise only because William E. Dodge, Jr., troubled by a technical problem and needing expert advice, asked a local firm of metal brokers to recommend a consultant. It was in response to their telegraphed invitation to meet with "an influential man" that Douglas boarded a train to New York. At the time he was superintendent of a none-too-robust copper smelter and refinery complex located in Phoenixville, Pennsylvania, and it was to supplement his meager income that he moonlighted as a consultant.[1]

What Phelps, Dodge & Company needed from Douglas was advice about whether the partners should build a copper smelter on Long Island Sound, or elsewhere, instead of shipping boatloads of ore across the North Atlantic to Swansea, Wales, at that time the world center of copper smelting and refining. The 43-year-old Douglas did not hesitate to tell Dodge that building the smelter he envisioned would be a waste of money; it would greatly reduce costs if ores were processed near the mines rather than shipped from the American West to reduction works on the East Coast or in the British Isles.

Douglas's intimate knowledge of mining and smelting greatly impressed Dodge, who soon steered their conversation to the subject of William Church's recent visit. He asked if Douglas knew anything about Church's Arizona properties. The consultant shared with Dodge what he had learned from his recent personal inspection of the Copper Queen, allowed that he had heard about the Longfellow Mine in the Clifton area, but claimed no firsthand knowledge of Church's own holdings. The following morning, after checking Douglas's references, Dodge hired him to serve as the partnership's eyes in distant Arizona Territory.

When Douglas agreed to investigate the Detroit Copper Mining Company, he made a decision with far-reaching consequences both for himself and for Phelps, Dodge & Company. In time, Church received his loan, and the firm's partners quickly came to value Douglas for his knowledge and honesty. Dodge wrote later in 1881 that he knew of no one he would trust more to make the examination. From 1881 until his

⤜ THE GOOD DR. DOUGLAS ⤚

*A*LTHOUGH IN LATER YEARS *he was universally known as Dr. Douglas, the one thing the first president of Phelps Dodge Corporation did not possess was a bona fide medical degree—unlike his father, a Scottish-born surgeon. Since childhood, young James had dreamed of becoming a surgeon himself, early amassing his own box of scalpels and hanging around the operating room in hopes of absorbing his father's medical skills, but before he completed medical training at Laval University, a family crisis shoved him headlong into a career in mining. In 1899, however, another Canadian university, McGill in Montreal, did award him an honorary doctorate of laws, and after that date he was often addressed as Doctor or Professor Douglas. In fact, Douglas later endowed a professorship in Canadian history at Queen's University.*

In time he became both a multimillionaire and a person greatly respected in the North American mining industry. Yet his youthful desire to alleviate the physical ills of humanity never left him. So strong was the physician in Douglas that during his final years he doggedly pursued his interest in uranium deposits in Colorado in hope of finding a low-cost way to extract radium to cure cancer. In fact, quite possibly his largest single bequest was a gift of 3.75 grams of radium, originally valued at $375,000, to what is today Memorial Sloan-Kettering Cancer Center in New York.

death in 1918, the fate and fortunes of James Douglas and Phelps Dodge remained so tightly entwined as to be inseparable.

This story of what turned out to be a pivotal event in Phelps Dodge history may seem improbable, but in its primary details it is as true as a whole series of improbable but factual events that led Douglas to Cliff Street in the first place, and Phelps Dodge to far-off Arizona. Nothing, absolutely nothing, in Douglas's idyllic childhood spent on his family's estate in rural Quebec, or in his undergraduate education at Queen's University in adjacent Ontario, or in three years' study at Scotland's University of Edinburgh that prepared him to become a Presbyterian minister, or in two years of medical training at Laval University back in Quebec suggested that this man, so obviously at home in an academic setting, would make his mark as a corporate executive. By any standard he was a remarkable person, a Renaissance man who made good.

Detours on a Winding Road to Success

DOUGLAS WAS A successful business leader not just because he gained money, status, or power. Those three measures of success were all the result of his being such a keen student of the world around him. During eighty active years he devoted himself not just to the study of history or metallurgy or chemistry or medicine or Presbyterian theology or geography or architecture or business, but to all of those subjects and many more. To Douglas this kind of broad education was a requisite for success in the complex business of mining. It was central to his personal angle of vision.

Remarkably, Douglas entered the world of copper originally intending only to rescue his imperious, stubborn, and erratic father from financial ruin. He later recalled that his father refused to trust his savings to ordinary depositories, "but he unhesitatingly invested them in enterprises of which he understood absolutely nothing." Long-term success for the younger Douglas in mining was as improbable as his initial meeting in 1881 with Phelps, Dodge & Company.[2]

More than a decade earlier, when reckless speculation in a Canadian copper mine had plunged the elder Douglas deeply into debt, his son redirected his intellectual energies at Laval University away from medicine and toward the study of mining and metallurgy. With Thomas Sterry Hunt, an eminent professor of chemistry, he collaborated during several months of laboratory experiments to develop what became known as the Hunt-Douglas electrolytic method of refining copper ore. Until then the only known way to purify copper was by fire, a refining method that lost most of the precious gold and silver by-products that the new Hunt-Douglas method recovered. This gave a major boost to copper ores from the Rocky Mountain West, which industry experts had earlier scorned as inferior to those from Michigan.

Although he failed to salvage his father's investment—virtually all of the family estate had to be sold to pay outstanding debts—his knowledge of copper gained the 38-year-old Douglas a much-needed job managing the Chemical Copper Company, a small smelting and refining works erected on the banks of the Schuylkill River above Philadelphia. After the complex opened in the mid-1870s it used the Hunt-Douglas process to extract copper from ores mined in the nearby mountains of Pennsylvania. In this way it produced some of the first electrolytically refined copper in the United States, but the enterprise itself never became a financial success.

With its chimneys, slag heaps, dumps, and idle railroad tracks, the industrial landscape of Phoenixville provided a depressing contrast to the comfortable environs the Douglas family had known back in Canada. Rising early each morning James went immediately to the refinery, where, with the exception of an hour off for lunch, he spent his day. He recalled that "the work was interesting, and my wife made the salary go so far that we could educate our large family, supply our larder, not with luxuries, but with sufficient healthy food, and clothe our-selves warmly if not fashionably." Family ties, Presbyterian piety, and hard work sustained Douglas during these difficult years, and perhaps one reason he later found kindred spirits in the Phelps Dodge partners was that an identical trilogy of values sustained them too during difficult days.

Douglas rented a house for his family on Tunnel Hill, a working-class section of Phoenixville closer to the Chemical Copper Company works than a more moneyed locale would have been, and hence much less expensive. Though sparsely furnished, the large stone house had an enclosed verandah in front where for lack of room elsewhere Douglas stored several mummies his father had acquired in Egypt during

one of his adventures there. Those ancient remains were among the few items salvaged from the family's Quebec estate, and later, with the return of prosperous days, the son donated them to New York's Metropolitan Museum of Art. Neighbors regarded the mummies with superstitious awe, some ill-educated people believing them to be Douglas's own departed ancestors whose bodies he carried with him for the supernatural protection they afforded. In any case, "our neighbours' houses," recalled Douglas, "were broken into, but the supposed corpses of our ancestors, which stood menacingly on guard on our front verandah, protected us."

While in Phoenixville, Douglas often returned to unfinished duties at the refinery after dinner, carefully picking his way along the busy tracks of the Philadelphia & Reading Railroad, and still later when he arrived home, he would sit at his desk, often until well past midnight, writing letters of advice on mining matters or articles on history and mining. When an essay was published it might earn him a much-needed sum of $10 to supplement his $150 monthly salary that had to stretch to provide for his wife, six children, and aging father. Writing provided a chance, he later recalled, "to clarify my own ideas." Because the family living room was his study, "I acquired also the invaluable habit of abstracting my thoughts and senses from surrounding influences."

During the most active years of his profes-sional life after he allied with Phelps, Dodge & Company, Douglas spent considerable time on long railway journeys. But that was when he did much of his writing, and also aboard commuter trains between his home and New York's Grand Central Station. Even when he enjoyed the

Humble Beginnings for a Mining Dynasty

RESIDENTS OF PHOENIXVILLE *grew accustomed to a number of strange sights, such as two of the Douglas children, little Walter and James, trudging off to school each day clad in kilts that their Scotland-loving father insisted they wear. Other children laughed at the brothers and often threw things at them. Their father required them to wear shoes too, rather than go barefoot, as was the neighborhood custom, but he was not so stern or rigid that he could not enjoy a refreshing swim with his boys in an old mill canal. Both sons eventually followed their father into the mining industry, and each left his own mark on the southwestern borderlands.*

The younger, James S., or "Rawhide Jimmy" as he was known to succeeding generations, supervised various mine and railway construction projects before personal investments in the United Verde Extension Mine at Jerome, Arizona, made him fabulously rich. He was by any standard a loud and colorful character. Less outgoing was Walter, about eighteen months older than his rambunctious kid brother and temperamentally much more like his father. Slender and quiet-spoken, Walter later served as general manager of the Copper Queen Mine and followed in his father's footsteps to become president of Phelps, Dodge & Company.

hard-won luxury of his own private railway car, the Nacozari, he never felt wealthy enough to waste a single minute of time, and so he seldom spent a day in travel or an evening at home without devoting precious moments to research or writing on historical subjects. Over the years, Douglas penned enough essays and books to fill a small library, including scholarly studies of Canadian independence, New England and New France, and the influence of railroads on industry in the United States and Canada. Two articles that dealt with surprisingly modern themes were titled "Conservation of Natural Resources" and "The Status of Women in New England and New France."

Among Douglas's early writings were articles published in 1873 and 1874 that detailed his homeland's struggle to build a transcontinental railroad, the Canadian Pacific. Who could have imagined during the hardscrabble 1870s that Douglas could later apply theoretical knowledge he gained this way to forge a railroad empire for Phelps Dodge across the American Southwest? When in more prosperous times he journeyed from New York to Bisbee aboard his own railway car, he often completed the trip behind one of the El Paso & Southwestern's sleek steam locomotives as it sprinted ninety miles per hour across the lonely desert country of southern New Mexico and Arizona. When the train slowed, he sometimes stood on the rear platform to wave to railroad employees, most of whom he knew personally by name.

In truth, none of his several careers before Phelps Dodge and none of his vast and varied fund of knowledge seemed inappropriate when Douglas shouldered the task of developing a new mining empire in frontier Arizona and New Mexico. In the austere but enchanting Sonoran and Chihuahuan deserts he fashioned a modern landscape where more than a thousand miles of railroad track connected copper and coal mines, concentrators, and smelters and linked them all with freight and passenger trains to distant refineries, markets, and metropolitan centers.

He was a town builder too, and when he fostered a new settlement, Douglas the would-be medical doctor insisted that it have a good hospital, while Douglas the scholar promoted libraries to remove the rough edges from frontier life. Douglas the theologian felt a moral obligation to care for his employees both on and off the job. Critics denounced Phelps Dodge's paternalism—at one time it was possible to be born in a Phelps Dodge hospital, to work for Phelps Dodge all your adult life, and to be buried in a Phelps Dodge cemetery in a coffin purchased from Phelps Dodge Mercantile—but to be called paternalistic would not have bothered Douglas the philanthropist in the least. Over the years, the founders of Phelps Dodge funded an enormous number of worthy causes, many of them for religious and educational purposes ("good work and good works" was the old partnership's hallmark), and Dr. Douglas continued and expanded this family tradition by leaving a considerable portion of his own fortune to museums, charitable insti-

tutions, and various beneficial medical projects. The average person's conception of a mighty captain of industry—bloated, arrogant, and worshiping only the almighty dollar, as depicted in the popular political cartoons of Thomas Nast— could not be stretched far enough to encompass this reserved, slightly built, and bookish man.

Dr. Douglas retired from the Phelps Dodge presidency in 1916 but continued as chairman of the board until his death two years later. He left a fortune estimated to be worth $40 million, most of which he amassed only after he turned forty. Some people might claim that Douglas was a "late bloomer" or that he was just lucky when he unearthed a fortune in copper for Phelps Dodge; yet it was always true that the harder he worked the luckier he got. The professor seemed to have an uncanny ability to locate ore. Many a miner claimed he could envision the hidden location of a rich mineral deposit, but Douglas's "nose for ore," as this ability "to see in the ground" was popularly phrased, was bolstered by his accumulated scientific knowledge. However, this is not to say that his judgment was infallible.

Even he admitted that on occasion he was simply "groping about for ore." On Douglas's recommendation, Phelps Dodge acquired several minor mining properties, or prospects, that the company later abandoned as unprofitable. On the other hand, he failed to recognize a couple of major mines. Still, he and Phelps Dodge were right often enough to rank the partnership among the Big Three United States copper producers by the early twentieth century. Douglas's true genius lay in forging landscapes of production that made Phelps Dodge's mining investments profitable. Yet for all his boldness in business, he never plunged recklessly into anything. Many longtime employees remembered that his motto was, "In time of prosperity always prepare for adversity—which is sure to come." That was sage advice in an industry noted for its pronounced boom-and-bust cycles, and whether succeeding generations of Phelps Dodge leaders were aware of Douglas's aphorism, they followed a similarly careful course.[3]

From Old Bisbee to New Morenci

VIEWING THE LOCAL LANDSCAPE of production from Dr. Douglas's desk in early 1997, I could easily see that mining activity in Bisbee was at a low ebb. Headframes for underground mines still loomed like long-forgotten gallows on the distant horizon, but their machinery was idle, and the Lavender Pit, once so productive, was mainly a tourist attraction. After more than a century of round-the-clock activity, the Copper Queen Branch was quiet. Uncannily so, some would say, for perhaps it was only resting and quietly gathering strength for the next big boom. There is, in fact, a chance that at some future date the Copper Queen Branch could actively mine the Cochise deposit to extract still more copper ore from beneath the ruddy surface of the Mule Mountains. Meanwhile Phelps

THE MORENCI CONCENTRATOR *used spiral classifiers to return coarse particles of copper ore to the ball mill for further crushing. The immensity of the plant probably would have stretched the imaginations of Dr. Douglas and other Phelps Dodge pioneers. Inconceivable too only a few years ago are the 320-ton haul trucks first regularly used in the Morenci mine in 1997.*

Dodge leaches a small amount of copper from stockpiles of low-grade ore it mined years earlier.

By contrast, Phelps Dodge Morenci was a hive of activity. That complex consists of four contiguous open-pit mines (Morenci, Metcalf, Northwest, and Southside), two concentrators, and three solution extraction facilities, and two electrowinning tank houses, or "Sexy EW"

plants. That's slang engineers sometimes used to describe the solution extraction–electrowinning process (SX/EW) that removes copper from low-grade ores and produces a finished product ready to ship to the rod mill by the almost magical application (to nontechnical minds) of chemistry and physics.

Visualize a maze of silvery pipes, blue liquids

carrying copper in solution, and a tank house the size of a football field that in the end will yield almost pure copper at an amazingly low cost and with virtually no air or water pollution, and you can understand why *sexy* in this case inspires awe. Like underground mining, SX/EW spawned a specialized lingo replete with technical terms like "pregnant leach solution" and "raffinate" and transformed what were once scorned as waste dumps into low-grade ore stockpiles. Though Phelps Dodge temporarily closed its Hidalgo smelter in 1999, pounding ball mills, trainloads of copper concentrate, and the fiery furnaces of smelters remain necessary to produce copper from ores that do not easily lend themselves to the solution extraction–electrowinning technology, but in contrast to the pyrotechnics of converter aisles at Chino and Hidalgo, the SX/EW process is quiet and subtle. The sound is mainly that of liquid in motion—not unlike the soft gurgle of a high mountain stream.

This marvelous technology is a primary reason for the average Phelps Dodge worker's producing 20 percent more copper per hour in the late 1990s than in the early years of the previous decade. Many people now claim that adoption of SX/EW technology saved Phelps Dodge during the dark days of the early 1980s when copper prices fell through the floor. That's partly true, but it is one of many affirmations waiting to be explored in depth on the pages that follow. In any case, beginning with a plant at Tyrone, Phelps Dodge installed additional SX/EW facilities at Morenci and Chino.

From the observation point high on the rim at Morenci, heavy equipment far below looks like mere toys, and the capacious mine itself seems almost empty. Actually it never sleeps. Haul trucks, crushers, conveyors, and processing plants continue without ceasing, as one shift follows another around the clock all year long. Several times each day sirens sound, traffic pauses in an area, and then come the shake and dull boom of an explosion created by a mixture of ammonium nitrate and diesel fuel. When the dust settles, the process of loading hundreds of tons of copper-containing rocks into haul trucks begins anew. Every twenty-four hours Phelps Dodge moves an average of 1.3 million tons of rock in its mines and 700 tons of fluid through the closed-circuit pipes of its SX/EW plants.

Imagine each day assembling a freight train 180 miles long consisting of thirteen thousand hopper cars, each carrying a hundred tons, and seven thousand tank cars, all pulled by dozens of diesel locomotives. Traveling at forty miles per hour the train would take four and one-half hours to pass an intersection. That gives some perspective to what Phelps Dodge does every day to produce copper. Fortunately, you'll never need to wait for such a long train because most movement of the rocks and fluids is self-contained on mine properties where it is both unobtrusive and governed by strict environmental regulations.

At Morenci, giant spiraling steps fifty feet high and more coil down into the depths of the mine. Workers blasted those benches out of solid rock and hauled their treasure of copper to crushers to begin the task of processing ore. In the late 1930s a railway line extended down the benches to the bottom of the ever-deepening pit, and electric and later diesel locomotives hauled trains of loaded mine cars up to the rim. Giant haul trucks joined the trains in 1986 and

OPPOSITE: THE PROBLEM OF *how best to portray the expansive landscape of production at Morenci has long challenged photographers. These men undoubtedly used their 1940s-era motion-picture cameras to record black-and-white panoramic images.*

completely supplanted them within the pit five
years later because they could get the job done
faster at lower costs. Although the spiral of tracks
down the pit walls is gone now, Phelps Dodge
regularly operates trains along its industrial railroad linking Morenci and Clifton, where carloads
of copper concentrate begin their journey to
company smelters in New Mexico, or where
bundles of finished copper sheets produced by

SX/EW facilities head off to market or the
rod mills. The red, white, and blue diesel
locomotives, each with the name Phelps Dodge
lettered in copper, bear the slogan America's
Leading Copper Producer. That is no idle boast.

During 1996, the Morenci mining operations
produced a record billion pounds of salable
copper—fully 56 percent of Phelps Dodge's
total production, 39 percent of all copper mined

LOADING ONE OF THE *massive haul trucks inside the open-pit mine at Morenci. This 1997 image reveals only a fraction of the landscape of production.*

annually in Arizona, and an impressive 27 percent of the total produced in the entire United States. Physically, Kennecott's Bingham Canyon mine in Utah is larger, but in the late 1990s only two or three mines in the world produced more copper each year than Morenci. Those would be found in Chile, the South American country that at the dawn of the new millennium ranked ahead of the United States as the top producer in the world.

At the Morenci complex in 1997 were several electric shovels with 56-cubic-yard dippers, or buckets. Each bucket was larger than the living room of an average house and could easily load a 240-ton "muck truck" in just three passes. Each shovel produced up to five hundred truckloads a day. One January morning that year I rode one of those behemoth trucks—its cab sits nearly three stories above the ground and when fully loaded it weighs as much as a Boeing 747 jetliner—as the driver, Rachel McCarthy, with more than two thousand horsepower at her command, skillfully eased us down mud-and-snow-covered roads toward the bottom of what is known as the Standard Pit. On a good day, the truck can proceed at speeds in excess of thirty miles an hour.

This 240-ton haul truck is one of seven that arrived from Caterpillar in 1992. About six months after my ride, Phelps Dodge added the first of fifteen Komatsu trucks to its fleet at Morenci. Each of the Komatsu HaulPaks carries a payload of 320 tons! A Caterpillar grader with a blade three feet high and twenty-four feet wide maintains the haul roads. The state-of-the-art truck shop that services the mammoth equipment is larger than a football field, and it sees lots of action. Phelps Dodge Morenci spends $1 million each month on new tires alone, and a like amount

on truck parts. The haul trucks of the 1930s cost about $16,000 apiece, or less than a single tire on one of the modern giants. Today the entire 240-ton truck costs about $1.5 million.

There was a time when writers seeking to give human dimensions to numbers too large to be easily grasped calculated how many Washington Monuments, for example, could be stacked atop one another to dramatize the depth of the Grand Canyon. Using that familiar measure, no fewer than five Washington Monuments (or two Empire State Buildings) would be needed to reach the twenty-five hundred feet from floor to rim of the open-pit mine! The word *canyon* seems appropriate because the Morenci excavation is so extensive that it is possible to see lightning flash through dark clouds that stretch along one rim, while a snowstorm shrouds the opposite rim in white. Down at the bottom the falling snow may have turned to rain and sleet.

Located high above the west rim of the pit is a control tower where computers, technicians, and a centralized truck-dispatching system send all fifty-eight haul trucks of various sizes shuttling ore between shovels and crushers in the pit and instruct drivers on which routes to follow to avoid long waits for loads. Aboard each truck a sophisticated global positioning system links it to satellites circling far overhead.

Landscapes Lost and Found

IF YOU ASK longtime residents of Morenci where they were born, many of them will simply point from the rim of the open-pit mine and

say, "Out there!" Some claim they were "born in space" because today there is nothing but open sky where old Morenci once clung precariously to steep slopes. How different everything looked when mining began here in the early 1870s. Until 1932, when the Great Depression temporarily halted activity, Morenci copper came entirely from underground mines.

Modern open-pit operations date from development activity begun in 1937. As workers enlarged the mine they sometimes cut into one of the old underground workings that spoke of another landscape and time when local ore regularly assayed around 2 percent copper in oxide mines, some early oxide bodies assaying as high as 20 percent copper or more. Ore mined at Morenci today averages only about 0.5 percent copper, and compared to early-day ore it is as much a study in contrasts as the landscape itself.

"Every time I go up there, I'll see that old place in my mind. It's strange," mused the retired Phelps Dodge Corporation vice-president for engineering Richard W. Rice, who grew up in old Morenci. His hometown was reborn at a location south of the pit, though residents of the original town often call the replacement community "new Morenci," with a tone of disparagement in their voices.[4]

Equally difficult for first-time visitors to visualize is the expansive system Phelps Dodge engineered to provide the water it needs to process Morenci ore. Beginning in the early 1940s, the corporation ultimately created one of the largest, most complex private systems of dams, reservoirs, and pump stations found anywhere in the United States. But it is not necessarily a highly visible one. I vividly recall a summer morning in 1997 when for more than fifteen minutes my pilot flew large figure-eight patterns through the cloudless skies of northern Arizona in search of one major component of that water system, the site of Blue Ridge Dam and lake. Even with modern charts and sophisticated aids to navigation, the big concrete structure remained difficult to see from a twin-engine Cessna.

The two aviators and two water lawyers who served as my guides intently studied their instruments and maps and carefully searched the terrain below until at last they spotted Blue Ridge Dam nestled low in the blue-green forest of ponderosa pines that cloaks the Mogollon Rim. Aerial reconnaissance helped keep things in perspective by reminding me that all talk of Phelps Dodge's oversize boot prints across the Arizona landscape is nothing more than poetic language. Even its open-pit mine at Morenci would easily fit into one tiny corner of the Grand Canyon. Flying along the remote Mogollon Rim to obtain a lofty view of Phelps Dodge's water system revealed how difficult envisioning the corporation's varied landscapes would have been had I remained deskbound—even at Dr. Douglas's evocative rolltop in Bisbee.

FOLLOWING OUR OVERVIEW of two of the twentieth-century mining landscapes that defined modern Phelps Dodge, let us return now to the corporation's chronological beginnings—to the day Mr. Phelps met Mr. Dodge, or more accurately, to the time Mr. William E. Dodge met Miss Melissa Phelps. The history of one of America's largest corporations begins, improbably enough, with a tale of romance in old New York.

Chapter Two

FAMILY TIES, *or the*
RELIGION *of* BUSINESS

O NCE UPON A TIME. It is tempting to begin with a familiar phrase that emphasizes the fairy-tale quality of early Phelps Dodge history, except that this tale of romance and heartbreak, success and setback, is all true. Once upon a time an orphan lad named Anson Greene Phelps became one of America's earliest millionaires, and once upon a time Anson and Olivia Egleston Phelps raised three attractive daughters. The ambitious young clerks of New York City who met weekly at the Phelps home at 32 Cliff Street for Sunday prayer meetings found it difficult to concentrate fully on weighty matters of Presbyterian theology. Melissa married William E. Dodge, the tall, fair-haired proprietor of a dry goods store on Pearl Street, in 1828; Elizabeth, the eldest Phelps daughter, married another young merchant, Daniel James, the following year; Caroline was betrothed to Josiah Stokes, Phelps's confidential clerk, when tragedy struck and nearly ruined the growing family clan. Until then, those were good years.

The prosperity of the late 1820s and early 1830s encouraged the partnership of Anson G. Phelps and Elisha Peck to expand their import-export business. After the Erie Canal opened in 1825 to link

---

IN NEW YORK *the road to success was constantly under construction. In this engraving from the early 1880s, workers extended a cable car line along Broadway. Cars powered by a moving cable beneath the streets offered a popular form of transportation before electricity.*

25

New York with the rapidly growing farms and towns of the Middle West and ship owners began offering regular service to Europe, the city sprinted ahead of Boston, Philadelphia, and Baltimore to become the nation's leading port on the Atlantic seaboard. During one of his usual morning strolls, Phelps could count the masts of more than five hundred ships in harbor at any one time, representing a tonnage greater than at any other maritime city in the world except London.

Perhaps he lingered to watch as brawny longshoremen hoisted freight of all types into or out of the cargo holds of ships. Typically they unloaded the heavy bales of cotton that arrived from Charleston or other Southern ports, then reloaded them aboard Liverpool-bound packets or set them aside for temporary storage in local warehouses. Wrapped in burlap and tightly bound, each bale contained four hundred pounds of cotton, more or less, pressed into a rectangular shape like a giant loaf of bread so as to fit easily into a vessel's hold.

King Cotton had launched the industrial revolution and caused thousands of new textile mills to be built in England. Those mills in turn wove American cotton into clothing that Peck regularly shipped back to New York where Phelps sold the imported merchandise to buyers from throughout the nation—including the South where the cotton trade originated. And what a trade it was. At one time, fashion dictated that well-dressed women of the antebellum South be clothed in no fewer than sixteen petticoats under a full skirt. Cotton, which remained America's premier export until the 1920s, was

without doubt the grand prize of international commerce.

When ships returned to New York from Liverpool, the British port that dominated the European side of the North Atlantic trade, they brought Phelps both cotton fabric and clothing and a variety of metal products that he supplied to factories and retail stores all over the United States. A typical cargo from England—the "workshop of the world" for nearly three-quarters of the nineteenth century—might include square wire for American umbrella makers, copper buttons, brass kettles, and heavy bundled sheets of brass, copper, and zinc. There were usually copper nails too, and brazier's copper, rivets, solder, lead, antimony, and tons of pig lead and iron.

Tin and tinplate ranked as Phelps's number one import, and that was true both for Phelps & Peck and the successor partnership of Phelps, Dodge & Company. What made tinplate so popular, apart from its use in cooking utensils or as roofing material, was that it could be made into cheap, clean, and nonabsorbent containers of every type. Long before cardboard and plastic, tinplate was used to package everything from cookies and crackers to motor oil.

To store their expanding quantity of merchandise, Phelps & Peck erected a six-story office-warehouse at the corner of Cliff and Fulton streets. The partnership insisted that builders make its exterior walls, support columns, and floor beams extra solid to carry the weight of tons of baled cotton and metal goods that they planned to store there. On the afternoon of May 4, 1832, an earthshaking crash startled New

York's mercantile and financial district. Running, waving, and shouting people soon filled the narrow streets as rumors passed from one incredulous person to another that Phelps & Peck's giant warehouse had collapsed. That seemed improbable—the massive building was still relatively new—but the combined weight of hundreds of heavy bales of cotton and tons of metal had exceeded the builders' calculations, pushed apart thick pillars supporting the floors, and caused the whole front section to peel away and collapse into the street. A fortunate few escaped, but beneath broken and twisted beams, bricks, and plaster lay trapped and dying clerks and warehousemen.

Phelps, who had briefly left his desk to attend an afternoon meeting of one of his many philanthropic boards, hurried back to the unbelievable scene, dropped to his knees, and tore away at the mountain of debris with his bare hands in a desperate search for his only son, Anson Phelps, Jr. The normally mild-mannered and imperturbable man was frantic. On the opposite side of the rubble, the son searched for his father. Under the glare of torches and oil lamps the work of recovery continued all through the night and into the next week.

The disaster shocked the elder Phelps. Material losses were bad enough, but the human tragedy was infinitely worse because among the seven dead was his clerk, Josiah Stokes, a silver pen reportedly still in his hand. He had been killed instantly by a falling beam. "Dear Josiah," Phelps confided to his diary, "on him I had begun to lean in hopes." The 51-year-old businessman had been grooming the amiable 23-year-old son

of a merchant to become his newest partner and heir apparent. Sad as that lost opportunity was, the death of Caroline Phelps's fiancé greatly added to the burden of sorrow that weighed heavily upon the family patriarch and suddenly made him feel very old and tired. His rags-to-riches saga seemed destined to end unhappily; the fairy tale had become a personal nightmare.[1]

How far Anson Greene Phelps had come since 1781. The son born that year to Lieutenant Thomas Phelps and Dorothy Lamb Woodbridge Phelps was named for his father's commanding idol of the Revolutionary War, General Nathaniel Greene, and for George Anson, the British admiral who had perfected ocean navigation in the mid-eighteenth century and whose memoirs became best-sellers. Orphaned at age fourteen, the boy grew up in modest circumstances in Simsbury, Connecticut. Five years later, in 1800, after a time of being "fed and clothed from place to place," Anson apprenticed to a saddler in nearby Hartford. Quickly mastering the trade, he established his own small shop where he made and sold saddles, trunks, and other leather goods. He sent some of his saddles to the South where they found favor among wealthy planters; and as he came to understand Southern agriculture better, he grew familiar with the growing transatlantic trade in cotton.

As early as 1805 the 24-year-old Phelps had joined the ranks of the vital middlemen in a commerce that linked Southern planters with British millowners who formed the vanguard of the new industrial revolution. He lent money to raise cotton, hefty bales of which he then shipped

THE RUINS OF PHELPS & PECKS STORE.

Fulton St. New York as they appeared in the morning after the Accident of Jan. the 1st.

to New York and sold to acquire imported iron-
and tinware, which he retailed to customers in
Connecticut. As the cotton trade boomed, young
Phelps's business prospered. By 1807, in addition
to running his store, he purchased several sloops
for a few hundred dollars each and entered
coastal shipping. Because roads linking the North
and South varied from poor to nonexistent,
Phelps's commitment to coastwise commerce
placed him at the cutting edge of transportation
developments in the new nation. In years to
come, transportation no less than mining
remained a key component of the history of
Phelps, Dodge & Company.

Phelps's ships and those of other New
England merchants helped catapult New York
ahead of Philadelphia as the nation's leading port
and financial center. Grasping the promise the
fast-growing city offered, Phelps in 1815
relocated his family from Connecticut to 178
Broadway, near Fulton Street, and opened his
"metal commission" business nearby at 29
Burling Slip on the East River. The Dodges, who
later figured so importantly in his family and
business affairs, were also transplanted Yankees
from Connecticut. That was a common pattern
of migration for the New Englanders who
emerged as New York's merchant elite.

Three years later, in 1818, a new shingle that
read Phelps & Peck swung in the brisk wind at
No. 181 Front Street. Phelps took Elisha Peck as
a junior partner because the young man possessed
energy, money, and family connections in
England. With Peck based in Liverpool and
Phelps in New York, the partners successfully
reconstructed transatlantic trade left in ruins by
the War of 1812. Their firm prospered and by

the 1820s ranked among the prominent
mercantile establishments in New York. Phelps
reinvested a portion of his share of the earnings
by purchasing real estate in Manhattan and shares
in half a dozen banks and insurance companies.
Until the abrupt collapse of the partnership's
warehouse in 1832, Phelps rode the wave of
prosperity that buoyed the growing city of
New York.

Following his tragic loss, Phelps found
that his enthusiasm for business seemed gone.
Somehow his son-in-law William E. Dodge in
due time rallied his spirits. Daily prayer was no
doubt a major part of the recovery. Stern
religious convictions sustained Anson Greene
Phelps during his season of troubles, both
personal and financial, and gradually he
reconstructed his life and work. With his sons-
in-law, Dodge and James, he formed two new
family businesses that superseded Phelps & Peck
(Peck having gone into business for himself).
In partnerships established in 1834—New
York–based Phelps, Dodge & Company and
Liverpool-based Phelps, James & Company—lay
the genesis of Phelps Dodge Corporation. In the
near term, though, the familiar pattern of coastal
and transatlantic trade continued much as before.

Presbyterian Copper

THOUGH A FULL generation separated Phelps
and Dodge, the two men were remarkably
similar in background and belief. Dodge's
forebears, like those of Phelps, sailed from the
mother country to New England in the 1600s
and eventually, after several generations, migrated

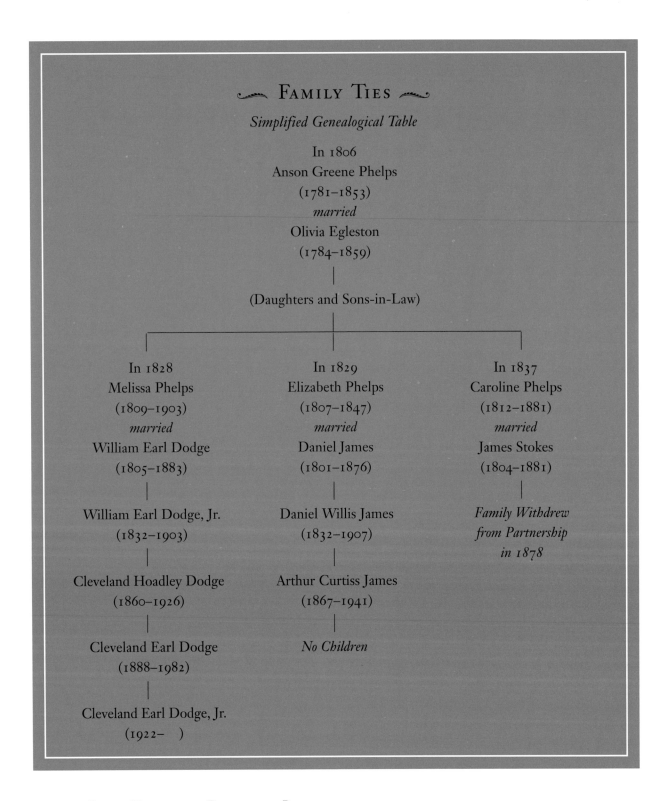

✎ FAMILY TIES ✎

Simplified Genealogical Table

In 1806
Anson Greene Phelps
(1781–1853)
married
Olivia Egleston
(1784–1859)

(Daughters and Sons-in-Law)

In 1828	In 1829	In 1837
Melissa Phelps	Elizabeth Phelps	Caroline Phelps
(1809–1903)	(1807–1847)	(1812–1881)
married	*married*	*married*
William Earl Dodge	Daniel James	James Stokes
(1805–1883)	(1801–1876)	(1804–1881)
William Earl Dodge, Jr.	Daniel Willis James	*Family Withdrew*
(1832–1903)	(1832–1907)	*from Partnership*
Cleveland Hoadley Dodge	Arthur Curtiss James	*in 1878*
(1860–1926)	(1867–1941)	
Cleveland Earl Dodge	*No Children*	
(1888–1982)		
Cleveland Earl Dodge, Jr.		
(1922–)		

✎ ANSON GREENE
PHELPS *(1781–1853), pious
patriarch of the family
partnerships known as Phelps,
Dodge & Company in the United
States and Phelps, James &
Company in England, both of
which dated from January 1,
1834.*

to the Hartford area, where they alternated between periods of prosperity and poverty. William Earl Dodge was born in Connecticut's capital city on September 4, 1805.

It was at church and prayer meetings in Hartford that Anson and Olivia Phelps first became acquainted with David and Sarah Cleveland Dodge. The restless Dodges moved between Connecticut and New York City several times before finally settling in the growing metropolis. David Low Dodge was a man of forceful religious convictions that he imparted to his son, William, along with a position as a junior clerk in a dry goods store that two kindhearted Quakers operated on Pearl Street near Peck's Slip. The boy was thirteen years old.

In later years, from 1827 until 1834, William E. Dodge continued the wholesale dry goods store his father had established on Pearl Street a decade earlier. New York—by the 1830s already a city of some 120,000 inhabitants—was a robust mercantile center, and Pearl Street was to the nation's dry goods trade what nearby Wall Street became to finance. Business on the two streets was in turn closely linked to the fleet of tall ships that brought commerce from around the world to South Street wharves on the East River, an area once known as the street of ships.

During young Dodge's formative years on Pearl Street, religion remained uppermost in his life, and it was religion that attracted him to Melissa Phelps, to whom he proposed marriage in June 1828. After their marriage, the bride's father confided to his diary, "I trust the Lord has led by His Providence to this union and hope He will greatly bless them in all their journey through this wilderness." Religion, however, was

far more than a bond that united Anson G. Phelps and his sons-in-law and business partners. The stern tradition of English Puritans formed the core values that influenced major business decisions no less than the most trivial personal matters.

The pious partners were inclined to see the hand of God at work in every conceivable phase of human life, from birth to death, and in every part of their business, both successes and setbacks. Rigorous religious beliefs conditioned them to work hard, live frugally, take care of their families, deal honestly with customers and employees, and to support numerous philanthropic causes.

A later and far more cynical generation condemned the partners' personal interest in the well-being of their employees as paternalistic, but their solicitude, however it may be defined, was among the important legacies that Anson G. Phelps and William E. Dodge bequeathed to the enterprise that carried their names through the twentieth century. So central was religion to the way Phelps Dodge Corporation did business that, profiling it in the early 1930s, *Fortune* magazine headlined its story with the words "Presbyterian Copper." Though the founding partners never formally articulated a mission statement, commitment to church and family was certainly central to the way they did business.

Forging a Partnership

WHEN PHELPS, DODGE & Company was formed in 1834, the sons-in-law Dodge and James received a 20 percent share each, and

Phelps's son, Anson, Jr., received 12.5 percent. The older Phelps retained the remainder. In terms of organization, Dodge chose to deal mainly with the domestic side of the business, while James sailed for England to handle trade there under the name Phelps, James & Company. As for Anson, Jr., until marriage in 1845 helped boost his interest in the business, he remained a man of restless temperament who seemed to prefer relaxing in a small boat on the East River to the mind-numbing drudgery of sitting at a high desk to scribble numbers in a ledger.

Although Phelps Dodge's domestic sales increasingly relied on imported metals, the firm continued to handle an array of merchandise that encompassed everything from pottery and spittoons to moleskins and feathers. In some ways the incredible variety of goods it offered for sale caused the offices of Phelps, Dodge & Company to resemble an enormous general store. A typical load of trade goods traveled by sailing ship from Phelps Dodge's New York warehouse to port cities along the Atlantic and Gulf coasts or by steamboat along the great rivers that linked Pittsburgh with Saint Louis and the frontier beyond.

The growth of Phelps, Dodge & Company, which by the late 1840s ranked among the great mercantile houses in the United States, as well as Phelps's personal investments in banks and other enterprises, made the firm's senior partner very rich. That was no small feat in a young republic that did not even chronicle its first bona fide millionaire until the late 1830s or early 1840s. According to one list of prominent New York citizens in 1842, Phelps's wealth amounted to $1 million, and his son and Daniel James were worth

$400,000 each. Dodge's wealth was not inventoried, although a similar compilation made only a few years later in midcentury tallied his wealth at $500,000 and that of the senior Phelps at five times as much.

Growing wealth did not divert Anson G. Phelps from his long-established personal routine. He arose early and began each new day with prayer. After communing fervently with his God, he dressed in a high choker, ruffled shirt, and swallowtail coat, as was the fashion of his day. He would stir the smoldering fire back to life, prepare his own breakfast, and spend the next hour or so writing in his diary as he pondered the day's events, which for Phelps often included examples of God's providence. Every day but Sunday he would stroll to the Battery at the southern tip of Manhattan to see if one of his cargo-laden ships might be sailing through the Narrows. After surveying the bustle along the East River wharves, which in the mist of a sunlit morning resembled a romantic painting with towering masts and fine homes and warehouses contrasting with dark green hills in the background, he would walk over to his countinghouse to attend to the day's business.

During the 1830s the young nation prospered as never before, but its economic boom rested upon an extremely flimsy foundation. Inflation and speculation grew worrisome and then rampant. When the bubble finally burst in 1837 it caused hundreds of banks to close their doors. That froze millions of dollars in deposits and dried up credit that was the lifeblood of many a business. Factories shut down, and for the next five years hard times stalked the land. Amidst all the financial gloom,

DANIEL JAMES (1801–76). Based in Liverpool, James fretted about his father-in-law's "speculations" in land, which he feared would make it hard for him to pay the firm's bills in England. "I had rather hoe corn and dig potatoes than to always be in the state of anxiety and perplexity I have been in," he once fussed.

VIEW OF WALL STREET IN 1850.

The only buildings which remain unchanged are the Sub-Treasury (with flag), the U. S. Assay Office adjoining, and Trinity Church.

the Phelps family enjoyed a moment of public happiness when Caroline, after five years of mourning for her fiancé following the collapse of her father's big warehouse, married Josiah's older brother, James Stokes.

The general business depression caused Phelps to confide to his diary that "the wedding was to me anything but joyful, and every day since, the gloom has been growing thicker and thicker, and not a day passes but a number of respectable houses fall. . . . Never did such gloom before cover my prospects both here and in England." In Liverpool, Daniel James could not sell his partners' American cotton for a profit, and therefore he could not buy any tinplate for them to sell in the United States. For a while in the late spring and summer of 1837, Phelps, Dodge & Company actually had to suspend payments, and to James it appeared that "we are struggling for existence like a drowning man."[2]

The Phelps Dodge partnership almost collapsed beneath a burden of debts that caused

⌒ From the Desk of Anson Greene Phelps ⌒

COMPARED TO *Dr. James Douglas's large Victorian rolltop desk in Bisbee, the one from which Anson Greene Phelps regularly conducted business is plain and austere, much like the interior furnishings in the early countinghouses of Phelps & Peck and subsequently Phelps, Dodge & Company. Fashioned from mahogany, white pine, and tulip wood, Phelps's desk is now preserved at the South Street Seaport Museum in New York. It offers a rare surviving example of the type of office furniture used during the first half of the nineteenth century.*

Though the desk's full story is not known—the Stokes family donated it to the museum several generations after severing business ties with the Phelps Dodge partnership—it certainly functioned as the New York nerve center of both of Phelps's trading firms. In the typical countinghouse individual offices were rare; a myriad of transactions centered at various desks and counters in large open rooms. Typically, a clerk would appear at No. 19 or 21 Cliff Street with an order beginning "Please deliver bearer," followed by a list of the metals desired. The senior partners of a firm usually placed their desks at the far end of the main room, often in a raised area comparable to the quarterdeck of a ship, from which they oversaw all the activity.

The ship comparison should not be surprising because New York's merchandise trade, which by the 1840s consisted of more than four hundred businesses engaged in foreign commerce, was concentrated within a few blocks of brick countinghouses located near the southern end of Manhattan Island and adjacent to some sixty wharves that lined the East River. Many early business practices came from the maritime world.

Though overshadowed by modern high-rise buildings, the quaint South Street neighborhood of stone-paved streets and four- and five-story brick countinghouses stirs memories of days long ago when ships from port cities of every ocean discharged mountains of

⌒ THE COUNTINGHOUSE DESK OF *Anson G. Phelps, where Phelps Dodge originated in 1834. Later the desk was used by James Stokes and by Anson G. Phelps Stokes, his son. Several generations later, the Stokes family donated the desk to New York's South Street Seaport Museum.*

goods packed solidly in wooden boxes, barrels, and bales. Prosperous merchant princes like Anson Phelps were distinguished by their black broadcloth coats and tall beaver hats as they paced the narrow lanes.

Like a captain directing his shipboard crew, a New York merchant believed it was good business to keep his countinghouse clerks always occupied and well disciplined. If no customers needed help, the clerks might be busied simply by moving merchandise from one corner of the store to another. Like a ship's captain, too, a good merchant was expected to be solicitous of the welfare of his clerks even after work, and so this form of paternalism was ingrained in Phelps Dodge and other countinghouses from the beginning.

The duties of a clerk in a New York countinghouse varied with seniority. The most junior person, often a mere boy, might be responsible for drawing water or starting a fire to heat the countingroom early each morning. He might also have to deliver goods. Next he might be assigned to copy letters, and after he could do that without error or blot, he was promoted to making duplicates of letters to go by the packet ships. Next he was promoted to copying accounts, and then he was trusted with the responsible duty of making those accounts himself. It was customary to work up through the ranks, as William E. Dodge did. The office staff was all male; and instead of the clatter of typewriters there was simply the soft scratching of writing pens as clerks atop high stools hunched over correspondence, accounts, bills of lading, and ships' manifests. Because carbon paper had not yet been invented, this was slow work, and a clerk who had particularly fine penmanship had to make out duplicate copies by hand.

Wholesalers or jobbers like Phelps, Dodge & Company rarely sent salesmen to call on their customers. Instead, shopkeepers from distant parts of the United States and Canada regularly traveled to New York, often by canal boat and sailing packet, to bargain face-to-face with merchants for the goods they wanted to fill their shelves for the coming year—and to catch up on news and gossip. A merchant often hesitated to keep an empty chair beside his desk because that might encourage idle chatter with customers whose dawdling interrupted a busy day's work. One can only guess what business deals were transacted at Anson Phelps's desk or what influences shaped his view of the world from there.

(Adapted from Robert Greenhalgh Albion, *The Rise of New York Port [1815–60]* [New York: Charles Scribner's Sons, 1939].)

Phelps to admit to a friend that this was the most trying time of his life, except the day "when our store fell and killed seven persons." Fortunately, and as if to emphasize the continuing fairy-tale quality of early Phelps Dodge history, the knight who rescued the struggling firm was none other than Phelps's newest son-in-law, James Stokes. He provided his partners a sizable loan they needed to meet financial obligations.

Although exporting, importing, and general merchandising remained the major business of Phelps, Dodge & Company during the greater part of the nineteenth century, the firm's American partners were ever alert to promising new fields for investment. The Old Man, as Phelps was sometimes referred to out of his hearing, was inclined toward financial risk, and he encouraged his partners to invest in banking, brass manufacturing, coal and iron mines, railroads, and real estate, among numerous other business ventures. Back in the early 1830s, for example, the Old Man had invested in banking in New Jersey and obtained a controlling interest in the Union Bank at Dover. All of his many business activities eventually turned out well, no doubt much to the relief of his partners and sons-in-law, who were legally bound to pay off any debts incurred by the firm or by one of its partners. By the time of his death in 1853, Anson Phelps had become one of the largest property holders in the state of New York, and one of its richest.

Far more cautious than Phelps were his three sons-in-law, especially Daniel James of Phelps, James & Company in Liverpool. His primary assignment was to turn a profit on the exchange of cotton from the United States for tinplate

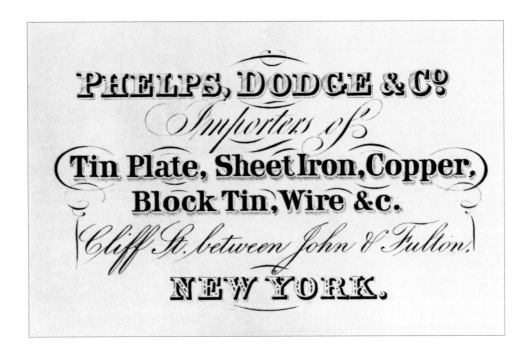

from England, and thus it dismayed him to see his New York partners spend money outside the firm's principal line of trade while he struggled to overcome an almost unceasing shortage of cash. James continually begged his New York partners to cease granting long-term credit to customers and, instead, to "keep as snug as possible!" In fact, James's protests had a salutary impact on the partner in "the middle," William E. Dodge, whose conversion to tight credit policy underscored the conservative character of Phelps Dodge, a trait that commentators still noted more than a century later.[3]

As American business gradually recovered following the panic of 1837, the Phelps Dodge partners in New York regained some of their former enthusiasm for investments, albeit conservative ones. Phelps was drawn to metals, among numerous other business ventures, and Dodge to railroads and lumber. Dodge became

ONE OF THE ORNATE BUSINESS *cards issued by Phelps, Dodge & Company emphasized the role that metals played in the firm's import-export trade. Phelps, Dodge & Company by 1874 met fully one-third of America's demand for tinplate by annually importing about fourteen million boxes of the metal.*

interested in the nation's vast timber resources, particularly the immense stands of virgin timber in northern Pennsylvania, and involved the partnership in a tentative way in the mid-1830s when he accepted a thousand acres of forest in rural Tioga County in partial payment of a debt. Pennsylvania, named for William Penn's woods, was once so densely forested that it was claimed a squirrel could easily cross the state on the closely entwined branches of white oak, chestnut, hickory, maple, sycamore, walnut, wild ash, wild plum, and white pine trees, the latter especially favored for tall ship masts and for framing houses.

A seemingly limitless expanse of virgin white pine—straight-grained, easily worked, and light but very strong—contributed to a building boom in early America and a bonanza for all who harvested and sold the timber. Dodge saw a chance to gain more of Pennsylvania's valuable timberland, both for himself and for his Phelps Dodge partners. The purchases, which included several sawmills, formed what was for a time the largest of many Phelps Dodge subsidiary enterprises.

After some thirty-five years, when their timber holdings in Pennsylvania were at last fully logged, the Phelps Dodge partners sold the land to other investors and searched elsewhere for other sources of timber, or "green gold." It was primarily Dodge who developed vast new holdings that included whole forests in Canada and the Great Lakes states of Michigan and Wisconsin before the Civil War and the Deep South states of Georgia and Texas afterwards. For several decades, investments in timber contributed approximately 10 percent a year to the profits of Phelps, Dodge & Company.

"Good Copper" and the Building of Young America

ESSENTIAL AS COPPER is to the modern Phelps Dodge Corporation, red metal early touched the life of Anson G. Phelps in curious ways that foreshadowed future developments and added yet another improbable dimension to his life story. Members of the Phelps clan sailed to New England in 1630—their family name was perhaps spelled Ffelpes or Guelph back in Britain—and following several years of frequent relocations, they settled in the village of Simsbury, about twelve miles northwest of Hartford, Connecticut, on the far western edge of the frontier. Scouts for the British navy who reconnoitered the area in 1643 reported fine timber and pitch, both needed for shipbuilding, and good pasturage for settlers.

The first landscape young Anson Phelps knew well was the verdant Farmington River valley and the gentle slopes of the low mountains that defined the horizon east of Simsbury. On the far northeastern edge of the village, near what is now the town of East Granby, stood a hill once known as Copper Mountain because of its mineral outcroppings. On it was located the first copper mine in the English colonies of North America and the first smelter for reduction of ores. Though tiny by today's standards, Connecticut's workshop in the wilderness yielded its first red metal in 1707 as the Simsbury Copper Mines Company. A part of its hard-won profits underwrote the cost of a local schoolmaster and provided support for Yale College.

Because strict British laws forbade smelting

copper ore outside the mother country, the colonists, who included some of Phelps's own ancestors, had to work in secret to make their mine productive. In 1737, for example, the proprietors ran afoul of crown officials for minting "Granby coppers," and the blacksmith responsible for the threepence pieces was quickly stopped. Although he struck no more, his original coins circulated through the American colonies for years. On one side was a stag and the motto "Value Me as You Please," and on the other were three sledgehammers surmounted by crowns and the motto "I Am Good Copper."[4]

During the Revolutionary War the onetime copper mine served Connecticut and the thirteen rebel colonies as Newgate Prison. General George Washington sent captured British soldiers there as early as 1775. One of Phelps's relatives supplied food to the inmates, and his father worked at the prison briefly before enlisting to fight with the rebel colonists.

How much of Newgate's history young Phelps knew is not certain. Both it and other small copper mines in New England continued to operate intermittently well into the twentieth century, but those often marginal mine properties apparently held little interest for him. However, as a young man he was quick to grasp moneymaking opportunities in a related field, Connecticut's emerging brass industry, which soon ranked first in the nation. Brass is an alloy of copper and zinc; bronze is an alloy mainly of copper and tin (though modern bronze alloys may contain other metals as well), but England had successfully prevented her American colonists from learning the "mysteries" of making brass. Even after the Revolutionary War, the new

United States still had to import brass from Britain. That monopoly could not last forever.

A group of enterprising Connecticut manufacturers, who were the new nation's leaders in fabricating tinware, combined Yankee ingenuity with timely knowledge of British trade secrets to turn brass into a promising new field of endeavor. During the first half of the nineteenth century they transformed the Naugatuck Valley into the heart of the nation's tin, copper, and brass industry. By the 1880s that small river valley produced 85 percent of all rolled brass and brass ware in the United States. Many of the industry's pioneers located their factories in Waterbury, where local businessmen were among Phelps's earliest buyers of English metal products; and they remained valued customers of the later Phelps, Dodge & Company. All together the manufactories of the Naugatuck Valley consumed about two thousand tons of copper annually by 1855.

Phelps himself joined with one of his customers in the 1830s to erect a factory in the Brass Valley, as that part of Connecticut became popularly known, to hammer out copper kettles by machine instead of casting them in the usual manner. This business venture, alas, remained unsuccessful until artisans developed new, more malleable alloys that kept copper plates from cracking under repeated hammer blows. The impatient Phelps meanwhile built factories of his own to produce a wide variety of copper and brass goods, including those favored for naval vessels.

As a result of his long-standing involvement in the Brass Valley, in 1844 Phelps established a village called Ansonia around a new rolling mill and there organized the Ansonia Brass and Battery Company to hammer out a steady stream of copper kettles. To power the overhead shafts in the mill and turn the broad belts that drove his clanking machinery, Phelps built a large reservoir and water delivery system. Each new wheel his water power set in motion meant larger-scale fabrication of metal products. True to his Puritan heritage, Phelps, the builder of manufactories and waterworks, also donated money to erect the First Congregational Church of Ansonia.

From Phelps's factories and those of his competitors who grasped an opportunity in the emerging copper city came a swelling stream of industrial and consumer goods that included brass sheet metal, brass or bronze rods, brass wire for the manufacture of pins, hooks, and eyes, brazed and seamless tubing, rivets, copper kettles, buttons, lamps, door railings, clocks and gongs, hinges, brass bedsteads, gas and kerosene chandeliers, polished brass library lamps, and many thousands of miles of soft copper wire for the fifty or more companies that were formed after the first successful demonstration of the telegraph in 1844. One increasingly popular item was brass tubing used to carry gas for indoor

~ THE ANSONIA BRASS AND *Copper Company located several mills in the area around Ansonia. In this image the artist took the liberty of combining them into a single landscape of production. Anson Phelps imported Irish labor to construct one of the large reservoirs needed to provide waterpower to his mills. To operate his factories, he relied on English craftsmen, machinery, and methods.*

illumination. Later, beginning in the 1880s, application of electric power to home and industry created another huge market for the Naugatuck Valley's copper and brass manufacturers.

Renamed the Ansonia Brass and Copper Company in 1869, and popularly called the A. B. and C. Co., the Connecticut firm consumed more copper annually than any other mill complex of its type in the United States. At the time the company boasted of five enormous factories—including brass, copper, and wire mills—located in the vicinity of Ansonia that employed as many as fifteen hundred hands. It met a total payroll of $900,000 per year. Forty-eight fiery furnaces melted twenty thousand pounds of metal each day, and the thirteen smoking chimneys featured prominently in products brochures were seen as a sure sign of prosperity.

The town of Ansonia itself—where the official seal still includes a likeness of Anson Phelps—incorporated in 1889 and matured into a modern city of twenty thousand inhabitants. Because copper from small mines in New England and New Jersey grew too scarce to meet the Brass Valley's expanding needs, a steady stream of copper arrived from Wales, and much of its passed through the New York warehouses of Phelps, Dodge & Company. For years the main offices of the Ansonia Brass and Copper Company and its salesrooms were located at 19 and 21 Cliff Street, the same address as Phelps Dodge. James Stokes and then William E. Dodge, Jr., served as president. Ansonia Brass & Copper remained an integral part of Phelps

Dodge for almost seventy years until the partners sold it to the American Brass Company, which emerged in 1899 as the largest fabricator of brass products in the world. Ironically, Phelps Dodge's longtime rival, Anaconda Copper, purchased American Brass in 1922 as part of its effort to integrate vertically from mine to consumer.

Connecticut's Brass Valley added to its fame when it became the center of clock manufacturing in the United States. It was there during the panic of 1837 that Chauncey Jerome invented a cheap brass clock with mass-produced metal gears to replace the wooden ones formerly made by time-consuming hand labor and which were influenced by dampness, a big drawback in the humid South. The metal timepieces quickly became so popular that Dodge reported, "Yankee clocks are ticking all over England." Rather than miss out on a new market for its imported metal, the Phelps Dodge partners joined in the manufacture of clocks too.

Their enterprise, formed in 1854 as a spin-off of Ansonia Brass and Copper, fabricated everything from large clocks for courthouse towers to all kinds of regulator clocks for home and office. Other items for the home ranged from cheap clocks for the kitchen to the "most expensive and artistic timekeepers, encased in onyx or gilded bronze." Many models featured angels, cupids, and draped ladies in the ornate style popular at the time. The Ansonia Clock Company's offices were located within those of Phelps, Dodge & Company, and among its six directors were William E. Dodge, Jr., and D. Willis James.[5]

If watches, and especially public clocks with

ANSONIA CLOCK BUILDING
destroyed by Fire Oct. 27th 1880.

DONALDSON BROTHERS, FIVE POINTS, N

their large and clearly visible dials and pendulums, represented the beating heart of the new industrial age, then railroads surely symbolized its veins and arteries, and an expanding network of telegraph wires formed its nervous system. After America's first telegraph line linked Baltimore with the United States Supreme Court chambers in Washington, start-up companies worked rapidly to stretch a network of iron and copper nerves across the nation. To the Phelps Dodge partners, railroad and telegraph companies presented both a market for their metal products and an intriguing investment opportunity. The moneymaking possibilities spelled out in dots and dashes of Morse code were just as alluring in the

1840s and 1850s as the bits and bytes of computer networks today.

Of all the partners, the elder Dodge had the strongest faith in railroads. When he took over his father's dry goods store on Pearl Street in 1827, not a single mile of regular railroad line was in operation in the United States; but by 1840 the nation recorded twenty-eight hundred miles, and by 1880 nearly one hundred thousand miles. Eventually the nation's railroad network would be more than double that size. Dodge's impassioned belief that railroads would in time surpass the nation's numerous canals and turnpikes in importance led to his being elected a director of the struggling New York & Erie

Railroad. In 1844, he played a leading role in a successful campaign to raise $3 million needed by the railroad to extend its tracks west from the Hudson River to the shores of Lake Erie. Seven years later, in 1851, he welcomed President Millard Fillmore, Senator Daniel Webster, and the other distinguished national guests who gathered to celebrate completion of New York's valuable new railroad link.

The Erie pioneered techniques of management that eventually became commonplace for big businesses, including Phelps, Dodge & Company after it outgrew the countinghouse and its simple management hierarchy derived from shipboard life. Alas, the Erie was also an American pioneer in one of the ugly aspects of the rise of big business. That was after a strange new man, Daniel Drew, joined its board of directors in 1853. He was the first of a breed of ruthless speculators who did not hesitate to bilk investors to enrich themselves. So brazenly did he manipulate Erie stock that the railroad became synonymous with chicanery and deceit.

For four years, Drew and Dodge sat on the Erie board together—Dodge striving to keep the line out of trouble, and the ferretlike Drew stirring up ever greater trouble that would weaken and ultimately bankrupt the railroad. Dodge abruptly resigned in 1857 after a split with his fellow directors. Outrageous as Drew's financial behavior was, Dodge did not leave the railroad because of that but rather because the Erie's management stubbornly insisted on running its trains on Sunday.

Dodge's Puritan New England heritage led him to oppose vigorously any plan the Erie had to operate its trains on Sunday—and subsequent events only bolstered his beliefs. Scarcely two years after Dodge sold his Erie stock at a good price, the railroad failed. The same scenario happened on the New Jersey Central Railroad, on the board of which Dodge served for fifteen years. Before he resigned he protested to his fellow directors that they should place a flag on all their locomotives bearing the legend "We break God's Law for a dividend." He disposed of his stock for $116 to $118 a share. Two years later, the New Jersey Central went bankrupt, and its stock sold for ten cents a share. Until his dying day, Dodge continued to protest "Sunday Railway Desecration."[6]

Both Phelps and Dodge bankrolled railroad expansion in several parts of the United States, including the Delaware, Lackawanna, & Western line close to home and the Houston & Texas Central out West. After the Civil War, Dodge and other New York investors took over the Houston & Texas Central Railroad and thereby gained some three million acres of Texas timberland. As valuable as that was, Dodge took special pride in the fact that as long as he was president of the railroad none of its trains operated on Sunday. Despite their numerous investments in pioneer railroad projects, including the Union Pacific when it formed the eastern part of the nation's first transcontinental railroad, by the time of Dodge's death in 1883, Phelps, Dodge & Company had virtually retired from the railroad field, or so it seemed at the moment.

The Next Generation

DURING THE FIRM'S new era of prosperity and expansion, the health of Anson G. Phelps began to fail. In 1853, at the age of seventy-two, he confessed in his diary that he was growing tired. That summer he and his wife traveled to Europe, where he suffered a stroke. He returned home that autumn in poor health. On the final day of November, surrounded by his wife and children, Anson Greene Phelps died. The orphan lad who trudged the dusty roads of Connecticut carrying all his earthly possessions in one small bundle had become one of the nation's foremost merchants and businessmen.

A decade and a half earlier, during the grim days of the panic of 1837, when survival of his firm seemed anything but certain, Phelps had confided in his diary that if God would spare his life and enable him to pay off his just debts he would never again call anything his own. His philanthropy during the remainder of his life and at his death was in keeping with that pledge. Over the years, Phelps made sizable contributions to the American Bible Society, the American Board of Commissioners for Foreign Missions, the American Home Missionary Society, the New York Institute for the Education of the Blind, and many other religious and benevolent organizations. The total of these and other bequests came to almost $600,000, the largest sum left for charitable purposes up to that time by any benefactor living in New York.

After the death of its founder, the thirty-year partnership of Phelps, Dodge & Company added two junior partners, Daniel Willis James and William Earl Dodge, Jr., sons of the original partners in Great Britain and the United States. All shares of the firm remained in the hands of the Stokes, James, Phelps, and Dodge families, although Anson G. Phelps, Jr., survived his father by only five years, dying in 1858 at the age of thirty-nine. Because of poor health, young Phelps found his chief interests in travel, music, and literature, not in business; but like his father, he gave liberally to support various causes and institutions, and chief among these was the Union Theological Seminary of New York, which sought to better the condition of African Americans. After 1858, the Phelps family name may have disappeared from the roster of active partners, but Phelps, Dodge & Company remained a close-knit family enterprise.

Upon the death of Anson G. Phelps, Jr., the surviving Phelps Dodge partners absorbed his interests in the company and also fell heir to most of the estate of his mother, Olivia, who died at almost the same time. Those two deaths required formation of yet another partnership, effected on January 1, 1859, to maintain a family firm whose owners now consisted of Anson G. Phelps's three sons-in-law, Daniel James, William E. Dodge, and James Stokes; and his two grandsons, William E. Dodge, Jr., and D. Willis James. Those two cousins, representing the partnership's third generation, guided Phelps Dodge when it explored the new frontiers of copper mining in Arizona after 1881, but that was a future still undreamed on the eve of the American Civil War.

At the beginning of the 1860s, when the bonds of nationhood were strained and soon would snap, Phelps, Dodge & Company continued with the largest business in metals of any mercantile house in the United States, and possibly the world. Its Ansonia Brass and Copper Company mills manufactured a variety of iron, tin, and brass products and used more copper than any other enterprise in America.

Just as the panic of 1837 severely tested the original firm, so too the American Civil War that began in 1861 encouraged Phelps, Dodge & Company to evolve in new directions in order to survive. The firm won a $5 million contract to sell sheet iron to the Russian government, and it arranged for American sewing machines valued at $1 million to be sold to a British customer. The partners tapped into new markets closer to home, such as when they sold imported English wire to builders of the first transcontinental telegraph line, completed to California in October 1861.

The Civil War encouraged Americans to seek out domestic sources of raw materials to supply their mills and factories. The Phelps Dodge partners acquired an iron mine in New Jersey, gained interests in steel mills and iron foundries in Illinois, and briefly entered the tin-mining business in California. The expansion cost money. Thus in 1864, at about the time General Ulysses S. Grant launched his final campaign to destroy the Confederate States of America, Phelps Dodge found it necessary to double its capital to $3 million.

America's economic boom did not end with the Civil War. Just two years later, in 1867, the brass and copper complex at Ansonia was handling more than a hundred tons of metal a month. A quarter century of prosperity, interrupted only by a brief period of hard times during the mid-1870s, caused the partners to pause periodically to increase their firm's capital and to admit new family members to the circle of owners. Individual partners maintained their large and profitable railroad, lumber, coal, and iron interests; and as in earlier years there was no clear line of demarcation between the business activities of partners acting as individuals and those of partners acting as representatives of the company, for usually the two were inseparable.

More and more, the third generation of family members, Daniel Willis James and William Earl Dodge, Jr., oversaw day-to-day activity at Phelps, Dodge & Company. Both men had joined the firm in 1854, shortly after their grandfather died, and worked side by side as junior clerks at the high desks of the family countinghouse. The senior William E. Dodge assumed the role of elder statesman. During a lifetime of activity, he had participated in many different business enterprises and given liberally of his time to serve on various civic and charitable boards. At the time of the Civil War he had raised funds for the benefit of Union regiments then preparing for combat in the South and to aid English factory workers thrown out of work because of the North's naval blockade of Southern cotton. He had opposed slavery, rum, war, and Tammany Hall, the corrupt New York political machine that in 1864 tried unsuccessfully to cheat him out of his election to Congress, where he was noted mainly for collecting pledges of total abstinence from alcohol from some fifty senators and congressmen.

Now in his twilight years, when he had

THE STATUE OF William E. Dodge surveys New Yorkers relaxing in Bryant Park.

become less acquisitive and even more pious, Dodge wanted to devote his remaining time to spending money on philanthropy. The list of worthy causes and institutions he supported was lengthy. His benefactions to education alone were large and numerous, and they included literally scores of schools, colleges, universities, and seminaries throughout the United States and abroad. Among these was helping to found the Syrian Protestant College at Beirut (later known as the American University of Beirut), which became distinguished as a center of Western learning and culture in the Arab world, or Holy Land as it was called in Dodge's time. He influenced the formation of the Young Men's Christian Association (YMCA). In addition to his support of New York's Metropolitan Museum of Art and the American Museum of Natural History, Dodge was widely recognized as a leading layman of the Presbyterian church.

Given his charitable inclinations, it should not surprise anyone that Dodge assumed personal responsibility for the religious, moral, and physical well-being of his firm's many employees; and not without reason did Phelps Dodge towns become known for their churches, schools, and comfortable homes for workers. It was said that Dodge gave away an average of a thousand dollars a day during his final years, and that was during an era when a thousand dollars was real money, or about twice the average annual income of the typical American worker.

Dodge's brother-in-law and business partner Daniel James, who had for decades presided over the company's affairs from Liverpool, died in 1876, and two years later, a second brother-in-law, James Stokes, and his two sons withdrew from the family firm to enter the investment banking business on Wall Street. Once again the business had to be reorganized, for the number of partners had decreased to three in 1879, and one of these, the aging William E. Dodge, no longer participated actively in company affairs. One day as the elder Dodge sat in the offices of the firm he had helped to found back in 1834 he carefully studied the portraits of the deceased members. He mused that he was almost the only old partner left. His face lighted up as he added, "But I am ready!" Dodge died on February 4, 1883, having engaged in charitable and church activities until just two days before his death. He was seventy-eight years old.

In Bryant Park today, located just west of Fifth Avenue behind the New York Public Library, stands a sizable statue of William E. Dodge keeping eternal watch over the city he loved and whose financial and civic well-being he did so much to cultivate. Incidentally, keeping him company in that popular green oasis in the midst of Manhattan's towering skyscrapers are three literary figures: William Cullen Bryant, Johann Wolfgang von Goethe, and Gertrude Stein. That odd juxtaposition, perhaps wholly unplanned, seems to symbolize the continuing commitment of Phelps Dodge Corporation to the arts, which its modern foundation liberally subsidizes.

Metals Matter

IN THE TIGHTLY woven tapestry that records details of Phelps Dodge's early history, the strands that stand out most prominently are the metal ones. Besides importing millions of boxes of tinplate and other metals from England, the partnership was actively involved in the copper and brass mills of the Naugatuck Valley. Even earlier there was the onetime copper mine near Anson Phelps's ancestral home in Simsbury, which seemed to foreshadow his future as a dealer in metals. Finally, there were the early metal mines that Phelps, Dodge & Company and the individual partners acquired.

About a decade before the Civil War, when Anson Phelps was engaged in the large-scale manufacture of brass products, he learned of amazing deposits of copper on the remote Michigan shore of Lake Superior. His memories of the War of 1812 and how its embargoes choked off metal imported from Europe made him conscious of the need for good sources of American copper to guarantee a steady supply for his Connecticut brass manufactories. That was why Phelps joined with other investors in 1849 to form the Waterbury and Detroit Copper Company. The new firm, financed largely with money from the Naugatuck Valley, supplied a large portion of the 650 tons of copper produced in the United States the following year. About the same time, Dodge became a major stockholder in a Lake Superior copper outfit called the Minesota Mining Company (officially spelled with only one *n*). Between 1852 and 1856,

PHELPS, DODGE & Co.,

Nos. 11 to 23 Cliff St.,

Bet. John & Fulton,   **NEW YORK,**

IMPORTERS AND DEALERS IN

TIN & ROOFING PLATES

Of all Sizes and Kinds,

PIG TIN,

Russia Sheet Iron, Charcoal and Common Sheet Iron,

LEAD, SHEET ZINC,

COPPER, SPELTER,

Solder, Antimony, etc.

MANUFACTURERS OF

COPPER, BRASS & WIRE.

the Michigan-based enterprise doubled its shareholders' money, and it remained extremely profitable for years to come.

To its copper investments in Michigan, Phelps, Dodge & Company added the Temescal tin mine in California (not far from the present-day city of Riverside). During the Civil War, young Anson Stokes traveled to the West Coast to sell the firm's tin mine, which disappointed its investors because the ore proved so sparse and scattered; but while on the Pacific Slope he grew so excited by copper claims along the lower Colorado River that he urged the Phelps Dodge partners to sell off their Lake Superior investments and buy copper claims in the Southwest.

The partners did eventually dispose of their Michigan mine holdings, but the time was not yet right for them to look West to the emerging copper mines of Arizona and New Mexico. When they finally did in 1881, a religious person might be tempted to describe the improbable series of events that followed as a case of "predestination." Perhaps, but investment in Arizona copper was only one logical outcome of a story that dated back to that once-upon-a-time event in 1815 when young Anson G. Phelps first advertised that

---

A PHELPS, DODGE & COMPANY *advertisement in the 1873 directory for New York City. During the post–Civil War decade the partnership supplied tinplate for the cans used to package kerosene, or coal oil, that the Cleveland merchant John D. Rockefeller sold to light American homes. Its allied Ansonia Brass and Copper Company manufactured a variety of ornate lamps that householders used to burn Standard Oil illuminants.*

Entered according to act of Congress in the year 1874 by Geo Degen, in the Office of the Librarian of Congress at Washington, D.C.

NEW YORK,

he had sixty-two boxes of imported tinplate available for sale at his new warehouse at 29 Burling Slip.

When William Church journeyed to New York sixty-six years later to discuss his Arizona properties with Phelps, Dodge & Company, few if any firms in the United States understood the metals business better—and especially copper. To be sure, the partners' loan to Church involved risk, but by 1881 they had learned much about the risks of metal mining; and because they earned steady profits from several large businesses in New York, Connecticut, and elsewhere, their investing several thousand dollars in Arizona copper did not place Phelps, Dodge & Company in any financial jeopardy. In fact, at least two more decades elapsed before most observers regarded the partners' modest investments in the mines of Morenci and Bisbee as the major turning point in Phelps Dodge history.

Chapter Three

FORGING *the*
COPPER KINGDOM

I
N 1882, AND little more than a year after the initial visits of William Church and James Douglas awakened Phelps, Dodge & Company to its destiny in Arizona copper, gangs of laborers employed by Thomas Edison started digging up cobblestone streets in the neighborhood surrounding the partnership's New York offices to lay fifteen miles of underground electric conduits. Clerks heading to and from work no doubt understood that the acclaimed genius of the incandescent light bulb expected to install his new type of artificial illumination in a few dozen firms in the financial district, among them J. P. Morgan's bank and the *New York Times*, and that only a few blocks away at 255–57 Pearl Street, the Edison Electric Illuminating Company struggled to complete the world's first central power plant.

On Pearl Street, just a few doors down from Edison's unfinished facility, young William E. Dodge had once operated his dry goods store. But that was history. On Pearl Street you could see the future too. Whether Phelps Dodge employees fully realized it or not, whenever they took time to watch Edison's progress along New York's lower East Side, they were really glimpsing the dawn of the

---

~ THIS MID-1880S IMAGE *looking up Broadway from Cortlandt Street in New York City illustrates the maze of overhead wires that caused Edison to place his electric lines underground. In time, all of the city's overhead wires would be relocated under the streets.*

electrical age. In the coming years, world electrification would alter the fortunes of Phelps Dodge and radically transform the enterprise, not just in terms of the increased amounts of copper it mined to produce electric conduits but also in its quest for better forms of wire and cable insulation. Phelps Dodge first joined the search in 1930 when it extended its reach from mine to market, and it continues today at Phelps Dodge Magnet Wire.

Back in 1882, when the work of sheathing bare electric lines was still more art than science, a droopy canopy of poorly insulated wires of all types hung above New York's narrow streets to carry telegraph and telephone messages, connect burglar alarms and stock tickers, and supply power to recently installed arc lights that brightened formerly dark corners. Because there were already so many overhead wires, notably the high-voltage lines that powered arc lights, it was hazardous to both utility linemen and pedestrians below for Edison to string any more. That was the primary reason he buried his thick copper mains underground like gas and water conduits, an innovative step for sure, but one that raised anew the nagging questions of how best to insulate them.

Bare telegraph wires had typically been attached to glass insulators, but their voltage was also low. The question of how to transmit large amounts of electric power, some of it at comparatively high voltages, had not yet been answered by cheap and effective insulation, especially in an underground installation. But

LAYING THE EDISON ELECTRIC MAINS –THE SERVICE BOXES AND EXPANSION JOINTS.
NEW ENTERPRISES IN NEW YORK CITY.

A WORKMAN LAYING *Edison's electric mains beneath the streets of New York's lower East Side. This image appeared in* Scientific American *in November 1881.*

ELECTRICITY, ALONG WITH *laughing gas, was once a compelling form of popular entertainment, as this advertisement from the early 1840s suggests. Even after inventors established electricity's many sensible uses, it remained a source of merriment, such as when fashionable women of the early 1880s, most notably Mrs. Cornelius Vanderbilt, wore evening dresses sequined with dozens of tiny electric bulbs that sparkled when they greeted their guests.*

Edison, with his usual can-do optimism, simply combined available materials in a way he thought ought to work. His employees wrapped two power conduits in Manila hemp and then inserted them into twenty-foot lengths of cast-iron pipe, which they packed in turn with a smelly homemade mixture of refined Trinidad asphalt boiled in oxidized linseed oil with paraffin and a touch of beeswax. The strange concoction hardened as it cooled.

Edison predicted that his insulated copper conduits would last more than half a century. For the most part he was right, although one short section developed air holes in the asphalt mixture that permitted current to leak out and energize a wet spot on the pavement of Fulton Street above. This caused plodding draft horses suddenly to rear up and spill their loads, much to the amusement of bystanders in the crowded market area.

Pearl Street had declined in prestige since the early decades of the nineteenth century when it was home to numerous fine taverns frequented by fashionable guests. But it was close enough to Wall Street that Edison hoped his lighting system would have a big impact in high financial circles. When all was ready, on the afternoon of September 4, 1882, he put on his Prince Albert coat, white cravat, starched shirtfront, and high-crowned derby. Together with a few trusted lieutenants he strode confidently into the somber banking offices of Drexel, Morgan and Company nearby at 23 Wall Street. Here Edison personally switched on the first lights, a dazzling array of 106 incandescent bulbs.

The display of "bottled sunlight" so

impressed J. P. Morgan that he bankrolled Edison enterprises that later combined as General Electric Company, which together with Westinghouse, another youthful giant of the industry, sold equipment to local utilities emerging all over the United States. Back on Pearl Street that historic September evening, commuters heading toward the waterfront and ferry connection to Brooklyn marveled at the brightly lit building interiors.

When the decade of the 1880s dawned, manufacturers had long recognized that copper was an excellent conductor of electricity. It was nonsparking and nonmagnetic; and it was so ductile that wires could easily be drawn from it. For nearly four decades, copper wires had transmitted the clicking dots and dashes of telegraph messages; and after Alexander Graham Bell demonstrated the first practical telephone at the Philadelphia Centennial Exposition of 1876, they carried the first scratchy voice messages too. But until Edison's plant on Pearl Street began operating, the market for copper to transmit electricity for light and power remained small.

Electric technology in 1882, like computer technology a century later, was still so new that no one could possibly imagine all its applications. For most people it was inconceivable that electricity would someday power elevators, mine trams, streetcars, looms, pumps, and who knew what else. Even Edison himself originally thought mainly of using it to light the incandescent bulbs he invented three years earlier.

When the Pearl Street station first provided electricity to light five thousand of Edison's bulbs, the total amount of copper required to wire and equip the relatively small installation totaled

128,793 pounds. Multiply that figure by the several thousand electric utilities that soon sprang up across urban America, and also in many cities and towns of other nations ranging from Russia to Australia, and no one will wonder why the electric industry created more demand for copper during the 1880s than had existed in any previous decade in world history. Production of copper in the United States totaled about seventy-two million pounds in 1881. Thirty years later domestic output skyrocketed to an amount more than a billion pounds *greater* than that in 1881, an increase of 1,500 percent!

Ironically, before the Pearl Street works rumbled to life, Edison's tireless critics had complained that there was "not enough copper in

the world" to permit him to extend his power system outside the immediate neighborhood, much less serve all of New York City. Apparently none of his detractors realized that Manhattan's main metal-importing firm, Phelps, Dodge & Company, had quietly begun to invest in Arizona copper properties. In future years the mines of Bisbee and Morenci would contribute to a flood of red metal available for all kinds of uses, most notably to foster electrification, the wonder technology that around the world became synonymous with civilization. It was the dawn of a new day for an old metal and a time of unprecedented opportunity for Phelps Dodge.

Looking West to the Future

By the late 1870s and even before the age of electricity, it was clear that the present and future prosperity of Phelps, Dodge & Company depended on its various metals businesses. Investments by the firm as well as by its individual partners in timber, real estate, and

This old woodcut provides an artist's perspective on underground mining in sixteenth-century Europe. It appeared in De Re Metallica, a mining and smelting bible published in 1556 by Georgius Agricola. Earlier, in the days of the Roman Empire, much copper came from the island of Cyprus, which in Latin was called Cyprium and from which is derived the English word copper. Romans identified copper with the beauty of Venus, both the goddess and the planet; and alchemists of the Middle Ages always represented copper by the astrological symbol for Venus, a circle with a cross attached below. That is the same symbol biologists now use to represent females, but in ancient times it equated copper with enduring life.

"Out of Whose Hills I Can Dig Copper"

WHY SHOULD MALACHITE, *a bright green mineral, yield a yellowish-red metal called copper? The first person to make that connection is lost to history; nonetheless, during the past five thousand years, humans on several different continents have put copper to work in many different ways. Along with iron and aluminum it easily ranks among civilization's most useful metals.*

Ancient Egyptians hammered copper into sheets out of which they fashioned ornaments and tools. In The Odyssey Homer wrote of a Greek ship setting sail for Cyprus carrying iron to barter for copper. For the Children of Israel in biblical times, the Promised Land was a place not only where milk and honey flowed, but also where "the rocks are iron and you can dig copper out of the hills" (Deuteronomy 8:9). With the Romans came an era of even more intensive copper usage. It soon appeared in everything from works of engineering and weapons of war to personal adornment and domestic articles—and its use spread wherever Rome's legions marched. From the Middle Ages to the late 1700s, European artists preferred thin sheets of copper to canvas and other materials for their oil paintings. Its smooth surface offered little resistance to brush strokes in detailed compositions.

The copper industry of the United States dates from 1664 when a foundry in Lynn, Massachusetts, manufactured copper and brass kettles. By the year 1725 a copper factory operated in Philadelphia, and in 1759 a plant in Waterbury, Connecticut, produced shiny knee and shoe buckles. By the time of the American Revolution, shipwrights had learned to sheath the lower hulls of wooden sailing vessels in hammered sheets of copper to protect the timbers against destructive shipworms common in warm seas, and builders roofed many homes as well as public buildings with shiny sheets of copper. The versatile metal eventually appeared in everything from nails, wires, seamless pipes, boilers, stills, downspouts, and gutters to engraving plates, lightning rods, writing pens, and cookware.

The rise of large-scale copper mining in the United States, as distinct from early fabrication, dates from the 1840s. For centuries,

railroads were all important, but the house of Phelps Dodge still rested firmly on a foundation of metal—and on imported tinplate in particular.

The popular packaging material produced a steady stream of profits for the partnership until the 1890s, when a meddlesome Congress imposed a protective tariff on imported tinplate to boost America's still young but increasingly robust iron and steel industry. Fortunately, during the closing two decades of the nineteenth century the nation's copper and brass manufactories grew rapidly too, and their long-term prosperity helped Phelps Dodge offset revenues lost as

metal imports dwindled. In the early 1880s, and well before the protective tariff ruined its trade in imported tinplate, Phelps, Dodge & Company wrestled with questions of how its brass and copper works in Ansonia, Connecticut, could most effectively capitalize on increased popular demand for its products.

Should the partners build a smelter on land they owned on Long Island Sound? That way the West's growing number of mines would not need to ship their mostly unprocessed copper across the Atlantic to Wales for smelting prior to final purification, or refining, back in the United

Indians had dug copper on Michigan's remote Keweenaw Peninsula in Lake Superior, but not until the 1840s did the outside world understand just how huge those deposits were. The Michigan copper rush from 1844 to 1847 was one of America's first great mining booms. By 1869 the Calumet and Hecla Mining Company had emerged as the world's single largest producer of copper, with an annual output of some twelve million pounds.

Major companies on the Keweenaw Peninsula—principally Calumet and Hecla and the Copper Range—dominated production in the United States until the 1870s and 1880s when prospectors unearthed vast deposits of copper in the northern Rocky Mountains. Montana emerged as America's leading producer of red metal by the end of the nineteenth century, a title it claimed from Michigan in 1887 and yielded to Arizona in 1910. Michigan nonetheless continued to be an important, though declining, producer until well after the Second World War.

Arizona remained the nation's number one copper state for most of the twentieth century—often outproducing all other states combined—although Utah, Nevada, and New Mexico held important ranks in America's copper kingdom. Unlike Montana, where deposits of copper were concentrated in the Butte area, Arizona had ore bodies scattered in a wide belt encompassing mountain, mesa, and range, often in a remote frontier setting.

The first mining company in the United States to produce more than fifty thousand tons of copper a year was Anaconda in Montana, and except for 1905–1907, when Calumet and Hecla again took the lead, and 1908–1909 when the Copper Queen surged ahead, Anaconda maintained its position as the greatest copper producer until 1920, and Butte could claim to be the world's greatest mining camp. Bisbee often ran a close second, and by 1900 it along with Clifton-Morenci, Globe, and Jerome dominated copper production in Arizona Territory.

States. But before proceeding, they needed expert counsel. That was when William E. Dodge, Jr., and D. Willis James, who had originally sought only technical advice from James Douglas, hired him to evaluate William Church's property in far-off Arizona Territory. As a result of that initial consultation, the partners increasingly looked to the West to anticipate their business future.

Just as serendipity had led to his first assignment for Phelps, Dodge & Company in early 1881, it was a series of serendipitous events that had introduced Douglas to the Arizona mining frontier in the first place and radically changed the course of his career. One was the handwritten letter he received from a speculator named Edward Reilly. The two men had initially met at the Philadelphia exposition of 1876, the world's fair where Alexander Graham Bell introduced his newfangled device called a telephone. Reilly told Douglas of a fabulous copper find beneath the reddish slopes of the Mule Mountains near Arizona's border with Mexico.

A short time later, two railcars loaded with ore from Reilly's Copper Queen Mine reached the smelter and refinery complex that Douglas managed in Phoenixville, Pennsylvania. Amazingly, when he refined the Arizona copper it proved of such exceptional quality that it sold for only half a cent less per pound than Lake Superior copper, the industry standard that had always brought top dollar. That represented a triumph for backers of the Copper Queen Mine and for the newly developed Hunt-Douglas electrolytic method of refining ore.

Unfortunately for Douglas, a pattern repeated in his early middle age was retreating one step for every two steps forward. This time it happened when a thief took four bars of refined Copper Queen copper from an unguarded car on a siding at the Phoenixville plant and cost Douglas future business from Reilly. The latest setback, disappointing as it was, did not sever the developing connection between Douglas and Arizona copper. In fact, in December 1880, a group of eastern investors hired him to look over several promising copper claims in mountainous terrain near future Jerome, high above the Verde River that meanders through central Arizona.

The inspection trip required a long and arduous journey to a remote and sparsely populated corner of the United States not yet easily reached by scheduled passenger trains from the East. It meant dangerous and grueling rides between unfinished rail lines aboard the stagecoaches that sped swinging and swaying along Arizona's dirt roads, but Douglas needed the money. He accepted the hazards and discomforts of frontier travel with poise and serenity, and only in later years would he ride in luxury across the desert landscape aboard the Nacozari, his private railway car.

When his stagecoach reached the end of the line at Prescott, the territorial capital, Douglas continued by saddle horse deep into the Black Hills, an outback of broken and uplifted ground. Modern mining in that area dated from 1876 when a local farmer and sometime prospector named M. A. Ruffner discovered a giant outcropping of copper-bearing rock. He located two claims, the Eureka and the Wade Hampton, but neither property much impressed Douglas. Whatever copper they contained seemed utterly valueless because the closest railroad line was still two hundred miles away at the opposite edge of the rugged terrain. He recommended against the risky purchase because his clients were not financially strong. That decision, logical though it was because of the primitive transportation technology of the time, later returned to haunt Douglas.

Had he been evaluating gold or silver deposits, he might have advised clients to take a chance and buy the claims. But for copper, even a whole wagonload was worth little compared to gold or silver, especially after lengthy and inefficient hauls to reach the nearest smelter trimmed away anticipated profits. The Southwest was landlocked, except for mines along the lower Colorado River where steamboats and barges offered connections, and until low-cost railroad transportation replaced the odd mix of packtrains, ox carts, buckboards, and even an occasional camel that hauled freight to distant railheads, only the very richest mineral deposits could be exploited successfully. The region's mining industry originated with precious metals; copper deposits created little excitement until advancing rails seemed almost magically to lift the heavy burden imposed by freight rates.

PORTRAIT OF *William E. Dodge, Jr. (1832–1903), a member of the third generation of Phelps Dodge partners.*

The Eureka and other claims forming what became known as the United Verde Copper Company changed hands several times after Douglas first evaluated them. In 1887, when the price of copper dropped and expenses mounted, disheartened investors again put the property up for sale. Douglas recommended without hesitation that Phelps, Dodge & Company buy a controlling interest in United Verde, though he had no solid evidence that it contained an ore body rich enough to earn any profit during times of low copper prices. However, this time around, the tracks of two transcontinental railroads had been spiked into place across Arizona Territory, and that made all the difference. The deal almost collapsed when the United Verde's eastern owners demanded that Phelps Dodge acquire complete control of the operation at a price $300,000 higher than they had originally agreed upon. The last-minute switch was unethical, and it so infuriated Douglas that he hesitated briefly before agreeing to the new terms, but by then it was too late.

The Montanan William Andrews Clark, sensing an opportunity everyone else had missed, rushed to purchase the holdings. Phelps Dodge did eventually acquire the United Verde property from Senator Clark's heirs in 1935, but only after the mine enabled Clark, who was a millionaire by age forty, to become a multimillionaire before age fifty. Copper riches that the United Verde property hid from everyone but Clark enabled his family to earn a total of some $70 million from the Jerome area, and Phelps Dodge to earn another $40 million before it closed the mine in 1953. Even now, the property of the United Verde Branch only lies dormant as exploratory drilling deep underground continues to probe for viable deposits of copper and zinc.

As was true for all properties it gained over the years, when Phelps Dodge purchased a mine or mining company, it acquired its history as well. In the case of the United Verde, that history was bittersweet. It served as a nagging reminder of the copper fortune that slipped through the partners' fingers and into the outstretched hands of William Andrews Clark. If nothing else, the United Verde story underscored the element of luck in mining. Fortunately, Lady Luck was kinder to Douglas and Phelps, Dodge & Company in other parts of the Southwest.

Putting Down Roots in Stubborn Soil

AT THE REQUEST of one of his clients, during his December 1880 reconnaissance through the Southwest that took him to the site of future Jerome, the indefatigable Douglas headed south from the Prescott area to Bisbee, a scruffy little camp high in the Mule Mountains, to inspect Edward Reilly's new Copper Queen Mine—the one that produced the high-quality bars Douglas had refined in Pennsylvania only months earlier. Again he crossed the Sonoran Desert by a wearisome combination of train, stagecoach, and horse before he reached Bisbee in early January 1881.

Travel facilities were much the same when Douglas returned to Arizona in March and April to inspect Church's claims for the Phelps Dodge partners. On his second trip he journeyed to Lordsburg in southwestern New Mexico, where

he met Church, apparently for the first time, and together they traveled up to Clifton, a camp located about eight miles northeast of the confluence of the Gila and San Francisco rivers, near the eastern border of Arizona. The main object of Douglas's visit was to examine Church's properties in nearby Morenci, located some thirteen hundred feet higher on the slopes above Clifton. That done, he traveled back to Bisbee to study further some claims that had earlier attracted his attention. During the long, bone-shaking journey he must have pondered carefully what he had observed in Clifton and Morenci and what he wanted to report to Phelps Dodge.

Church's operation at Morenci had not been much to see. But something about the copper deposits there greatly impressed Douglas because upon his return to New York, he provided a favorable report that encouraged D. Willis James and William E. Dodge, Jr., to grant Church his hoped-for loan. In return, Phelps, Dodge &

Company received a substantial block of Detroit Copper Mining Company stock on May 21, 1881, and a contract to serve as the exclusive sales agent for its copper. At the same time that he reported on Church's prospects, Douglas mentioned to the Phelps Dodge partners that a claim in Bisbee called the Atlanta was for sale. They were interested enough to buy it. Thus on June 21, 1881, exactly one month after they inked their loan agreement with Church and the Detroit Copper Mining Company, Phelps, Dodge & Company obtained a modest mine property in Bisbee for $40,000 cash.

During the previous fifty years, the New York mercantile firm had weathered several financial panics and the dislocations of the Civil War. That was perhaps good preparation for the many challenges it faced mining copper in the Southwest. Henceforth, an increasing portion of Phelps Dodge profits depended on nonferrous metal deposits in two remote and largely

～ COPPER QUEEN OR COPPER DOG? ～

ARIZONA'S ROSTER OF *mines is lengthy. The mines number in the thousands, and their names are as varied as the sometimes addled imaginations of the prospectors who located them. Some became world famous, like the Copper Queen of Bisbee and the Longfellow of the Clifton-Morenci district, but others were quickly forgotten. Scattered across the desert landscape like so many tumbleweeds are the Crazy Swede, Devil's Cash Box, Dog Water, Sure Thing, and Fool's Folly. There are also the Copper King, Copper Princess, Copper Duke, Copper Belle, and Copper Dog. A century ago, only time would tell if Phelps Dodge had invested in a sure thing or a copper dog. To know for certain required luck, faith, and an abundance of money.*

unknown corners of Arizona Territory, where it sank its first several thousand dollars in 1881. For a time, after the first flush of optimism passed, it looked as if *sank* was indeed the most appropriate word to describe the firm's investment in a region where risk was a fact of everyday life. For good reason the New York Stock Exchange, during the years from 1881 to 1899, refused to admit any mining stocks because it believed they were too speculative.

Copper mining was assuredly not new to the Southwest when Douglas first evaluated Church's claims early in 1881; and Douglas's future fame derived not from discovering any major new deposits but rather from his organizing various landscapes of production to make deposits of varying quality profitable. That was no simple task. Back in the 1860s and 1870s when prospectors from the United States first located many Arizona copper deposits that in time became world famous, the Sonoran and Chihuahuan deserts still formed some of North America's most forbidding terrain. Even in the 1880s and 1890s the harshness of the environment continued to discourage investors and settlers alike.

In Arizona Territory, as elsewhere in the West, development of copper deposits tended to follow a common pattern. The road to riches in copper was far harder to follow than that in gold and silver, and its beginning sections were often completely obscured. Precious metal like gold dust and nuggets recovered from California's placer deposits of the late 1840s, not copper, made ordinary men rich. The intrinsic value of gold and silver, together with the fact that their fortunate finder could often develop gold or silver claims without outside capital, made them a real prize for prospectors. That was not the case with deposits of base metals. Often, however, it was a lone prospector searching for precious metals who stumbled across a promising outcrop of copper ore.

Development of the Southwest's early copper

deposits, including those at Bisbee and Morenci, typically began with one such lucky find. At that point the discoverer usually sold out and pocketed some money, perhaps even a small fortune. Next came the speculators who hoped to acquire and then sell the still largely undeveloped property to other buyers at a handsome profit. The Herculean task of actually developing a copper deposit and then mining it profitably was not for a person faint of heart or short of capital. Unlike Michigan's extremely high grade ore that scarcely needed smelting to make the copper

valuable, ore in a western deposit typically was not as rich, and it required an enormous investment in technology before any metal it contained could be successfully mined, smelted, refined, and marketed.

Because of the veritable mountain of money required to launch large-scale copper mining, a handful of giants emerged to dominate the industry in the United States, notably Anaconda in Montana, Kennecott in Utah, Phelps Dodge in Arizona, and American Smelting and Refining (ASARCO) in several different locations. By 1950

SITUATED DEEP IN *gorges carved by the San Francisco River and its tributary Chase Creek, early Clifton appears walled in by high hills of granite, basalt, and rhyolite.*

these four eastern-based companies produced 90 percent of America's copper, and their subsidiary plants processed 65 percent of all copper coming from their mines. That said, however, there was nothing inevitable about the rise to prominence of any of the nation's copper giants. During the first half of the twentieth century, these four were the principal survivors from among many enterprises that collectively forged the copper kingdom of the American West.

Putting roots down in Arizona posed a formidable challenge to Phelps, Dodge & Company. All of its early mine properties were located in a desert outback accurately described as America's bloody borderland because violence so often erupted there. As late as the 1870s and 1880s, the Southwest remained a frontier with little surface water, fierce climate, difficult topography, and no direct rail service. Added to that litany of difficulties were warring Apaches who ranged across large portions of Arizona and New Mexico and frightened away any would-be settlers or forced them to take extraordinary precautions for fear of attack. Without the lure of metal mining, why any outsider would venture into these arid lands is difficult to imagine.

The Southwest borderland was still dark and bloody ground when Douglas arrived in 1881 to inspect Church's claims for the Phelps Dodge partners. When describing the property to William E. Dodge, Jr., in May 1881, Douglas noted that Indians had recently raided the district, killed teamsters engaged in hauling ore to the tiny smelter, and even besieged the smelter itself, where armed workers held off them off. Not without cause did the younger Dodge later

write from New York to exhort Douglas, then on another reconnaissance trip for the partners, to avoid the risk of an Indian attack while traveling through Arizona.

Hard-won Copper from Clifton and Morenci

TODAY THE NAME Phelps Dodge is synonymous with copper mining in the Clifton-Morenci area. There it stands without peer. However, in that part of Arizona its historical roots spread through many different mine properties, the two most important ones being Church's Detroit Copper Mining Company and its onetime rival, the Arizona Copper Company. A similar process of acquisition occurred in Bisbee where Phelps Dodge merged its original Atlanta property with the adjacent Copper Queen to form the Copper Queen Consolidated Mining Company in 1885, which became an increasingly robust producer after the partnership won full control a short time later. With the ever more visible successes of the Copper Queen helping to publicize the underground riches of Bisbee, several formidable competitors arose at the turn of the century, most notably the Calumet and Arizona Mining Company that Phelps Dodge acquired in 1931 to enlarge its holdings. The two communities and their four big mine companies (in addition to many smaller properties) had much in common, yet each tells a distinctive story that illustrates why early-day copper was such a hard-won commodity.

No one understood that better than Henry

Lesinsky, whose dogged quest for successful production of copper in an unforgiving frontier environment accounted for many firsts in Arizona mining history and formed the basis for the Arizona Copper Company. His saga illustrates the numerous day-to-day hardships and the technical challenges early producers overcame.

Lesinsky did not come to the Southwest looking for copper. By the time he opened his general store in Las Cruces, New Mexico, in 1868 he was already seasoned by an unproductive quest for gold in Australia and silver in Nevada. He would correctly label his earlier search for gold "a lottery." Maybe that was why he took another chance in the early 1870s by acquiring some claims in the Copper Mountain area, a remote site located about seven miles north of Clifton, Arizona Territory, in what later became the town of Metcalf (now part of Phelps Dodge's big Morenci mine complex).[1]

The Longfellow Copper Mining Company was born in the back of Lesinsky's store in 1874. None of the original investors could have foreseen that the Longfellow claim alone would eventually produce some twenty million pounds of copper, or that during the first ten years of production the ore averaged 20 percent copper, or about forty times greater than the copper content of ore Phelps Dodge mined at Morenci in the 1990s. All that Lesinsky knew for certain in his first years as a would-be mining magnate was unremitting hardship.

Copper Mountain was so remote that Lesinsky took five armed men with him when he first traveled the primitive trail that wound through forests of cedar and piñon pine common to the uplands of New Mexico and Arizona to

investigate the claims. Even so, the adventurers still lost their horses, blankets, and provisions to stealthy Apache raiders. An even more formidable challenge was that among the handful of original investors in the Longfellow Copper Mining Company not one person knew a thing about copper mining or smelting, and none owned a single piece of machinery. That did not lessen Lesinsky's enthusiasm.

Because Arizona Territory was so isolated in the mid-1870s (the nearest railroad was in Kansas), it would cost a small fortune to ship even the highest-grade copper ore halfway around the world to a smelter in Swansea, Wales, but there were no comparable reduction works close at hand. A person could always build a smelter of his own to increase the value of the shipped product, and as crazy as the idea seemed to many people, that is exactly what Lesinsky proposed to do.

He journeyed to El Paso and Juárez where he hired Mexican artisans who understood the basics of smelting. Drawing upon a mining heritage that stretched back several generations to Spain, workmen constructed a crude but workable woodburning adobe smelter for Lesinsky. They turned local mesquite trees into the charcoal they needed for fuel. The tiny furnace, not much bigger than an outhouse, processed copper at the rate of about six hundred or seven hundred pounds per day, but usually less because its great heat kept melting the adobe walls. Located on Chase Creek below the Longfellow Mine, it was Arizona's first smelter.

Lesinsky "blew in," or began production at, a bigger and better smelter in Clifton in 1874, where the current of the San Francisco River

⤳ Baby-Gauge Wonders ⤳

TO TRANSPORT ORE *more efficiently from the base of their Longfellow Mine to the charcoal-fired smelter in Clifton, the ever-inventive Henry Lesinsky extended a "baby-gauge" railway (twenty inches between the rails) down the canyon in 1878. Mules hauled empty ore cars to the base of a sharp incline running to the mine itself. When workmen had loaded all the cars except the last one, both the driver and his mule boarded the empty car and coasted down four and a half miles of track back to the smelter.*

The union of mule power and gravity worked so well that Lesinsky decided he could do even better with steam. In 1880 he purchased a diminutive locomotive from H. K. Porter in Pittsburgh and had it shipped around Cape Horn to provide added power for his home-built mine tramway. This first narrow-gauge railway engine in Arizona Territory became a source of wonder and admiration. The tramway cost about $50,000. Its rails, however, did not extend outside the Clifton area. From the smelter, Lesinsky still shipped his output of copper to distant railheads by plodding bull teams and wagons; and those same wagons returned heavily loaded with merchandise to stock his store shelves.

powered a homemade bellows fashioned from cowhide. He had the new plant built entirely from local stone, and it required all of five days to erect. It processed a ton of copper ore per day. But like its predecessors, it suffered from serious technical problems: poor Lesinsky soon discovered that when the stones became hot they tended to explode, and so he returned to adobe for his building material. Alas, he learned too that any copper from his little smelter cost him about four times as much to process as it was worth on the market.

There are numberless ways *not to build* smelters, and long-suffering Henry Lesinsky seemed determined to test every one. There was, for example, the smelter painstakingly designed

by a German metallurgist whom he lured to Arizona from New York City, which burned down less than a day after it began operation. After eight months' work and an investment of $20,000, all Lesinsky had to show for his time and trouble was a heap of fire-blackened bricks. Only the income from his store continued to finance his mining and smelting misadventures.

Friends once described Lesinsky as a jovial man. However, with each failure he grew less happy. Finally, having sunk at least $60,000 of family money into the unremunerative mine and smelter complex, he hovered at the brink of a nervous breakdown. But he would not fail next time, Lesinsky's gambler mind reasoned.

He tried importing special fire bricks from

⤳ TRAINS OF THE *diminutive Coronado Railroad hauled ore from mines in Metcalf to a smelter in Clifton in the 1890s. For most jobs, except where smoke presented a problem in underground mines, steam was far more efficient than animal power.*

Germany, but when he finally did get a smelter design that worked reasonably well, the price of copper unexpectedly tumbled and halved any profit he hoped to earn. The good news was that ore from his Longfellow Mine was so rich that four tons of it yielded a ton of copper. Further, he had no taxes to pay, and local water for power and cooling and wood for construction and charcoal were free for the taking. By the time Douglas arrived early in 1881, Lesinsky had struggled for eight long and frustrating years to produce copper from the Longfellow Mine, and in his quest he had fashioned the first of many industrial landscapes in the Clifton-Morenci area.

Competition: The Detroit Copper Mining Company

THE OTHER PERSON who played a major role in shaping the early industrial landscape of the Clifton-Morenci area was William Church. His quest for success was no less arduous than that of Henry Lesinsky, but at least he could draw upon his personal experience as a mining engineer. He arrived by prairie schooner from Georgetown, Colorado, in 1872, the same year Douglas traveled to Georgetown to inspect some mines there, though no evidence suggests that the two men actually met at the time. After Church reached Arizona (and at a time when Lesinsky still battled to build a successful copper smelter), he spent three months carefully investigating the surrounding terrain. He eventually acquired four contiguous claims that seemed most promising—the Arizona Central,

Copper Mountain, Montezuma, and Yankie—near the camp of Clifton.

Needing money to develop the properties, Church turned to Captain Eber Brock Ward, a hard-driving Detroit industrialist and investor always on the lookout for a good buy. The Detroit Copper Mining Company was incorporated in its namesake city in 1872, and Ward became by far its largest stockholder. Detroit, with ties to Michigan mining in the state's upper peninsula, was at the time a center for copper processing. The newly formed company's claims in Arizona were located at Joy's Camp, a rudimentary settlement named for Captain Miles Joy, who served as Ward's eyes and ears in far-off Arizona. Joy's Camp later became Morenci.

Little development work took place at the remote holdings of the Detroit Copper Mining Company before Captain Ward dropped dead on a Detroit street in January 1875. At the time he was reported to be Michigan's wealthiest resident. Church, determined to push ahead with his Arizona adventure, served as the new firm's president and general manager and in that capacity dominated industrial life in Morenci for the next twenty years.

With his loan from Phelps Dodge in 1881, Church enlarged his mine and erected a smelter on the San Francisco River about three miles below Clifton. It consisted of two furnaces, thirty-six inches in diameter, a type widely used in the Southwest. A blower driven by water power supplied the blast of air necessary to reduce copper ore. During the two years it operated, Church's smelter complex converted about sixteen thousand tons of ore into black

EARLY MORENCI WAS A
*typical urban outpost on the
Arizona mining frontier. The
source of its name remains a
mystery. Some claim that Morenci
was the stage name of an actress
William Church knew in the
dance halls of Denver; others say
that it derived from a town in
southern Michigan. Perhaps it
comes from Canada, where E. B.
Ward, the Detroit Copper Mining
Company's largest stockholder, was
born and where the first Roman
Catholic bishop was Montmorency-
Laval (for whom Quebec's Laval
University, which James Douglas
attended, was named).*

copper and earned the company a tidy profit. At
the smelter site too was located a combination
store and boardinghouse. In those early days the
Detroit Copper Mining Company employed
some 250 men who did all drilling and ore
breaking by hand. Workers used wheelbarrows to
transport broken ore from mines to animal-
drawn wagons that teamsters guided down seven
miles of winding mountain roads to the smelter.

There was a major problem, however.
Indian attacks, real and rumored, frightened the
teamsters needed to haul ore and drove up wages
and transportation costs. Soon Church had to
decide whether to leave Detroit Copper's
reduction works at the river's edge and build a
new mine railway for hauling ore downhill or to
move the reduction works closer to the mines.
That meant relocating the smelter uphill and
pumping thousands of gallons of water a day up
a steep mountainside.

Always there were more problems to solve,
but this time Church gained a valuable ally in
James Douglas, who provided the technical
assistance needed to install a pumping system to
lift water to Morenci, where the smelter was
relocated in 1884. Four years later, in March
1888, the Detroit Copper Mining Company
"blew in" a 120-ton furnace, which indicated
that its mines were then in full production.
The following month, William E. Dodge, Jr.,
accompanied James Douglas to Morenci. That
is the first recorded visit of one of the Phelps
Dodge partners to the little town and its copper
deposits that were to become indispensable to the
modern corporation.

In time the reserves of high-grade ore
dwindled. In those days there was no simple
process available to extract copper from lower
grade ores. Profits shrank, and discouraged
citizens of Clifton and Morenci grew restless.
That was when Lesinsky and other shareholders
in the Longfellow Copper Mining Company sold
out to an Edinburgh syndicate of investors in
1882. In all, the Scotsmen paid them the then
enormous sum of $1.2 million. Henry Lesinsky
joyfully recalled, "Here was a change of affairs.

Ten years of agony were forgotten. . . . We had reached the promised land at last." In addition to the mine property, the business-minded Scots bought the Clifton reduction works with its three furnaces and a company store.[3]

The Scottish syndicate easily capitalized its new enterprise under the name Arizona Copper Company, Ltd. They quickly raised $4 million because their ores typically ran 20 to 40 percent copper. With a fistful of money, they tried to modernize their Arizona properties virtually overnight. They spent lavishly to run a railroad from Southern Pacific tracks at Lordsburg to Clifton and built a bigger smelter, but the cost of improvements when combined with local production troubles almost bankrupted the young enterprise. To save it, the Arizona Copper Company had to be reorganized in 1884.

Nearly forty years later, in 1921, Phelps Dodge added the Scottish-owned enterprise to its growing list of mine holdings. It combined the Arizona Copper Company with the Detroit Copper Mining Company it had purchased in full from William Church in 1897 and later renamed the Morenci Branch. Together they formed the basis for its subsequent open-pit operation.

How the Copper Queen Became Consolidated

Fifty years before Morenci emerged as Phelps Dodge's "flagship" operation, the company's major success story in Arizona was its Copper Queen Consolidated Mining Company in Bisbee. If it was Church's Detroit Copper Mining Company that first caused Phelps Dodge to look far west to Clifton and Morenci, it was the mines of Bisbee that made the partners their first fortune in southwestern copper.

Significant copper finds in Bisbee occurred at least a decade after those in the Clifton-Morenci area. In 1877, when Henry Lesinsky still struggled with his trouble-plagued smelters and William Church suffered a chronic lack of development capital, an Irish-born army scout named Jack Dunn paused during his search for Apaches to take a closer look at some rust-stained outcroppings at the site of future Bisbee. He was a hundred miles from his base at Fort Bowie at a place local Mexicans called Paso de las Mulas, so named because the two granite peaks looked like mule ears to people below. When Dunn and his fellow soldiers drank from one of the springs, they grumbled that the water tasted somewhat like iron, and some of the men developed stomachaches. Dunn noticed another spring nearby that bubbled from a towering mass of limestone. Here he picked up some pieces of green malachite, but copper, unlike gold and silver, held no interest for him. It was the possibility of a silver strike that thrilled Dunn, and so some weeks later he and two other men from Fort Bowie rode out to file a claim in the Mule Mountains.

The real excitement in 1877 lay twenty miles northwest of the Mule Mountains where rich silver deposits drew prospectors like flies to honey. One searcher was Ed Schieffelin, who named his mine Tombstone, having been warned by soldiers that the only thing he would find in the dangerous area would be his own tombstone.

The name of his rich strike became that of the booming settlement. Tombstone attracted an army of prospectors who fanned out from there into nearby mountains—the Dragoons, the Whetstones, the Huachucas, the Dos Cabezas, the Mules—where they staked thousands of claims.

One prospector was Edward Reilly, who abandoned a successful law practice and a comfortable home in Lancaster, Pennsylvania, to arrive in southern Arizona half-crazed with grief after the death of his only son in a railroad accident. For three years he tramped the lonely border country without making a successful strike. Discouraged, he was headed back East when he learned of promising new finds in the Mule Mountains. There he acquired the Copper

Queen property for $20,000, a small fortune in 1880.

Douglas later described Reilly as "an able but erratic man." The sellers of the Copper Queen property no doubt considered him easy to fool. The prevailing wisdom was that because so much Bisbee ore was found in limestone formations its copper deposits were of little value. Reilly actually had no money of his own to purchase the Copper Queen. He planned to interest others in the property and have them advance the necessary capital while he retained half ownership as a type of finder's fee. With that goal in mind, he borrowed enough money to reach San Francisco where he successfully sold a half interest in the Copper Queen to two railway contractors, William Martin and John Ballard.

I N THE GAME *of winners and losers played out in Bisbee in the 1880s and 1890s, the biggest loser was George Warren, who despite a problem with chronic laziness, somehow managed to acquire a one-ninth ownership of the Copper Queen Mine. When he drunkenly bet a friend that he could run to a post some fifty yards away and return faster than the other man could cover the same distance on horseback, he staked his interest in the Copper Queen on the outcome.*

Warren supposedly ran and lost his race before a large crowd on July 4, 1880, in the town of Charleston, which was located west of Bisbee on the banks of the San Pedro River. Warren's drunken boast cost him a fortune worth perhaps as much as several million dollars. He sold the rest of his property for $9,925 and used the money to drink himself insane. For a time he seems to have wandered from jail to jail along both sides of the border. He ended his days as Bisbee's village bum, sweeping saloon floors and cleaning cuspidors in return for a drink of whiskey. His name lives on today in the area's Warren mining district.

In August 1880 the Copper Queen Mine was capitalized for $2.5 million. Its owners renamed the camp of Mule Gulch in honor of their San Francisco attorney and business associate, Judge DeWitt Bisbee.

According to legend, Judge Bisbee never visited his namesake. Initially, he would not have missed much. When Douglas arrived early in 1881, the place was still a frontier outpost of some four hundred inhabitants, most of them single males. It consisted of a post office, two saloons, a brewery, three boardinghouses, and a general store, all facing a single narrow street. The only certainty was that miners, cowpunchers, and teamsters crowded the saloons at all hours. Everything else was tentative.

Mining in Bisbee began in earnest in 1880 as William Martin personally oversaw early development work at the Copper Queen. Ben Williams, tall and powerfully built, handled the mining activity, while his brother Lew, a short, dark-eyed, and rough-hewn man, prepared to make copper bullion from ore treated in a small water-jacketed furnace of his own design. For the next fifteen years Lew Williams remained in charge of the smelter, while during much of the same time his brother served as superintendent of the Copper Queen Mine. Thanks to the industry of the Williams brothers, who brought the carefully guarded secrets of their craft from Wales to Bisbee by way of the copper mines of Michigan, operation of the mine and smelter

complex was an unqualified success, at least in the long run.

During his first trip to Bisbee in January 1881, James Douglas accompanied Ben Williams out to the Copper Queen Mine, which was then basically the glory hole in the hillside still visible from old Bisbee today. Douglas learned that the ore coming from the open cut averaged 20 to 25 percent copper. After careful study of the property, he estimated that ore worth at least another $600,000 was still in sight. But negotiations for the sale of the mine to a Boston family were then in progress, and Douglas returned to the East Coast, unaware that he would return to Arizona Territory only two

months later to investigate Church's mine properties for Phelps Dodge.

Following yet a third inspection trip to Bisbee in June of that same year, Douglas reported to Phelps Dodge that purchase of the Atlanta claim, an undeveloped property adjacent to the Copper Queen, involved risks "too great to be taken by a purchaser who was not able and prepared to lose all that he invested." The partners nonetheless decided to take that risk, though no ore showed on the surface. Douglas's hunch that a vast store of minerals existed belowground was based solely on what he knew of the Copper Queen ore body. Phelps, Dodge & Company offered Douglas the choice of his usual consultant's fee or a 10 percent interest in the Atlanta Mine instead. He did not hesitate a moment, for as he later wrote, "though I sorely needed the money, I took the interest." He was willing to share in the risky venture.[3]

During the following months, Phelps Dodge's new mine in Bisbee seemed anything but promising, and the run of bad luck that dogged Douglas's earlier business ventures in the East seemed to dog him still. The Atlanta was soon mired in all kinds of troubles. Questionable legal claims were filed against the property, including one by a Bisbee merchant who claimed the disputed ground after noting a survey error. Far more serious, the men of the Atlanta spent three years running crosscuts and drifts in every direction and found little more than mere streaks of ore. Even after spending $60,000 on exploration, Phelps Dodge had not recovered even one carload of ore.

That disheartened D. Willis James and William E. Dodge, Jr., who summoned Douglas

to New York to review the mine's bleak prospects. Everyone was glum. Douglas did admit to one glimmer of hope: a streak of ore, or joker as it was commonly called, ran across the property line into the Copper Queen, but he added that he needed another $15,000 to reach the joker. The Phelps Dodge partners, who hated to throw still more money into the project, reluctantly advanced him the required sum. "But not another penny, not one single penny more."[4]

The spring of 1884 was a critical period for the copper mines of Bisbee. The Atlanta had yet to produce a significant amount of ore, and at about the same time that Phelps Dodge committed its final dollars to the mine, the adjacent ore body that the Copper Queen had worked since 1880 suddenly pinched out. Ben Williams had not so much as a single pound of ore to ship to his brother at the smelter. That was about when Douglas returned from New York and prepared to drill a final time at the Atlanta. If he failed, he would likely be job hunting once again at age fifty.

Well before the miners reached as deep as Douglas proposed to go, the gloom suddenly vanished. At 210 feet below the surface their shaft ran out of barren limestone and into an enormous body of copper-rich malachite. No longer was the Atlanta Mine, as a cynic might define such things, simply a hole in the ground into which the Phelps Dodge partners threw their money.

Amazingly, before anyone had time for serious celebration, a crosscut from the Copper Queen broke into the same rich ore body. Now the question became, who could claim the grand prize? Mining law provided no easy answer.

Fortunately, both Douglas and Ben Williams recognized the danger of protracted litigation. Instead of fighting one another in court and enriching only the lawyers, they decided to settle their claims amicably. After several months of negotiations, the San Francisco owners of the Copper Queen property agreed to combine it with the Atlanta and other claims to form the new Copper Queen Consolidated Mining Company. The Phelps Dodge partners and Douglas held two-sevenths of the new enterprise, while Edward Reilly and his associates held the rest.

In August 1885 Phelps, Dodge & Company began large-scale mining under the name Copper Queen Consolidated. William Martin was elected the firm's first president, but when he resigned two months later to retire to his native California, directors by unanimous vote conferred that office on Douglas. A short time later, Phelps Dodge and Douglas increased their two-sevenths ownership to about nine-tenths when most of the other investors parted with their shares at deep discount because of an unexpected dip in the price of copper.

For Douglas, whose stock in the company, if it became successful, assured him eventual investment income in addition to his salary as president, the long struggle for a secure livelihood appeared to be over, or nearly so. For both Douglas and Phelps, Dodge & Company a new era of prosperity appeared to be at hand. *Appeared* was certainly the operative word, for it was their good fortune that new copper discoveries in Bisbee and Morenci coincided with the dawn of the electrical age, which created a huge new market for red metal. But though the

TRAMMING IN THE murky depths beneath Bisbee. In the typical copper mine of old, if a candle remained lit in an underground passageway, many superintendents believed there was ventilation enough to continue work. In those days, the average mine foreman or shift boss was a hard-boiled and intimidating man.

price of copper should have begun to climb steadily along with rising demand, ironically in the mid-1880s it dropped steadily instead, because so many mines rushed into production. So, unfortunately, even with formation of the Copper Queen Consolidated Mining Company, Douglas and Phelps Dodge had not yet succeeded in turning the corner on their Arizona copper investments. It appeared that prosperity would have to wait a while longer.

As always, so much depended on the price of copper. When the Copper Queen finally did prosper later in the 1880s, it was because a French attempt to corner the world supply of copper temporarily drove up the metal's price. The strange financial venture known as the Secrétan Syndicate was ill conceived and ultimately unsuccessful, but for almost three years it provided the Copper Queen a boost during a critical time. When genuine prosperity finally did arrive in the early 1890s, Dr. Douglas and Phelps, Dodge & Company for the first time looked west to their future with real confidence.

Chapter Four

A LANDSCAPE DESIGNED
for PRODUCTION

IN LATE NOVEMVER 1912, only months after Arizona became the forty-eighth state, Walter Douglas (Dr. Douglas's elder son and general manager for Phelps, Dodge & Company) and other officials of the expansion-minded El Paso & Southwestern Railroad system inaugurated regular train service from their Tucson passenger station on West Congress Street. Tourists and business travelers alike hurried to claim their baggage or purchase tickets for far-off Bisbee or El Paso. More than eighty years later the refurbished Tucson station is again a grand edifice, and people fill its large waiting rooms, though not in anticipation of one of the El Paso & Southwestern's long-vanished passenger trains. They're all hungry diners at Carlos Murphy's Restaurant.

The station's imposing size and its many costly architectural embellishments graphically suggest how executives of the Phelps Dodge–affiliated railroad once aspired not merely to knit together scattered segments of Phelps Dodge's growing copper kingdom in the Southwest but ultimately to forge a major new transportation link for trains between Chicago and Los Angeles. To that end, they chose a classical design unusual in the desert Southwest to give Tucson a landmark station that featured an imposing rotunda topped by a stained-glass dome. In every detail—including the parklike grounds

---

INSIDE AN EARLY-DAY *smelter in Bisbee. Smelters remained a highly visible part of Bisbee's landscape until the new Copper Queen and Calumet and Arizona smelters opened in Douglas in the early twentieth century and those in Bisbee closed.*

79

outside, where a large fountain and hundreds of imported trees and plants created a verdant oasis—the building and its gardens symbolized the wealth that copper generated.

On November 11, 1924, shortly after Phelps Dodge sold its controlling interest in the El Paso & Southwestern to the Southern Pacific Railroad for a total of $64 million in cash and stock, the new owner closed the twelve-year-old passenger terminal and transferred all traffic to its own station on Toole Avenue. The former El Paso & Southwestern station during the succeeding decades was home to a tuberculosis sanatorium, a railroad hospital, and even a model railroad club. A local developer bought the worse-for-wear property in 1976, restored the building, including its original stained-glass dome, and converted it into a popular restaurant. The architectural monument that copper built is today honored by the Tucson–Pima County Historical Commission and the Arizona Historical Society.[1]

Tucson's El Paso & Southwestern station is a reminder that good rail connections to developed parts of the United States were essential to the Southwest's early copper mines. That was the primary reason Phelps Dodge scribed its bold signature in steel rails across the dazzling but often harsh terrain that westering settlers from the United States had formerly shunned as the Great American Desert. The region's

copper industry and its early railroads needed each other because large-scale production meant transporting huge tonnages of ore efficiently from mines to processing plants and the resulting metal to market, and the railroads profited from the copper traffic.

Processing Copper
on the Mining Frontier

THE EL PASO & Southwestern system did not result from any grand design on the part of Phelps, Dodge & Company to build itself a railroad empire. Rather, it represented the product of numerous small and often uncoordinated decisions. Over a span of several years, the copper trail that stretched from mine to market benefited from improvements in those two related technologies, such as when larger, more productive, and thus more profitable concentrators, smelters, and refineries went hand in hand with progress in rail transportation that lowered freight costs and boosted operating efficiencies. When large smelters outstripped local sources of wood for fuel, it became essential

COPPER CONNECTIONS

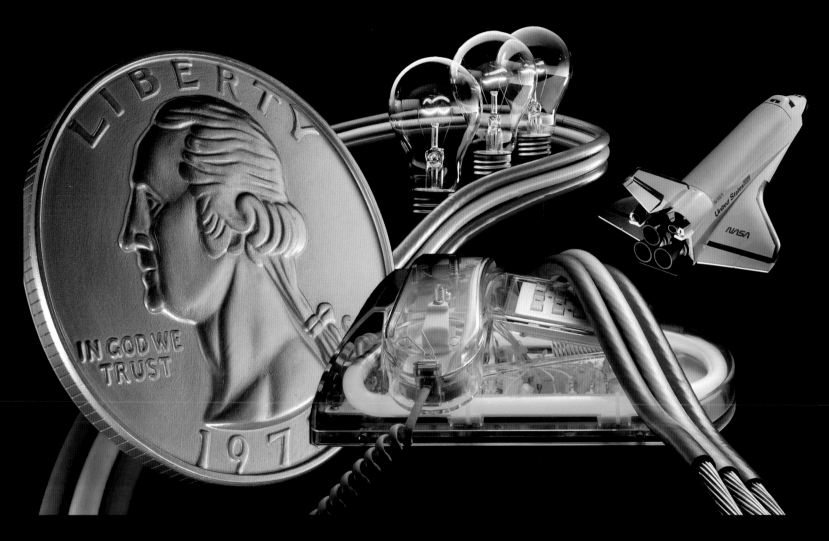

From more than *two tons of copper used in a space shuttle to less then one gram in a light bulb, one of the world's most versatile metals appears in an incredible variety of places. Even the common American quarter and half-dollar coins are more than 50 percent copper.*

THIS PAINTING OF *the Morenci Mine pit, which hangs in the boardroom of Phelps Dodge Corporation, was painted by noted Southwest artist Merrill Mahaffey. The massive landscape he portrays illustrates the impressive feat of engineering which represents the initial link in a chain of copper connections that Phelps Dodge forged from mine to market.*

...COPPER IN SOLUTION at Phelps Dodge's first SX/EW facility in Tyrone. The bulk of cathodes produced by SX/EW facilities in Tyrone, Morenci, and Chino will be melted at El Paso and Norwich, Connecticut, mills to produce continuous-cast rod.

~ HERDS OF CATTLE *continue to graze on the vast expanse of agricultural land that surrounds Phelps Dodge's Hidalgo smelter in New Mexico.*

~ THE FINAL *stage in the purification of copper. Inside the tankhouse at the El Paso refinery, where employees earned the 1996 Chairman's Safety Award for Best Safety Performance: U.S. Operations.*

THE CONTROL ROOM at the Norwich plant monitors the process of turning cathodes into an endless stream of continuous-cast rod. The New England facility won the 1989 Chairman's Safety Award for Most Improved Safety Performance.

ANOTHER EXAMPLE OF where mining meets the market: inside the continuous-cast rod mill at Phelps Dodge Philippines, one of several overseas plants of Phelps Dodge International Corporation. Phelps Dodge Philippines was a 1995 winner of the Chairman's Safety Award for Most Improved

FINISHED MAGNET WIRE *in the El Paso plant of Phelps Dodge Magnet Wire Company. After it acquired the plant in 1994, the corporation modernized the facility. One improvement was the Automatic Guided Vehicle, which has a body like a forklift but is driverless. It obtains directions from a computerized roadway formed by wires buried in the concrete floor that interconnect ten manufacturing locations and the warehouse where the robot takes pallets of newly processed wire to be stored.*

A LARGE CABLING MACHINE *at the El Salvador plant of CONELCA, part of Phelps Dodge International Corporation. CONELCA employees earned the 1997 Chairman's Safety Award for Best Safety Performance,*

load waiting freight cars. Where in 1905 there were thirteen shafts hoisting ore by cars in the Warren mining district, a decade later there were only five shafts, yet far more ore reached the surface than ever before.

Less visible to casual onlookers were changes in the landscape of production underground, where pneumatic rock drills supplanted an army of hand drillers. Even copper itself, in the form of wire for electricity, increasingly returned to the mines from where it originated. The Copper Queen utilized electricity belowground for the first time in 1887, and it soon extended power lines from the mines to the railroad roundhouse and depot, the company store, and the town's library. Until 1912, nearly all of the Calumet and Arizona Mining Company's producing shafts used steam-powered pumps to remove underground water, but after that date they rapidly converted to electricity. In Bisbee, as in all other communities, electrification symbolized prosperity and progress.

A pioneer user of electric haulage in underground passageways was the Arizona Copper Company, which in 1905 replaced mules at its Humboldt Mine in Morenci with trolley locomotives. That meant visible changes on the surface as well. Whenever companies switched to electric haulage, they typically strung miles of copper trolley line underground and added a power plant to the growing complex of surface buildings. Power plants and ever larger smelters in turn required an abundance of low-cost coal and coke to raise steam or provide heat high enough to melt copper ore. In consequence, Phelps, Dodge & Company expanded into both railroading and coal mining.

that increasing amounts of coal and coke be hauled inexpensively from distant mines.

Technology required to process copper ore was always a highly visible part of the local landscape of production, a landscape that steadily evolved if a mining company hoped to remain competitive. The same was true for all transportation in and around the mines. Shortly after the beginning of the twentieth century at Bisbee's Copper Queen Consolidated Mining Company, where workmen had customarily hoisted loaded ore cars to the surface at a number of different shafts, run them off elevator cages, and dumped their contents into railroad cars, the new superintendent Gerald Sherman saw how he could increase efficiency and save money by converting to one central shaft, hoisting ore to the surface in skips, or moving bins, rather than in diminutive cars, and using conveyor belts to

Smelter technology too changed dramatically over the years. Pyrometallurgy, or the application of heat to break the bond between copper and other elements, was another name for smelting, which dated back to ancient times. Early smelting installations often consisted of little more than a bowl-shaped hearth dug into the ground and surrounded above by a stone enclosure. In a later technology, crushed ores were fed into a charcoal-fired furnace and there reduced to metallic copper globules. Far more efficient were the coal-burning reverberatory furnaces that dated from the early 1600s in the British Isles. From Wales to the United States came a process that employed heat to extract copper, first by roasting the ore, then smelting it in a reverberatory furnace to obtain copper matte, and finally by burning the matte together with green logs to remove slag, iron, sulfur, oxygen, and other impurities from copper. From continental Europe came the blast furnace.

Because blast furnaces were somewhat less expensive than reverberatory furnaces to operate, they continued to be used in the United States until World War I, when they were rapidly abandoned for "reverbs," which worked much better with finely ground products resulting from the new flotation processes used in concentrators. But reverbs were also smoky, and that caused Phelps Dodge to phase out its last ones during the environmentally conscious 1980s in favor of its modern, highly efficient, and very low emission Hidalgo and Chino smelters in New Mexico.

Pyrometallurgy, even when greatly simplified as it is here, may sound dull to the nonspecialist, but the process of smelting copper on the American mining frontier was Herculean labor even under the best of circumstances. When James Colquhoun recalled his trouble-filled early days as manager of the Arizona Copper Company in the Clifton-Morenci area, he tried to put a human face on smelter technology, observing that "these little furnaces might well be called 'baby' furnaces, for they required the most unremitting care and attention. They scaffolded and froze on the slightest provocation and often we had to sit up for half the night nursing them through infantile troubles which would never have appeared in a grown-up furnace. Furnacemen talk of hard taps, but if they have never driven a bar through a taphole filled with frozen copper, they do not really know what a hard tap means."[2]

"To the casual onlooker," continued the onetime head of the Arizona Copper Company, "black copper smelting in small furnaces looked easy. I did not find it so. Even when the furnace was running nicely, there was always that nightmare—the loss of copper in the slags. At times we were so nervous about this loss that we were afraid to ask the assayer!" Colquhoun confessed that he envied his predecessor, Henry Lesinsky, "who had no assayer." Similar problems bedeviled other early smelter operators.

After it left the smelter, copper had to be further purified either by fire or by electricity, the latter being the more modern technology. In the 1880s nearly half of America's copper output went to Europe to be fire refined, usually to Swansea, Wales. Among the early electrolytic refineries was the one in Phoenixville, Pennsylvania, which James Douglas supervised between 1875 and 1883, and most of its copper came from Arizona's early smelters in "blister"

WORKMEN ONCE POURED *heavy copper anodes by hand at the United Verde smelter in Jerome. Workers in today's Phelps Dodge smelters in New Mexico use efficient casting wheels.*

form, so called because the slabs of smelted copper were covered with blisters. Additional treatment at the smelters made it possible to cast smelted copper into anodes, the large slabs with the big ears at one end. Today, most copper anodes travel from Phelps Dodge's New Mexico smelters to its electrolytic refinery in El Paso for final purification.

William Church of the Detroit Copper Mining Company once warned his staff there were only two types of rock in his mines—smelting ore and waste rock. In actuality, life was never so simple. The two most abundant forms of copper ore of all grades are sulfides and oxides, and each of those has its own distinct characteristics and each requires different types of processing technology. Because oxide ore occurred nearer the surface than sulfide ore, it was mined first. Only during early years of production, when copper oxide ores from the mines of Bisbee and the Clifton-Morenci area were rich in red metal, could they go directly to a smelter. With progressively leaner ores, concentrators became necessary additions to the local landscape of production. After Church realized that the Detroit Copper Mining Company was running out of high-grade oxide ore on its Morenci property, he constructed the first copper concentrator in Arizona Territory in May 1886. During the next decade, the area's two big copper companies also developed pioneer concentrators to treat the once-scorned (but abundant) sulfide ores too.

With concentrators came various machines required to crush and grind ore, a necessary first step in the complex process of eliminating as much noncopper-bearing material as possible.

Copper was often so minutely distributed through a deposit that in many cases ore had to be ground nearly as fine as face power before it entered the concentrator. Before the 1870s, men often crushed it with sledgehammers and then with mechanical jigs and stamp mills. The crusher used to feed Church's concentrator was a crude machine that employed rollers and a revolving screen to grind 6 percent copper oxide ore into particles the size of grains of sand. The innovative device eliminated much seemingly worthless rock and transformed what remained into concentrate that sometimes assayed as high as 23 percent copper, an impressive metal content that was equal to high-grade ore then mined in Bisbee. By the end of the century, innovative ball mills and jaw and gyratory crushers permitted the fine grinding required to separate particles of lead, zinc, and copper from sulfide ores.

Like crushers, grinders, smelters, and refineries, concentrators changed considerably over the years. The concentrator's basic task was to take finely ground ore, mix it with water to create a pulpy mass, and remove as much waste rock, or gangue, as possible without losing too much valuable metal, which in the early days was still quite a lot. Its most primitive form involved long and laborious processes of hand sorting, hand shoveling, manual stirring, and settling. A major step forward came in 1875 with the riffled shaking table, its most popular version being the Wilfley table, which separated various mineral particles by a combination of vibration and gravity. It was a bit like the oldtime prospector who used a ribbed pan and a swirling motion to separate sand and gold in a cold mountain stream.

CONCENTRATOR No. 6 of the Arizona Copper Company was once the largest such processing facility in the Clifton-Morenci district. Seen here in 1914, it was constructed in 1906 to handle the lower-grade sulfide ore from the Humboldt Mine. The concentrator reached a capacity of five thousand tons per day and continued to operate until 1932 when all underground mining in the area ceased.

The flotation process of concentration, introduced as a more efficient successor to the old clumsy gravity methods, was applied to both sulfide and oxide ores, though it is now primarily associated with the sulfide ores, which compose the greater part of the world's copper resources that are worked today. The story of the discovery of the flotation process recalls a lead miner's wife washing her husband's grimy overalls and noticing that particles of galena, a mineral of lead, always clung to the soap bubbles—but the incident is perhaps only mining industry folklore because it has been claimed so often around the world.

In any case, flotation, first used successfully in the early twentieth century, accomplishes separation by utilizing the chemical affinity of oils and fatty substances for mineral particles. In simple terms, the process consists of adding water and suitable chemicals and agitating the mixture to produce bubbles, to which fine particles of the minerals cling. The air bubbles, with the mineral particles adhering, rise to the surface, where they are skimmed off. Thus enhanced, the sludgy greenish-brown copper concentrate, containing about 20 percent copper, was shipped to smelters, while the "waste rock" was discarded as tailings. In recent years, however, concentrator tailings of early times have proved to be anything but waste because modern technologies are able to extract valuable copper from them.

LANDSCAPE OF PRODUCTION: *a maze of railway lines and trams links mines and processing works in the heart of old Morenci around 1912. The Joy Shaft seen here was the main entrance to the Humboldt Mine, which operated until 1932.*

⤳ A MULE HAULS
a string of ore cars at the top of
the Shannon Incline close by
turn-of-the-century Morenci.

Mule Trains and Mine Trams

I N ALL BUT the earliest and most primitive settings, railroad transportation was required to shuttle ores, concentrates, matte, and kindred products among the mines, concentrators, smelters, and refineries that formed Phelps Dodge's landscape of production, as well as to ship copper to fabricating plants and distant markets. Well before Phelps Dodge's rail empire developed into a thousand-mile-long system— before heavy freights of the El Paso & Southwestern steamed ponderously across the Chihuahuan and Sonoran deserts to link mines, mills, and smelters in Bisbee, Douglas, Nacozari, and Morenci to sources of fuel in northern New Mexico and to transcontinental railroad

connections in El Paso and Tucson—local managers grappled with problems of how best to move ore from mines to distant smelters and refineries.

Before steam and railroads, they used animal power—horses, mules, burros, and oxen—to pull large freight wagons. These four beasts of burden, each with its distinctive strengths and weaknesses, did most of the heavy-duty hauling and carrying on the West's mining frontier, and they remained hardworking servants until locomotives and motorized vehicles later replaced them. Back in the 1860s, for example, workers shoveled high-grade copper ore mined from shallow surface deposits at Ajo (now Phelps Dodge's New Cornelia Branch located west of Tucson) into oxcarts and sent the teams plodding across three hundred miles of desert to San Diego. From there, ore continued aboard tall sailing ships around Cape Horn to reach smelters in Wales. Oxen likewise hauled heavy freight in the Bisbee and Clifton-Morenci areas. Early residents of Clifton came to believe, in fact, that one reason their local beef was so tough was that it came from oxen once used to pull heavy wagons. For Sunday dinners this tough meat was beaten into a pulp on a wooden block, fried on a stove top, and served as veal cutlets, or so some residents claimed.

In the Clifton-Morenci area, the first mule teams hauled copper bars twelve hundred miles to the Kansas City area, the nearest railhead, where the metal was sold. Wagons traveled northeast across New Mexico and along the Santa Fe Trail. If all went well and the copper was safely delivered, the wagons were loaded with

MULE SCHOOL

At one time there were five schools in Bisbee—four for children and one for mules. The mule school, located on a hillside above Brewery Gulch, trained young animals for careers in underground mining. That facility consisted of circular track, one switch, and about thirty feet of straight track, all laid on a flat portion of the hill. Teaching aids included an ore chute where mules were trained to wait while cars were loaded. Most students at mule school learned quickly. By the end of their training, mules could direct cars by kicking switch points along the track with their hooves. They also learned to work without bridles or reins and to follow voice commands. By the time mules finished school, many of them understood commands in more than one language, a valuable skill in multiethnic Bisbee.

After many months of work in the semidarkness of the mines of Bisbee or Clifton-Morenci, a mule would be brought to the surface, but only at night. The animal would be put in a dark room with only candlelight. Slowly, each day, the door to the room was cracked open a little bit more until the mule's eyes adjusted to daylight. Retired mules were turned loose on the tailings area to live out their lives in leisure.

(Adapted from *Phelps Dodge Today*, June 1982, p. 12.)

supplies and hauled back to the Clifton-Morenci area to repeat the months-long cycle.

In mining towns themselves, mules and burros also met local transportation needs. After the Detroit Copper company store in Morenci advertised in 1901 that it would offer for the first time home delivery of goods by burro or mule as a modern convenience to its customers, its business expanded dramatically as a result. Before water was piped in, burros hauled it from door to door: each animal carried about forty gallons in two canvas bags that sold for as much as fifty cents a load.

Beginning in 1907, mules were used for hauling in the underground mines of Bisbee.

Prior to that time, men moved all ore cars by hand, but where a man could tram only a single loaded car, strong mules could move five. If a miner attempted to overwork an animal by attaching six of the three-quarter-ton ore cars, most mules were so intelligent that they felt the difference and refused to budge until he removed the extra car.

The Search for Better Connections

Arizona's first narrow-gauge carrier was the Coronado Railroad, the baby-gauge tram that Henry Lesinsky once used to haul ore to the

⚬ SEVEN INCLINES OF *varying lengths were once part of the mining landscape in the Morenci area. This photograph shows the base of the Coronado Incline near Metcalf, a settlement located on Chase Creek. Workmen lowered loaded ore cars down the 68 percent grade. When Morenci residents traveled to Metcalf or Clifton in the 1880s they often rode the cars up or down the inclines despite the efforts of mine managers to stop the dangerous practice.*

Longfellow smelter in Clifton. Apart from a web of tracks in and around their mines and smelters, Arizona's early-day copper companies needed better long-distance transportation too. Among the first such lines was the Arizona & New Mexico Railway, a subsidiary of the Arizona Copper Company, which extended narrow-gauge track (any track narrower than four feet, eight and one-half inches between the rails was narrow gauge; the Arizona & New Mexico's was three feet) from the newly laid Southern Pacific main line at Lordsburg into the Clifton-Morenci area in 1884. Unfortunately, construction of the seventy-mile Arizona & New Mexico proved far more difficult and costly than most people anticipated. Along one twelve-mile section over the low hills dividing Clifton and Guthrie the company spent a fortune to maintain an acceptable grade as it gained elevation by looping the track back over itself.

In time a spiderweb-like network of rails of various gauges crisscrossed the mining landscape of the Clifton-Morenci area. To take advantage of Arizona Copper's new railroad link to Lordsburg, which offered the Detroit Copper Mining Company a convenient outlet, the Phelps Dodge affiliate incorporated the Morenci Southern Railroad and extended a line down the mountain to a connection at Guthrie, south of Clifton. Despite numerous major engineering hurdles, including the need to construct three large looping trestles to gain eighteen hundred feet of elevation, workmen completed the line on December 31, 1901, by driving a special copper spike. Not without reason was the Morenci Southern line nicknamed the corkscrew railroad of America.

Over the years the network of railroads serving the Clifton-Morenci area continued to evolve along with the local landscape of mining, concentrating, and smelting. In the early 1920s, for example, Phelps Dodge halted service on the Morenci Southern, which had operated at a loss during most of its brief lifespan. Today the area's few remaining rails have all been widened to standard gauge, and freight cars travel directly from Phelps Dodge facilities at Morenci down the company's industrial railroad to Clifton, where they make a connection to Lordsburg.

Bisbee and the Copper Queen's quest for better rail connections involved far less arduous feats of engineering than was typical for the Morenci area, but in some ways it was a longer and far more complicated story. It began in the early 1880s when Southern Pacific tracklayers headed east from Tucson to El Paso by way of Benson and thus bypassed the Copper Queen by seventy miles. Bisbee's early copper ore had been so phenomenally rich that it easily covered the cost of shipping it by mule-drawn wagons north across the empty landscape of Cochise County to Benson, from where it continued by train to Pennsylvania for smelting and refining. But plodding mule trains could not handle the rapidly growing freight needs of the Copper Queen Consolidated Mining Company.

One way to reduce transportation costs was to smelt copper ore on site, and the Copper Queen depended upon tree-covered slopes of the Mule Mountains and the nearby Huachuca Mountains to supply the thousands of tons of wood and charcoal it needed each year for smelting operations. But by the end of the 1880s,

ten years of intensive logging, forest fires, and soil erosion had diminished the local fuel supply until it was no longer able to meet the needs of the Warren district. Wood was also required for mine timbers and for frame homes and business buildings as well. Fuel for Bisbee's hungry smelters had to come from coal mines in southern Colorado and northern New Mexico, and building lumber and mine timbers from the forests of Oregon. All had to be hauled by wagon from Southern Pacific tracks in Benson.

For a time it seemed that a rival railroad, the Santa Fe, would build a line to Bisbee and relieve the Copper Queen's transportation headaches, but in 1882 that route too missed the mines of both Tombstone and Bisbee. In truth, Douglas found neither the Southern Pacific nor the Santa Fe eager for Copper Queen business. He recalled the time when he approached Santa Fe officials to request that they build a branch to Bisbee (about forty miles from their tracks at Fairbank) and they treated him with "supreme indifference"— despite the fact that their railroad hauled a hundred tons of freight for the Copper Queen each day. The rebuff caused Douglas to ponder how the Copper Queen might extend its own rail link to Fairbank, but that seemed too expensive a solution to the transportation problem, especially when his company was long on ore reserves but short on the cash it needed for mine and smelter expansion.[3]

As production topped a million pounds of black (semiprocessed) copper a month in 1887, the need to move mountains of freight by wagon between Bisbee and the nearest rail siding grew ever more burdensome. The question was,

LOCOMOTIVE No. 1 *of the Arizona & South Eastern line, which connected the mines of Bisbee with the outside world. The railroad originally cost $476,420, about twice the sum Phelps Dodge had projected, but the investment quickly paid for itself.*

would Phelps, Dodge & Company in New York approve spending hard-earned dollars to build a railroad link of its own? If the partners assented, at least the new departure was in the tradition of the railroad building activities of the firm's founders. William E. Dodge, Jr., traveled to Bisbee during the spring of 1888 to survey the situation for himself before committing the partnership to underwrite the heavy expenses involved in building a railroad.

The resulting enterprise was incorporated on May 24, 1888, as the Arizona & South Eastern Rail Road Company. Construction began the following month. The first official run over the thirty-six-mile line between Fairbank and Bisbee was on February 1, 1889, and it touched off a public festival and holiday that lasted until the early hours of the next morning. The railroad dropped freight rates between Bisbee and Fairbank from $6 a ton that muleteers charged to $1 by rail, and that encouraged shipment of much larger amounts of freight than wagon trains alone could possibly have carried.

Apart from reducing the cost of hauling freight, the Arizona & South Eastern Rail Road reshaped everyday life in Bisbee by making widely available a variety of consumer goods that residents had once considered unattainable luxuries. Rails offered reliable and efficient transportation to all parts of the United States and thus proved to be the catalyst that ignited growth of both the mines and the community, which rapidly matured from a mining camp into one of Arizona's major urban centers. Bisbee's population grew to more than seven thousand residents in 1900 and to approximately twenty-five thousand by 1919.

Across the Border

DURING THE YEARS from 1881 to 1906, Phelps, Dodge & Company operated several small mines in the Prescott area, such as the Copper Basin property in which it first acquired an interest after William E. Dodge, Jr., visited the area in 1888. That was in addition to its primary holdings in Morenci and Bisbee. Individual partners also gained control of the Old Dominion Copper and Smelting Company in Globe, Arizona, but until that firm ceased business in the 1940s it always remained legally separate from Phelps Dodge itself. Of far greater significance was the Moctezuma property that Phelps Dodge acquired approximately seventy-five miles south of the international boundary in Sonora, Mexico, where two large iron-and-copper-stained pillars, los Pilares de Nacozari, provided a place name. The company's first international mining venture resulted from concern that the high quality of its copper ores in both Bisbee and Morenci was declining and that Phelps Dodge would soon require a new source of red metal.

Phelps, Dodge & Company hired a young consulting engineer in 1895 to survey the mineral resources of northern Sonora in anticipation of acquiring mine properties in Mexico. During his search the unassuming but brilliant metallurgist Louis Davidson Ricketts learned that Meyer Guggenheim and his seven sons regarded the several old mines they had earlier purchased at Nacozari as little more than minor cogs in an immense industrial machine. The Guggenheims—better known for their later involvement with American Smelting and Refining Company—were lace merchants who made a fortune investing in smelters in the United States and Mexico. Phelps Dodge, Ricketts believed, could purchase the largely undeveloped property at a bargain price. Indeed, when the Guggenheims realized how much money they needed to spend to turn a profit on their Moctezuma operation, they gladly sold it to Phelps Dodge in 1897 for roughly the amount they had already lost on the property.

Ricketts—destined to become a legendary figure in American mining because of his many innovative mill designs—held a doctorate from Princeton University. The miners with whom he worked considered him "common as an old shirt" and fondly addressed him as "Doc." The Mexican workers good-naturedly dubbed him El Polo Seco, "the Dry Pole," because of his Lincolnesque stature. To Phelps Dodge, he looked like the right man to oversee its development activity in Mexico. Besides, both William E. Dodge, Jr., and Douglas knew and liked Ricketts.

In 1897 Phelps, Dodge & Company hired him as general manager of the newly acquired Moctezuma Copper Company and assigned him the task of launching a major development program to transform a semi-wilderness area into one of the largest and most active mining centers in northern Sonora. To that end, Ricketts supervised an army of geologists, engineers, and construction workers who busied themselves updating the local landscape of production. One improvement project involved building a narrow-gauge rail-road line approximately five miles long to link

the expanded mine complex at Pilares with the large modern concentrator that replaced the small Guggenheim mill and smelter. Moctezuma Copper also constructed a well-planned community along the river about six miles north of the old town, Nacozari Viejo, which it abandoned. Reborn Nacozari was a picturesque settlement surrounded by mountains, except for the narrow canyon through which the river ran: it featured Spanish and English schools, a general store, hospital, free library, parks and recreational facilities that included a clubhouse where all residents met on an equal footing, modern employees' houses with electric lights, and clean running water.

In short order, Phelps Dodge's Mexico property proved to be an excellent investment because the mine and mill complex yielded copper at an average cost of nine cents a pound compared to a market price of fourteen cents a pound, a difference that added up to steady profits. Production, incidentally, derived from 3 percent copper ore at a time when mining men damned anything under 4 percent as waste.

Succeeding Ricketts as general manager of the Moctezuma Copper Company in 1901 was young James Douglas, who recently had served an eight-year "apprenticeship" at the Senator, Big Bug, and Bumble Bee, all small mines owned by Phelps Dodge in Yavapai County, Arizona, near Prescott. He was universally known as "Rawhide Jimmy," perhaps because of his hard-driving and abrasive ways, but especially because of his habit of patching together broken equipment with bits of wood, wire, and rawhide. At the concentrator at Nacozari he once protected rollers of an incline with rawhide to keep a cable from

The Hero of Nacozari

BEFORE "RAWHIDE JIMMY" DOUGLAS *left Nacozari in 1909, he witnessed a tragedy that created a new national hero for Mexico: a young railroad engineer named Jesus Garcia sacrificed his life to save the town of Nacozari and its mining operations. When Garcia was seventeen, the Moctezuma Copper Company hired him as a railroad water boy. He proved such a diligent worker that within three years he had advanced to become a locomotive engineer. On the afternoon of November 7, 1907, while aboard the cab of a steam engine pulling a trainload of explosives and other mining supplies, Garcia had to choose between his own life and that of the population of Nacozari.*

His train was four cars long. The two nearest the engine were open and loaded with blasting powder; the two behind were heavily loaded with baled hay. The trainmen, concerned that some hay might topple off the moving cars, placed two bales atop the load of powder. A spark from the engine landed on one of the bales, set it ablaze, and threatened to ignite the explosives.

After a desperate and unsuccessful attempt to extinguish the rapidly spreading fire, the 23-year-old engineer ordered his fellow crewmembers to jump from the train, which was then chugging up a steep grade just outside Nacozari. Had he abandoned his locomotive, the train would have stalled and rolled backward into the crowded heart of town. The ensuing explosion likely would have leveled most of the community of five thousand people. Instead, Garcia grabbed the throttle, opened it wide, and steamed away from town as rapidly as possible. When the two cars of blasting powder blew up, Garcia perished, but his quick action had saved Nacozari.

Garcia had been a frequent guest at the Douglas home and a companion of Douglas's son, Lew. Almost half a century later, when Lewis Douglas was the United States ambassador to Great Britain, a dinner with Winston Churchill and Field Marshal Bernard Montgomery turned to a discussion of heroism. They asked Douglas who his favorite hero was. Without hesitation he replied, "My favorite hero is Jesus Garcia."

(Adapted from *Phelps Dodge, a Copper Centennial* [1981], pp. 111–112.)

MONUMENT TO THE *railroad engineer Jesus Garcia as photographed in Nacozari in 1947. The president of Phelps Dodge, James Douglas, gave the main address at its dedication in 1909. All over Mexico, as well as in Cuba, Guatemala, England, and Germany, other monuments were raised to the young hero whose action saved many lives, but the grandest of all was this one in Nacozari's central plaza.*

damaging them. During Douglas's nine years there, total output of red metal at the Nacozari complex rose to an impressive twenty-six million pounds a year, a figure nearly equal to Phelps Dodge's production at Morenci and about a third of that at Bisbee. By the time Phelps, Dodge & Company incorporated in 1908, its Nacozari operations had become its second-largest producer of copper.

Smelter City

Bisbee's first bona fide copper smelter dated from 1880. It consisted of a water-jacketed furnace thirty-six inches in diameter built below the open cut from which Copper Queen ore was mined. Woodchoppers gathered fuel in the nearby hills and hauled it to town by burros. The Copper Queen built newer and larger smelters in 1887 and again in 1893, the latter facility in response to the rapidly growing volume of sulfide ores it mined.

In the early 1890s its miners had encountered large bodies of sulfide ores at lower levels. Initially the Copper Queen extracted only the incredibly rich oxide ores containing 20 to 25 percent copper, but the sulfide ores rapidly became the company's main source of profit. Yet as oxides graded into sulfides, the smelting process grew more complex (as, for example, converters were added); in the mid-1890s, the Bisbee smelter had to be enlarged once again. Further mine development at the turn of the century—as exemplified by the new Spray and Lowell shafts—required still additional smelter production. That proved a difficult feat. Smoke

was long a problem in Bisbee, but processing the swelling volume of sulfide ores had greatly worsened it. Pungent fumes sometimes collected in the canyons, but as was true in all nineteenth-century industrial centers, residents tolerated smoke as an indicator of their community's economic well-being.

Additional problems included little water and limited space for new facilities as well as transportation access to the works in Tombstone Canyon in the crowded heart of Bisbee. These caused managers of the Copper Queen Consolidated Mining Company to weigh constructing an entirely new plant at a more convenient site. After it bought the Moctezuma Copper Company south of the border, Phelps Dodge's desire for a new smelter conveniently located to process ores from both Nacozari and Bisbee only increased.

Dr. Douglas found his ideal site about twenty miles east of Bisbee, due north of the mines of Nacozari, and next to the international boundary. Here on a grass-covered plain at the southern end of the Sulphur Springs Valley, well-water land was available at a reasonable price. In the late 1890s the site was suitably located not just close to the Phelps Dodge properties in Bisbee and Nacozari but also to expanding copper mines across the Southwest and northern Mexico.

Initially, a hodgepodge of tents lined makeshift streets that consisted of little more than packed ground or ankle-deep dust. An early restaurant was fashioned largely from rough-hewn railroad ties. Fortunately, this phase of town growth soon passed, and Douglas, Arizona, rapidly became a community noted for its broad avenues, fine business and residential blocks, and luxurious Gadsden Hotel, with its well-stocked bar and New York–trained chef. It was one of the town's five respectable lodging places. In addition, there were shaded public parks, a handsome YMCA building, and a Phelps Dodge Mercantile store that opened for business in 1902. Some people correctly forecast a new street railway, the ultimate symbol of modernity at that time. Douglas, which incorporated in 1905, was by 1914 a city of five thousand people. But the Copper Queen neither controlled Douglas nor sought to make it a true company town. Despite setting aside an entire block for churches, Dr. Douglas could not prevent a thriving red light district, several squalid saloons, and gambling dens of all description from combining to mar the landscape he had hoped would serve as a model of urban beauty and modernity.

Phelps Dodge designed its Douglas smelter to be the most modern facility of its type in the world. It was also one of the largest. In contrast to Bisbee's early smelters, the new Douglas works covered some three hundred acres and required more than fifteen miles of standard-gauge railroad track just to link its five blast furnaces, powerhouses, machine shops, and foundry. Costing the huge sum of $2.5 million, the facility was built to handle an annual production of more than a hundred million pounds of copper, or about three times that of the old Bisbee plant.

Not only did the Copper Queen construct a sprawling smelter complex in Douglas, but the rival Calumet and Arizona Mining Company (one of several highly successful mining enterprises formed in turn-of-the-century Bisbee) announced plans for a smelter works of its own in the new border settlement. In fact, the Calumet and Arizona furnace was "blown in" (that is, produced its first copper bars) on November 15, 1902. The Copper Queen smelter commenced production in March 1904 and with periodic modifications served Phelps Dodge well for nearly three decades, until the merger of the two companies

⌐ INDUSTRY AT ITS heaviest: a panoramic image of the sprawling Douglas Reduction Works in 1928. The plant "blown in" in 1904 was originally a blast furnace smelter to which the Copper Queen added reverberatory furnaces in 1912. Further modernization of the smelter complex took place in the mid-1920s.

completed its rail line north from Nacozari to Douglas, ending the need to haul freight by mule teams over a primitive, twisting road and to smelt ores in Mexico.

Phelps Dodge's Lengthening Railroad Signature

IN ITS CONTINUING quest for cheap transportation, Phelps Dodge commenced to write its signature in iron and steel rails across the Southwest, tentatively at first with the Arizona & South Eastern Rail Road in the late 1880s, and then with bold and impressive strokes that extended to more distant horizons. It did so because the Copper Queen continued to experience problems with the Southern Pacific and Santa Fe over freight rates, and until after World War I there were no real alternatives to railroads in landlocked portions of the West. Except in those few places where two or more lines offered competition, or in port towns, railroads formed powerful monopolies, and not until the early twentieth century did the federal government seek to moderate their power to set arbitrary rates.

Time and again in the 1880s and 1890s the Copper Queen and the area's two big railroads butted heads. Even before Phelps Dodge extended tracks of its own from Bisbee to the newly planned "smelter city," its officials approached Southern Pacific management to urge the big railroad to run a branch line southwestward from the Lordsburg area to serve Douglas. The Southern Pacific refused. Further, its leaders only scoffed at the suggestion that

⌒ LOADING PROCESSED COPPER aboard boxcars at the Copper Queen smelter in Douglas. For many years the Copper Queen Mine in nearby Bisbee yielded a steady supply of 6 percent ore, which was a primary reason the Phelps Dodge smelter in Douglas in 1917 (the approximate date of this image) turned out nearly 12 percent of all copper used in the United States and almost 7 percent of that used in the world.

in the early 1930s. Ores from Bisbee's new Shattuck and Denn mines also traveled to the Calumet and Arizona and Copper Queen smelters in Douglas. Not without reason did Douglas become popularly known as the smelter city. In an age that equated heavily smoking stacks with jobs and prosperity, the growing industrial community took pride in its nickname.

As for the old Bisbee smelter, it closed in 1904 and was scrapped a few months later. Relocation of smelter operations to Douglas noticeably improved the air quality of Bisbee and vicinity, and that may have accounted for the construction of substantial homes on Quality Hill and at locations farther up the Tombstone Canyon. That same year, 1904, Phelps Dodge

Phelps Dodge's Arizona & South Eastern might be compelled to build an extension of its own all the way to El Paso. With a shrug of his shoulders, the railroad president Charles M. Hays had sighed, "You cannot very well expect the Southern Pacific to go out of its way just because you have seventy miles of railroad down there." That proved a costly rebuff.[4]

Phelps Dodge through its Copper Queen subsidiary decided to teach the obdurate Southern Pacific a lesson in proper treatment of important customers. To that end, some of its directors filed legal papers on October 19, 1900, to incorporate the Southwestern Railroad of Arizona to extend a line of tracks from Don Luis (near Bisbee) to Douglas and thence eastward "to some point on the boundary line of Arizona." Because of a recently enacted federal law, the mining company itself could not formally own a common-carrier railroad, but because James Douglas, William E. Dodge, Jr., Cleveland H. Dodge, and D. Willis James served on both corporate boards, Phelps Dodge retained effective control of a railroad empire to be known as the El Paso & Southwestern system after 1901. Both enterprises had their offices at a familiar New York address: 99 John Street.

As tracklaying progressed, officials of the newly formed railroad wondered why they should halt at Douglas, or even at the Arizona boundary. With copper selling at nineteen cents a pound when it cost only eleven cents to produce, and Phelps Dodge mines in Bisbee and Nacozari capable of large expansion, the company felt muscular enough to shoulder the risk of pushing a line of tracks farther east to gain convenient and competitive connections. The most

ambitious thing would be to reach 215 miles across the sparse country of southern New Mexico all the way to El Paso, where several different railroads beckoned.

Building a line from Douglas to El Paso seemed easy enough on paper. Not only did location engineers report that an excellent low-grade railroad could be quickly spiked into place parallel to the international border, but a confident El Paso & Southwestern purchased land in the west Texas city in anticipation of building passenger and freight stations, repair shops, and switching facilities. None of this was public knowledge, but numerous rumors combined with the news of gangs of men grading a right-of-way east from Douglas toward the New Mexico border made Southern Pacific

THE EARLY INTERIOR of Phelps Dodge's offices in Douglas, the city where Western Operations was headquartered until 1982. Those who served as general manager, the legendary top authority out West between 1920 and 1982, were Percy Gordon Beckett (1920–37), Harrison M. Lavender (1937–52), Charles R. Kuzell (1952–58), Walter C. Lawson (1958–69), John A. Lentz (1969–73), William W. Little (1973–76), and Arthur H. Kinneberg (1976–82).

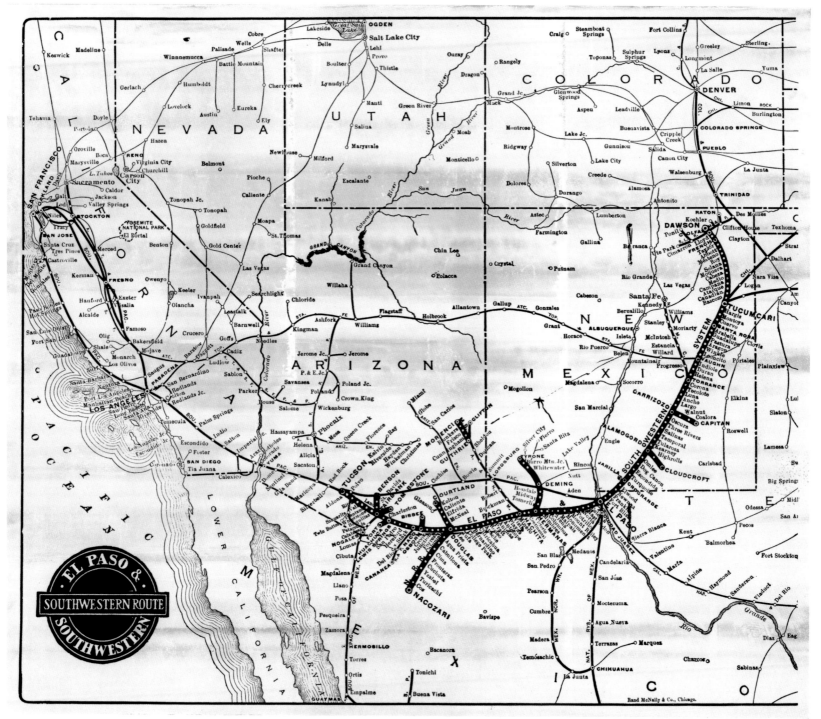

A MAP OF THE *El Paso & Southwestern system. Added in 1921 as part of Phelps Dodge's purchase of the Arizona Copper Company was the Arizona & New Mexico line between Guthrie and Hachita by way of Lordsburg.*

officials angry. The Octopus, as critics blasted the big railroad, fabricated one obstacle after another to prevent the El Paso & Southwestern from reaching its goal, and each time the plucky upstart prevailed.

In early 1902, when the first train steamed slowly over the new section of line from Bisbee to Deming, a triumphant Dr. Douglas was on board. Edward H. Harriman, the new master of the Southern Pacific, did not suffer defeat passively. The egotistical virtuoso of corporate intrigue simply bided his time, lurking quietly in the shadows like a spider until just the right moment to ensnare Douglas. His chance came just west of El Paso where a location engineer working for Douglas found a better route than the Southern Pacific followed from New Mexico into west Texas, but he inexplicably failed to file the legal papers needed to secure the right-of-way. The delay allowed Harriman to hurry his own civil engineers out to file claim to the new El Paso route, and he then invited El Paso & Southwestern officials to sue him.

Douglas, who detested litigation whether over disputed mining claims or railroad access, chose instead to extend his tracks into El Paso along yet another but vastly inferior and much more expensive line. The engineer's error probably cost Phelps Dodge a million dollars, but on June 20, 1903, the company's railroad line was open to El Paso, where it gained three additional connections to eastern destinations for its copper from Arizona and Mexico.

Rather than see rival railroads gain an advantage, Harriman, one of America's quintessential rail barons, finally made a rare concession. He relaxed the Southern Pacific's grip on the southwestern borderlands and proposed to Douglas that they now forget past differences and work together in future harmony. Not so fast, the Copper Queen president responded. Douglas declined to extend his hand in friendship until a suitably chastened Harriman came to his office bearing a check for the difference between the fourteen cents per ton-mile the Southern Pacific charged him for construction materials and the lower rate that had previously been in effect. Also, as Douglas reminded Harriman, there was the matter of additional costs the El Paso & Southwestern incurred when the Southern Pacific had blocked easy access to El Paso on a legal technicality. Only upon receipt of a generous settlement did he relent. His David-and-Goliath victory over the Southern Pacific was complete.

Quest for Coal

PART OF DOUGLAS'S willingness to battle toe to toe with the Southern Pacific came from his ongoing quest for mine timber and cheap fuel for smelter operations. Fuel was no small concern because by the late 1890s, for every ton of copper outbound, one to two tons of soft coal and about two tons of coke headed back into Bisbee. Much of it came from John D. Rockefeller's Colorado Fuel and Iron Company, with which the Copper Queen had signed a five-year contract in 1901. However, after enduring a series of sporadic strikes in Rockefeller's Colorado coal mines and occasional production-threatening shortages of railroad cars, Douglas

⌐ WHEN LUXURY RODE THE RAILS ⌐

*T*HE EL PASO *& Southwestern not only hauled a considerable volume of freight, much of it mining related, but it operated a first-rate passenger service too. By 1916, the year that marked the high point of America's railroad mileage, the Phelps Dodge–allied railroad had teamed up with two industry giants, the Southern Pacific and Rock Island lines, to run luxury trains from Chicago and Saint Louis to Los Angeles and San Francisco by way of El Paso, Douglas, and the graceful new station in Tucson. The desert was still forbidding, but passengers could cross it in style and comfort aboard sleeping cars and in the "Library-Buffet-Observation Car" that brought up the rear of the posh Golden State Limited. Handling the railroad's growing express business was Wells Fargo & Company.*

came to believe that Phelps Dodge needed its own secure supply of coal and coke hauled in its own railroad cars.

The mines that interested him most were located at Dawson, in northern New Mexico near Raton. Unexpectedly, as part of a deal it concluded for those mines on July 1, 1905, Phelps Dodge gained not only a coal company but an additional railroad, the El Paso & Northeastern, a ramshackle line then in the hands of the promoter and speculator Charles B. Eddy, who controlled both properties. He was eager to sell the nearly bankrupt railroad system that extended a total of 460 miles from El Paso across the plains of New Mexico to the coal-rich mines of Dawson.

Though much of the line between El Paso and Dawson had to be rebuilt to higher standards, the $16 million purchase gained Phelps Dodge

rail access to coal mines in northern New Mexico and to timber in the Sacramento Mountains above Alamagordo, northeast of El Paso. In that way too the expansion-minded El Paso & Southwestern forged a link with the Chicago, Rock Island & Pacific line. The Rock Island joined Phelps Dodge tracks in New Mexico with the nation's rail hub in Chicago, and thereby gave the copper company additional leverage when it sought to negotiate favorable railroad rates. The added track across eastern New Mexico, moreover, made Phelps Dodge the second-largest industrial owner of railways in the nation (after United States Steel).

Operated under the name Stag Cañon Fuel Company after July 1905, Dawson's coal mines and coke ovens supplied fuel to the Douglas smelter until 1924 when the reduction works switched to oil, which became more readily

available at the time, and then in 1930 to natural gas piped from New Mexico and Texas. At its peak, Dawson could boast of a total of ten underground coal mines and nearly a thousand beehive-shaped coke ovens; and like Bisbee and Morenci earlier, it thrived as a Phelps Dodge community. During the boom times before World War I it had as many as five thousand residents.

Phelps Dodge: From Partnership to Corporation

ALTHOUGH DIRECTORS OF Phelps Dodge always kept the El Paso & Southwestern and its several subsidiary railroads legally separate from their mining and smelting enterprises, the network of tracks formed a highly visible part of the signature that Phelps Dodge extended across the borderlands of the United States and Mexico in the early twentieth century. How much had changed since the Copper Queen spiked down those first thirty-six miles of track between Bisbee and Fairbank! The same thing could be said of Phelps, Dodge & Company itself.

Even as its landscape of production across the Southwest expanded and grew more impressive with each passing year, the old New York–based partnership evolved in a dramatic new direction. Because of the growth of domestic iron and copper production and congressional passage of the McKinley Tariff in 1890 to levy a prohibitive duty on foreign tinplate, the Phelps Dodge partners watched as their once huge trade in imported metals dwindled to a fraction of its former importance. In 1871, for example,

imported tinplate from Wales had totaled 82,969 tons; by 1890 it amounted to 829,000 tons; but following the McKinley Tariff it shrank rapidly to a negligible amount. Conversely, the domestic tinplate industry, which was negligible before 1890, grew rapidly to become the world's largest. Rather than mourn the decline of the imported metal, Phelps Dodge built a mill of its own in New Castle, Pennsylvania, and in that way the firm continued to market tinplate until it sold the mill to the American Sheet and Tinplate Company, a subsidiary of United States Steel.

In addition, Phelps, Dodge & Company owned the Ansonia Brass and Copper Company mills, which still ranked among the nation's largest fabricators of brass and copper wire, sheet copper, and rolled brass in the early 1890s. The Connecticut enterprise paid Phelps Dodge a steady dividend as well as offering a welcome outlet for copper from its mines in Arizona. But when the partners chose to focus their attention on copper from the Southwest after the mid-1890s, they sold their brass-fabricating affiliate. That was about the same time that Phelps, Dodge & Company commenced production from its first foreign mine at the Moctezuma Copper Company in Nacozari, Mexico.

During the first decade of the twentieth century the conservative partnership changed even more radically. In mid-February 1906, shortly after it received its last tiny shipment of tin from Britain, Phelps, Dodge & Company publicly announced that it was closing its mercantile and metals business on the Atlantic Coast "owing to the great increase in our Copper and Rail Road business in the West." Nor was that the only major transition taking place at

⌒ A LOOK INSIDE AN *early-day power plant at the Copper Queen reduction works illustrates why Phelps Dodge acquired the coal mines of Dawson in northern New Mexico: the mines provided fuel for both electric power and smelting operations in Douglas.*

Phelps Dodge. Within four years of each other, the last of the third generation of partners died— William Earl Dodge, Jr., in 1903 and his cousin, Daniel Willis James, in 1907. Both men had served Phelps Dodge for more than half a century. It was they who dispatched Dr. Douglas to southern Arizona in 1881 and risked precious capital on the Atlanta and Copper Queen mines in Bisbee and the Detroit Copper Mining Company in Morenci. It was those two partners who effectively led Phelps, Dodge & Company out of the cozy era of merchant princes into that of modern capitalism.[5]

Through it all, both men remained faithful to the original principles of the partnership by refusing to borrow money for stock market speculation or to pursue a policy of reckless overexpansion that might endanger the firm's rock solid financial reputation. All the while, they continued the company's tradition of giving generously to support churches, universities, hospitals, missionary enterprises, and philanthropies of every kind.

The time seemed favorable after the deaths of Dodge and James to dissolve the old family partnership and forge a modern corporation. The new arrangement would permit the firm to raise capital by selling its shares on the open market, allow convenient sale of any deceased partner's interest, and limit the exposure to financial risk by any individual or family member connected with Phelps Dodge, which within the boundaries of its copper kingdom astride the international border in the Southwest seemed to grow larger and more complex every day.

With these legal and financial advantages in mind, the seventy-year-old partnership of Phelps, Dodge & Company was incorporated under the same name in December 1908 and capitalized for $45 million. Still based at 99 John Street, the holding company took charge of the stock of its several subsidiary firms. At the time those were the Copper Queen Consolidated Mining Company in Bisbee, still its flagship property, the Moctezuma Copper Company in Nacozari, the Detroit Copper Mining Company of Arizona in Morenci, and the Stag Cañon Fuel Company in Dawson. A total of five hundred shares of common stock was distributed equally among Cleveland H. Dodge, Arthur Curtiss James, James McLean, George Notman, and James Douglas, who served as the first president of the new corporation. The El Paso & Southwestern Railroad remained a separate company in order to comply with federal law, but Douglas continued as president of that enterprise as well, and its controlling stockholders were members of the Phelps Dodge family.

From 1908 to 1915 Douglas continued to make frequent visits to Arizona, often four times a year, in order to inspect company properties and consult with various managers and chief engineers. No doubt on such occasions he savored the view from his private railway car as it glided past southwestern panoramas he had done so much to create. In addition to hulking industrial structures, there were the modern urban landscapes defined by the schools, libraries, churches, and hospitals that Douglas and Phelps Dodge took a personal interest in nurturing.

THE RIVER OF COPPER running through Phelps Dodge history took many impressive forms, including this once familiar drama photographed in 1938 at the Copper Queen Branch, Smelter Division. Later named the Douglas Reduction Works, it poured its last anodes in 1987. Today the smelter building is gone, and the complicated landscape of production can best be recalled through photographs and personal reminiscences.

⁓ *Chapter Five* ⁓

COPPER'S URBAN OUTPOSTS

I N THE AMERICAN SOUTHWEST, success in copper demanded more than simply mining or processing red metal, or even extending railroad lines to interconnect landscapes of production. It required that Phelps Dodge and other big copper producers support the building of homes, churches, and schools in the urban outposts that arose near major ore deposits, places that invariably were located in desert or mountainous areas remote from already existing population centers. The extent of copper company town-building activities "amazes the outsider unfamiliar with these peculiar and distinctive communities," noted an observer in the *Saturday Evening Post*, one of the nation's most popular magazines in 1923.[1]

Buildings that survive from King Copper's original urban landscape in frontier Arizona and New Mexico typically are visually arresting and historically engaging. Individual structures, no less than entire copper towns, memorialize a bygone era. One good example is the Phelps Dodge guesthouse in Clifton. The company retreat on the banks of the San Francisco River recalls the days when huge amounts of money, some of it coming from investors as far away as Scotland, circulated through remote corners of Arizona. This showplace set in a spacious lawn shaded by Italian cypresses and palm trees invites visitors to imagine the soirees of former times, when Japanese lanterns festooned the grounds

⁓ PHELPS DODGE'S BEAUTIFUL *guesthouse in Clifton, Arizona, in June 1998.*

107

and cast a soft glow over men dressed in tuxedos and women in silky evening gowns.

When frugal Scots of the Arizona Copper Company built the mansion shortly before World War I, reportedly at the almost unbelievable cost of a million prewar dollars, their apparent intent was to keep some of their hard-won earnings out of the grasp of British tax collectors. At the time the Arizona Copper Company paid annual dividends that sometimes ranged as high as 80 percent of earnings, and the money its executives invested in the Clifton estate reflected their smug prosperity. Today, the Phelps Dodge guesthouse contains six spacious bedrooms (each with its own bath and some with private verandahs), a formal dining room, a large and airy living room, wainscoted library-study, wet bar, grand staircase, and several imposing fireplaces. Ironically, the first general manager to occupy the mansion was Norman Carmichael, a bachelor from Scotland.

For many years Phelps Dodge used the home as living quarters for general managers of its Morenci Branch. The last occupant was Lyle M. Barker, who retired in 1959. When his successor, John A. Lentz, moved to Morenci to be nearer the company's mine and smelter complex, the wood-and-stucco manor in Clifton became the guesthouse. Today, no one who spends a night there, whether company executives or their guests, can forget that this is the mansion that mining built. As a result of recent renovations it is tastefully decorated throughout with mementos that include a geologist's pick, miners' carbide lanterns, boxes that once contained blasting caps or dynamite from Atlas, Hercules, and Apache Powder ("High Explosives—Dangerous"), and several immense

STANDARD MINES METCLAF ARIZONA

photographs that recall the history of Morenci and Clifton.

Among various structures that form the modern urban landscape of Phelps Dodge, this must surely rank as the crown jewel. Historically, however, the corporation's landscapes included not just buildings like the guesthouse in Clifton but also complete towns intended to be as beautiful as they were functional. Perhaps no business enterprise ever built a more resplendent company town than Phelps Dodge did in Tyrone, New Mexico.

A photograph of the Standard Mines Company community of Metcalf, near Morenci, clearly shows why deposits of copper in out-of-the-way places required mining companies to become pioneer town builders. A century later, in the 1970s, this was where Phelps Dodge developed the open-pit mine that is now an integral part of its sprawling Morenci complex.

The Impact of the Deacons
of 99 John Street

IN NEARLY A dozen outposts—copper, coal, and smelter towns—Phelps Dodge and the various mining companies it gained after 1881 played a major role in shaping the urban landscape. When it shouldered that responsibility (and it was enlightened self-interest that caused Phelps Dodge to desire a clean, comfortable, and uplifting environment for employees), it followed closely in the footsteps of the founding partners, religious men all, who combined good business with good works.

The frontier outposts of Bisbee, Clifton, Morenci, and Dawson existed before Phelps Dodge invested its first dollars in their copper or coal mines. Just the same, the company visibly modernized all those urban landscapes. The chaotic, free-for-all nature of early Bisbee's growth so appalled James Douglas that he and other company officials actively worked to remove the roughest edges from daily life by funding libraries, workers' clubhouses, and hospitals—even if that meant a quixotic quest to reshape town life according to their own lofty moral vision. To some observers, that moral vision was best exemplified by a sign posted in a public dance pavilion in Bisbee during the 1920s that read: "Do not shimmy, do not dance cheek to cheek." A writer in the *Saturday Evening Post* commented in 1923 that "the Phelps Dodge Corporation has been described as paternalistic, English and old-fashioned in its attitude. Its leading directors are usually known as the deacons of 99 John Street," a reference to the

century-old location of its New York headquarters.[2]

The best example of the exalted ambitions of the "deacons" as town builders was Tyrone, New Mexico, a showplace that attracted favorable comment from professional architects around the United States after Phelps Dodge began its construction in 1915. The project seemed exotic at the time, though in reality the building of Tyrone only continued a long-standing company commitment. Back in the mid-nineteenth century, the founding partners built settlements to benefit loggers and sawmill workers isolated in the forests of Pennsylvania and Georgia. In the last third of the twentieth century, the commitment continued when Phelps Dodge Corporation planned and built new Morenci and the smelter settlement of Playas in the boot heel of New Mexico southeast of Lordsburg.

Two visually arresting examples of former company towns associated with Phelps Dodge are Ajo and Clarkdale, Arizona. Both began as planned communities in the second decade of the twentieth century, though neither one was built originally by Phelps Dodge. The corporation acquired Ajo, a Sonoran Desert oasis defined by its charming Mediterranean-style architecture, together with the New Cornelia Mine when it bought the Calumet and Arizona Mining Company in 1931; and it gained Clarkdale and related properties in the Verde Valley when it purchased the late Senator William Andrews Clark's United Verde Mining Company in 1935.

At its new United Verde Branch, Phelps Dodge became a major presence in, but not the owner of, the adjacent settlement of Jerome. Like Bisbee and Clifton, Jerome was never a true

company town built and owned by a single enterprise. During the final years of the nineteenth century, all three outposts had expanded haphazardly to fill all available space within their confining locations. The main streets of Bisbee and Clifton stretched along narrow and flood-prone gulches, and those of Jerome clung precariously to a steep and unstable mountainside. Gravity, not long-range planning, determined the location of everything.

The corporation's many urban landscapes

were not limited to small, out-of-the-way settlements—Phelps Dodge still refines copper in El Paso, and it once did so at Laurel Hill in Queens, a borough of the nation's largest city— but it was in one-industry towns of the American West that the deacons of 99 John Street left their most visible imprint, even if that imprint has grown faint in some locations. Original Morenci simply vanished as its buildings disappeared into the open-pit mine at its doorstep. The most visible remains of the model company town of

Bisbee and named for one of the town's Brewery Gulch prostitutes. Dr. Douglas declined to purchase the Irish Mag property for Phelps Dodge because its seller—a "dangerous, mad man" who always carried firearms, slept with a rifle beside his bed, and "would sooner or later commit murder"—had once threatened to kill the Copper Queen's valued superintendent, Ben Williams. The Irish Mag claim passed instead to the Calumet and Arizona Mining Company, a newly organized enterprise that in 1902 discovered a mammoth body of ore calculated to be at least 325 feet thick and to assay as high as 30 percent copper. Even more amazing, the find lay beneath a tiny plot of land not much larger than twenty acres.

To relate the engaging details of each urban outpost is simply not possible in a general history of Phelps Dodge, and that includes details of the colorful saga of Bisbee's Irish Mag claim. Yet, a quick look at the evolution of early Bisbee, Clifton, Morenci, and Tyrone (and Dawson too in the following chapter) illustrates how Phelps Dodge became a pioneer town builder in the southwestern borderlands.

Tyrone are two red tile–roofed adobe buildings (a small Protestant chapel and the Hall of Justice), both of which provide for records storage and are now almost completely surrounded by the bustle of open-pit mining. Once-lively Dawson consists of little more than a large community cemetery filled with row upon row of silvery-gray iron crosses. Most markers bear the death dates of 1913 or 1923 (and therein lies a story told in the next chapter).

Phelps Dodge's long-vanished urban landscapes are no less instructive than its modern, highly visible landscapes of production. Each community maintained a unique relationship with the company, and each had its special history enriched by colorful local personalities and events. One particularly good example is the tale of the Irish Mag claim, located on Sacramento Hill approximately two miles southeast of old

Phelps Dodge and Mining Camp Modernization

B ISBEE, LIKE CLIFTON and Morenci, antedated the arrival of James Douglas and Phelps Dodge in 1881, and yet for more than a century whatever Phelps Dodge did in those three places invariably affected community life. From the beginning, most homes in Bisbee belonged to private owners; and when the growing settlement formed a municipal government, police and fire departments, and chamber of commerce, it did so more or less independently of the company. Yet for much of the twentieth century, whenever Bisbee residents spoke of "the company" they usually meant Phelps Dodge, or PD as it was familiarly known.

Early Bisbee appeared about as ragged as some of its recently arrived inhabitants. Brewery Gulch harbored the camp's cheapest saloons and dance halls as well as one of the most squalid red-light districts on the border; Tombstone Canyon was home to the better-class saloons and gambling houses, with a few general stores scattered among them. Both of Bisbee's primary commercial streets sank deep in mud during the rainy season or choked on dust during the rest of the year. Day or night, Brewery Gulch and Tombstone Canyon in the 1880s and 1890s teemed with half-drunken miners, mule teams hauling charcoal to the Copper Queen smelter, and Mexican peddlers leading strings of little burros to deliver wood, water, and supplies to customers living along the hillsides.

Both Brewery Gulch and Tombstone Canyon formed natural watercourses of the Mule Mountains, and when early arrivals stripped the hillsides of their juniper, oak, and manzanita for mining and domestic needs, and sulfur smoke from smelters stunted or killed every green plant, little vegetation remained to check rainwater as it cascaded down the slopes. That increased the problem of flash floods. Every now and then a summer cloudburst sent torrents of water and mud as much as four feet deep racing through the streets of Bisbee. Despite the destruction and occasional loss of life, the floods had their good points, or so some people claimed, because they

provided the only real street cleaning ever done in early Bisbee. But between storms the stench of sewage and garbage in the gulches grew oppressive once again.

The built environment of both Clifton and Morenci in the late nineteenth century was no fancier than in Bisbee, and daily life was no less crude. Morenci evolved from an unkempt mining camp of six hundred residents in 1891 into a rough-and-ready town of nearly three thousand by 1897. No one dared call it a beautiful place, not with its many crude, unpainted board shacks

By any measure turn-of-the-century Bisbee was a somewhat shabby place, still more of a mining camp than the modern city it soon became with help and prodding from Phelps Dodge.

opposite: One of early Bisbee's many cleansing floods. This one occurred in 1895, a time when many restaurant owners still threw garbage out their back doors, carcasses of dogs and burros occasionally littered the town's two main streets, and residents utilized open cesspools that contributed to the problems of pervasive stench, flies, and mosquitoes.

and the numerous saloons, gambling joints, and houses of prostitution that lined the main road leading into town.

In their earliest days, mining camps of the American West were literally only encampments or temporary settlements slapped together from local materials. Most never amounted to much, nor were they expected to last very long. But copper mining, unlike the recovery of placer gold, involved high costs and long-term investment, not the short boom and bust typical of gold diggings, and over time, as buildings of brick and stone replaced those of wood, places like Bisbee and old Morenci took on the appearance of modern industrial centers.

The only problem was that as those settlements grew older and larger—as they matured from primitive camps into real towns and cities—a growing number of substantial structures, often including railroad stations, large hotels, churches, schools, two or three banks, dozens of stores, and company offices and mine buildings, still squeezed into the original confining spaces. Residents of Morenci and Bisbee stacked their houses one above the other in tiers that extended up the hillsides, and neither place enjoyed the benefit of formal design or planning. They were both basically hodgepodges of buildings jammed together in seemingly whimsical ways, even if Bisbee's dense settlement, winding streets, and hilly location charmed early tourists and earned it the sobriquet Little San Francisco.[3]

Because mines, mills, and smelters crowded between commercial blocks and residential areas, these towns long endured an all-pervasive problem of dust, smoke, and fumes. At one time,

when the Arizona Copper Company located its smelter in Clifton at the confluence of Chase Creek and the San Francisco River, it simply directed the smoke into a drift in a nearby mountainside up which it rose and finally exited through a short stack at the top. Residents teased visitors by telling them that the peak towering above Clifton with smoke belching from its pinnacle was an active volcano.

Citizens living below still breathed sulfur fumes, and James Colquhoun, who headed the Arizona Copper Company, made the best of the situation by asserting that sulfur smoke had health-giving qualities. Actually, that was once a common belief. Some residents of Bisbee firmly maintained that when smelter smoke and fumes thickened the air of their enclosed valley it protected them from the ravages of typhoid fever. Even if their claims were suspect, few people doubted that smelter stacks belching smoke were sure signs of prosperity.

During copper's boom years of the early twentieth century, Bisbee's population increased to an all-time high of approximately twenty-five thousand residents in 1919, ranking it as one of Arizona's largest cities. Hundreds of businesses lined Tombstone Canyon and Brewery Gulch, and its streets teemed with new faces. Bisbee formed a chamber of commerce in October 1905 to sell itself to a world that knew little about the community except for its mines. Local boosters pointed with pride to the splendid new Muheim Building, which boasted a stock exchange complete with the latest market reports from New York City. These were years, too, when Phelps Dodge devoted a growing portion of its business resources to helping make Bisbee and

Morenci two of the most modern mining communities in the United States.

Phelps Dodge long played a leading role in the modernization of Bisbee through its Copper Queen Consolidated Mining Company. Among its many contributions to the betterment of town life were a school (1883), a library (1887), and a company hospital (1890). These first three and a growing number of similar institutions encouraged greater social stability by offering Bisbee residents an alternative to saloon life, the most common and often most unsavory place of recreation in any mining town. The company dramatically increased its commitment to modernization after 1900.

After the architect Frederick C. Hurst arrived in 1902, he drew plans for at least fourteen imposing new buildings in downtown Bisbee, among these a new hospital for the Copper Queen Consolidated Mining Company. Two stories high and steam heated, the colonial-style building accommodated as many as fifty patients at a time and featured the latest medical equipment. It was by far the grandest hospital in Bisbee to date. Also for the benefit of its employees, the Copper Queen in 1904 established the Gymnasium Club, which featured a large exercise room, shooting range, and later a hand-dug swimming pool.

The mining company built the Copper Queen Hotel, a luxurious seventy-five-room Mediterranean-style structure completed in 1902, to meet the lodging and dining needs of mining experts, purchasers of copper, company officials, and any other business folk who came to Bisbee. It quickly became the city's social center, and even today its red tile roof makes it a landmark in the urban landscape of Bisbee. The Copper Queen Hotel stood just behind the general offices of the Copper Queen Consolidated Mining Company. On the opposite side of the building was the company store, called the Copper Queen Mercantile after its founding in 1886. It was one of Bisbee's major commercial emporiums, although many privately owned enterprises competed directly with it.

Dr. Douglas was not alone in his early efforts to improve Bisbee and other Phelps Dodge settlements. Members of the partnership built the First Presbyterian Church in Bisbee, which contained a handsome pipe organ donated by Arthur Curtiss James and his wife. James and other Phelps Dodge directors liberally supported the Young Men's Christian Association, organized for company-community service in Bisbee in 1906, and the Young Women's Christian Association. To that organization, Phelps Dodge donated a modern and comfortable building, completed in 1916 and intended as a lasting memorial to Grace Dodge, the social-welfare-minded daughter of William Earl Dodge, Jr., who at the time of her death was national president of the YWCA.

After incorporation of the Bisbee Improvement Company in 1900, Phelps, Dodge & Company joined with other mineowners as well as local bankers and railroad men to extend electrical, natural gas, and telephone service to the town's growing population. The Bisbee Improvement Company assisted with community betterment projects such as street lighting, flood control, modern sewers, and road improvements. Even more impressive, in 1906, shortly after it relocated its smelter from Bisbee to Douglas,

Phelps Dodge joined with the Calumet and Arizona Mining Company and others to develop the new Warren townsite, named for that legendary local prospector George Warren, on relatively flat ground below original Bisbee.

As envisioned by the Calumet and Arizona, the project's primary sponsor, Warren was to be a beautiful and thoroughly modern community that would enable Bisbee mining companies to attract and retain their "best class of employees" at a time when booming copper production created a housing shortage. To that end, officials of the Warren Company carefully subdivided the land for individual homes and insisted that every lot be connected to water pipes, sewers, and electric lines, which they bragged would make this "the most sanitary town in the country." Warren took the shape of a large fan that flanked an impressive park, called the Vista, a public oasis six blocks long. Two of the region's leading architects, Frederick C. Hurst and Henry C. Trost, designed a range of housing types, but all residents, even the owners of modestly priced homes, still enjoyed large and landscaped yards on broad and tree-shaded streets. The most expensive homes tended to be located along the Vista, and the single most elegant one was the Italianate mansion of Phelps Dodge's Walter Douglas, strategically placed to command a sweeping view of Warren and the surrounding plain.

After Phelps Dodge bought the Warren Company in 1917, Calumet and Arizona abandoned much of the corporate paternalism of its earlier years and left its rival to shape Bisbee's future urban growth. Because Phelps Dodge actually owned few business and residential structures, the community maintained a split

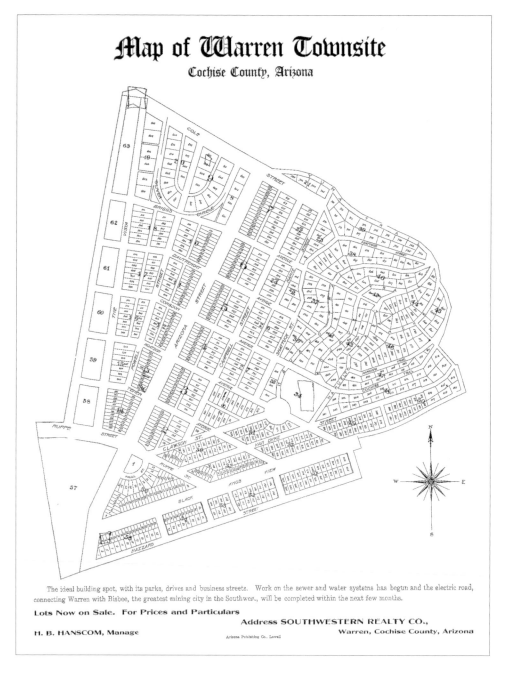

A MAP OF THE *Warren townsite, the Bisbee neighborhood that epitomized a well-engineered company town of the Southwest. It provided schools, parks, stores, company offices, a baseball diamond, water and sewer connections, and substantial homes in the bungalow, mission revival, and neoclassical styles. Water from nearby mines made Warren a green oasis surrounded by a desert dotted with ocotillo, agave, and other native shrubs.*

HOME DELIVERY BY *mule in early twentieth-century Morenci. Not until 1915, when a wagon road was at last graded into town, did automobiles serve local transportation needs. Before then nearly all travel was by foot, horse, or rail.*

personality as part company town and part boomtown.

During those same turn-of-the century years, Clifton and Morenci benefited too from company-sponsored municipal improvements. Among the most impressive new public buildings in Morenci was Phelps Dodge's Detroit Copper Mining Company store. It dated from March 1901, at which time the four-story structure was the largest commercial emporium in all Arizona. Fashioned from local stone placed over a framework of steel, it was reputed to have cost the munificent sum of $161,000. The D.C. Store, as it was popularly known, carried such luxury items as fresh vegetables and ice, both of which were scarce and difficult to purchase in isolated mining camps, and it stocked the latest styles of clothing and furniture from New York and San Francisco as well as imported European linens and laces.

Next door was the Hotel Morenci, a three-

story building designed by the architect Henry Trost, whose work was also represented in Bisbee. It was a turn-of-the-century example of the Moorish–mission revival style popularized by the California Building at Chicago's Columbian Exposition of 1893. Its large and elegant dining room was very popular, and together with the Morenci Social Club across the plaza, it gave a new air of refinement to community life.

The Morenci Social Club, the third of the town's trio of impressive buildings, offered yet another example of the popular Moorish–mission revival style. The three-story structure contained a two-lane bowling alley, clubrooms, billiard and game rooms, gymnasium and public baths (with seven showers, 165 lockers, and ample hot water on tap at all times), a library and reading room, and a south-facing verandah. To enjoy its many amenities, members paid dues of $1.25 a month, or $13.00 a year. In all, it represented quite an improvement over the town that originated as an unplanned camp, and in the eyes of mining company executives, it was infinitely superior to local saloons. A second club, this one for the town's many Hispanic residents, also met the community's recreation needs.

Morenci was home to five thousand people in 1910, while down the hill in the San Francisco River valley lived another eight thousand in Clifton. Because of its cramped location, Clifton, much like Bisbee in its basin in the Mule Mountains, suffered numerous flash floods. And like Morenci and Bisbee at the turn of the century, Clifton was a picturesque and lively town, albeit an unkempt one. It enjoyed the greatest building boom in its history in 1912 and 1913. That was when Arizona Copper built the

One showpiece of Morenci's turn-of-the-century modernization was the Morenci Hotel.

"million-dollar mansion" that now serves as the Phelps Dodge guesthouse.

The Search for a Model Company Town

WHEN IT PROMOTED municipal improvements in Bisbee and the Clifton-Morenci area, Phelps Dodge pretty much had to work with the cards dealt it by geography and history, but at Tyrone it owned the entire deck and would play the cards as it alone saw fit. By any measure, the planned community located in the Burro Mountains about fifteen miles from Silver City, New Mexico, was the most impressive Phelps Dodge company town ever built.

There have actually been three Tyrones. The earliest one already existed when Phelps Dodge began purchasing claims in the area in 1909; the second was the model town designed by the New York architect Bertram Goodhue and built between 1915 and 1918; and the third was modern Tyrone, which dates from 1966, when Phelps Dodge announced plans to develop the Tyrone open-pit copper mine. Since most of Goodhue's Tyrone was situated atop the proposed mine, and its buildings had grown decrepit and unsightly by the mid-1960s, mine construction crews tore it down before they began stripping away the overburden.

Copper mining in the Tyrone area, which straddles the Continental Divide, easily ranks among the oldest mining in the Americas. More recently, in the late nineteenth century, several different companies tried their luck in the Burro Mountains, but none left an indelible mark. In 1904 the Leopold brothers of Chicago acquired the property of the Southwestern Copper Company and reorganized it as the Burro Mountain Copper Company. The enterprise erected a small concentrator, and nearby a collection of shacks took the name Leopold. Perhaps as many as two hundred men worked in its mines and mill. Also close by was original Tyrone, comprising a hospital, a two-story frame hotel, a store, and several tent-and-clapboard residences.

Phelps Dodge purchased the Burro Mountain Copper Company from Nathan F. Leopold, Sr., of Chicago. It was his son, incidentally, who was convicted of the infamous Leopold and Loeb thrill murder of young Bobby Franks in 1924. The acquisition inaugurated Phelps Dodge's continuing involvement in the New Mexico mining industry. The company rapidly acquired several hundred more claims in the area, though its stated intention was to hold them as a large reserve. In truth, the ore at

Tyrone averaged only 2 percent copper in contrast to the 12 percent ore coming from the Copper Queen. Underscoring its Phelps Dodge connection, the Burro Mountain Copper Company, with James Douglas as president, was headquartered at a familiar address: New York's 99 John Street.

Development work commenced in 1910, and by 1916, after Phelps Dodge had acquired control of virtually the entire district and copper prices climbed sky-high, it was at last ready to commence underground mining of Tyrone's highest-grade ore. To prepare for production, it completed the Burro Mountain Railroad in 1914, a small link in the El Paso & Southwestern system that extended from Tyrone to connections at Burro Mountain Junction, a distance of about ten miles.

Phelps Dodge launched its great experiment in town planning and development on July 14, 1915, when Walter Douglas announced plans to build a model company town in Grant County. In contrast to the typical mining camp of the West, which grew like a tumbleweed and disappeared almost as fast, Tyrone would be carefully planned and constructed. Phelps Dodge hired one of the best architects in the United States to design the commercial buildings and employees' houses. He was Bertram Grosvenor Goodhue of New York City, who delighted in fanciful and romantic creations, the designs of which were stimulated by trips he made to Mexico in 1892 and to Iran and India ten years later. He made the latter journey as the guest of a wealthy client from Montecito, an upscale suburb of Santa Barbara, who wanted the architect to design Persian-style gardens to complement his mansion. Among his many commissions were Saint Bartholomew's Cathedral (1919), which is located adjacent to New York's posh Waldorf-Astoria Hotel and diagonally across Park Avenue from Phelps Dodge's onetime headquarters in the Colgate-Palmolive Building, a campus plan for the California Institute of Technology in Pasadena, and the initial design for Nebraska's skyscraper capitol in Lincoln.

Just before his Tyrone commission, Goodhue achieved lasting fame by designing the Spanish colonial–style buildings of San Diego's Panama-California Exposition, held in Balboa Park in 1915 to celebrate the opening of the Panama Canal. They are still a major tourist attraction. By that time, Goodhue was probably the world's foremost authority on Spanish colonial architecture, and it was no accident that buildings in Tyrone resembled those in Balboa Park.

In the California resort community of Santa Barbara, where the Walter Douglas family kept a summer home to escape the desert heat of southern Arizona, the connection between Phelps Dodge and Goodhue was apparently first forged. Legend states that two influential women goaded Walter Douglas into supporting construction of Tyrone: his wife, Margaret, the inspiration behind the station gardens and other forms of beautification along the tracks of the El Paso & Southwestern Railroad; and Grace Parish Dodge, wife of Cleveland Hoadley Dodge, a company director and son of William E. Dodge, Jr. Mrs. Dodge, who was awed by the beauty of the New Mexico site, strongly urged that the run-down shacks and company tents of old Tyrone and Leopold be replaced by an attractive and comfortable town that would improve the lives of

workers at the Burro Mountain Copper Company.

Phelps Dodge management itself viewed Goodhue's Tyrone as something of a social experiment. A company town with an ideal living and working environment might deter the kind of labor unrest becoming so common in industrial regions of the United States, and it provided good public relations. Goodhue's Tyrone permitted Phelps Dodge to distance itself from shabby settlements that all too often typified company towns elsewhere in the nation. In addition, the building of a model community was entirely consistent with the prevailing social philosophy at Phelps Dodge, particularly the Presbyterian philanthropic commitment that had led four generations of partners to fund schools, churches, hospitals, and decent affordable housing for their employees. Town construction along Niagara Arroyo began in 1915 and continued for three more years.

Goodhue's Tyrone

In his search for a suitable design, Goodhue wandered around New Mexico with his camera in hand to record images of early colonial adobe houses and Indian pueblos. As a result, his plans for a model company town adapted Spanish colonial architecture to the public and mining company buildings and arranged these around an attractively landscaped plaza. At its center were a marble fountain and a bandstand. Tyrone as designed by Goodhue embodied his idealized image of a Mexican village.

To prevent smoke and noise from disrupting

The RECENTLY *completed administration building of the Burro Mountain Copper Company in Tyrone in 1915. The architect Bertram Goodhue devoted special attention to its details, and sculptors from Italy ensured that the cornices were done right. Tile designs once used in Spain's famous Alhambra, along with large murals, adorned its interior walls, and wrought iron ornaments and fancy chandeliers contributed to the building's elegance.*

and multiple-unit dwellings to accommodate 235 families.

Plans were in place for other buildings, but before work could begin, the copper boom ended unexpectedly. Even without additions, the existing ensemble of Mediterranean-style buildings with red tile roofs and stuccoed exterior walls stood in colorful contrast to the juniper- and oak-covered terrain. When completed, Goodhue's Tyrone rated favorable coverage in the *Architecture Review*, the leading publication of his profession. It was said to have cost $1 million, but Phelps Dodge probably spent only about half that sum. Even so, the results were still stunning.

Not only was it beautiful, but Tyrone was also a model of social order—at least as Walter Douglas and his father envisioned it. The settlement contained no saloons and no brothels. Phelps Dodge sold no land to outsiders, yet it encouraged retail competition by leasing space to merchants on the plaza opposite the Mercantile. Further, the corporation charged low rents because it hoped to encourage mine and mill workers to remain loyal employees for years to come.

By early 1917, amidst the hammering and sawing of construction crews, new Tyrone emerged as a community of thirty-five hundred residents. As was typical in all corners of the Southwest at the time, townspeople clustered in ethnically and economically segregated residential districts, where nearly everyone shared the language and customs. The Anglos (Americans as they were then called), who were most often employed in management jobs or as skilled miners, occupied the high ground southwest of the central plaza; Mexicans and

the cleanliness and order of the main plaza, Phelps Dodge officials directed that locomotive engineers back their passenger trains into the depot, which Goodhue consciously modeled after the Santa Fe Railway station in San Diego. It featured Moorish-style arches, mahogany panels, blue glazed tiles, and long shaded corridors.

All down the line, Phelps Dodge spared no expense to create beautiful public buildings for Tyrone. For example, the Thomas S. Parker Hospital reportedly cost $100,000 to construct, and it featured sunken tubs appropriate to a fine resort hotel. In addition, there was a nineteen-room public school, the Tyrone Union Church (which embraced all Protestant faiths), a small justice of the peace court (which originally cost $3,200 to build and still stands), a clubhouse, restaurant, garage, recreation hall, and morgue. On nearby hillsides, workmen constructed single

Mexican Americans, who generally worked as unskilled laborers, settled below in Encinal and Pinal canyons. Such housing arrangements made visible the geography of power that was central to the layout of every company town, and it reflected the racial and ethnic segregation then so much a part of everyday life in the American Southwest.

On the other hand, Goodhue intended the central plaza to integrate residents of all income levels and ethnic backgrounds. On a typical Sunday, and following a custom popular in Mexico, two lines of boys and girls promenaded past one another so that youthful eyes could meet and romances flourish. Serenading the young lovers and the watching crowds were Mexican bands. In the central plaza, Phelps Dodge posted no "Keep Off the Grass" signs because it intended the plot for public use; it also provided residents a baseball park, tennis court, golf course, and rifle range. Farther down the Niagara valley, the Burro Mountain Copper Company erected a concentrator with the capacity to process two thousand tons of ore per day. Although industrial structures bulked large in the landscape and the rumble of crushing reverberated along the arroyo, Phelps Dodge assured that the noise and dust from milling and mining operations had little impact on the town itself. That was in contrast to other parts of industrial America where workers' hovels often clustered around grimy mills or factories.

The mines at Tyrone operated from April 1916 until May 1919, when slumping copper prices after World War I combined with relatively high mining costs to force Phelps Dodge to suspend underground mining there. Development work and mill expansion continued

for a time longer. In fact, mine production resumed in August 1920 only to cease again eight months later. Then, Tyrone was the seventh-largest city in New Mexico. The Burro Mountain Branch produced an average of 10.5 million pounds of copper each year (ranging from a high of 17 million pounds in 1918 to a low of 4.5 million pounds in 1921), but its annual output paled in comparison to that of Phelps Dodge properties in Arizona, and its profits were not impressive. The corporation anticipated that Tyrone would make a profit as long as the price of copper dropped no lower than 15 cents a pound. Thus, after the price fell to 13 cents a pound in 1921, the town's future looked doubtful.

In an emotional meeting held in the plaza with director Cleveland H. Dodge, miners offered to trim their pay by 25 percent to keep working, but there was nothing Phelps Dodge could do. The El Paso & Southwestern provided special trains for all workers who wanted to return to Mexico, while the rest of the town's

An "American house" built in Tyrone in 1916. Arched openings and tinted stucco enhanced the austere cubes that ranged in size from three to five rooms. Each one cost Phelps Dodge from $1,500 to $2,800 to build and rented for $25 to $30 a month. By the time construction halted—when the United States joined the First World War in April 1917—a total of 77 of these company houses and 124 more spartan ones for Mexican workers had been completed.

population simply drifted away. After operations ceased a second time, Tyrone shrank from seven thousand residents to a few hundred. Goodhue's Tyrone lasted a scant six years.

Instead of selling the property, Phelps Dodge held on to its Burro Mountain Branch and continued a small-scale copper-leaching operation. Even that ended in 1928 when the company removed the pumps and allowed groundwater to flood the tunnels, thus effectively ending hopes for resumption of underground mining. Tyrone remained a virtual ghost town for the next forty-five years—perhaps the nation's most beautiful one. During those years it was never completely depopulated, but certainly it was a ghost of its former self. Gradually some buildings fell into decay, and Phelps Dodge dismantled or relocated others. It sold the power plants, pulled up the tracks on the railroad line in 1935, and removed the freight depot,

foreman's house, and laborers' quarters. However, it kept the elegant passenger depot on Tyrone's main plaza standing until 1967.

A portion of the old property generated modest income as a vacation resort called Rancho de los Pinos. Among the health seekers and vacationers were Jonas Salk, later famous for his polio vaccine, and professors and their families from Yale University and the University of Chicago. All came to ride horseback in the Burro Mountains, play tennis and swim, and enjoy outdoor and indoor games and recreations. Phelps Dodge executives and their families frequented Tyrone too.

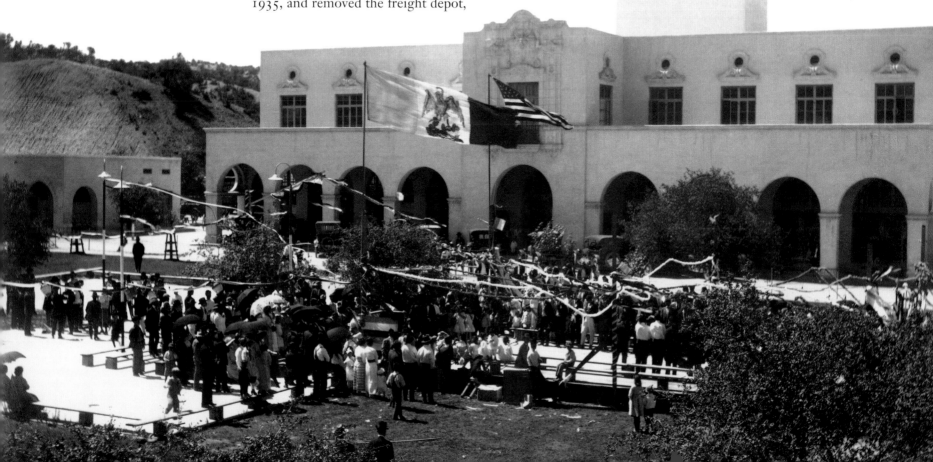

Except for some development work and intermittent copper leaching, the mining area remained dormant until a drilling program initiated in 1949 delineated the limits of the present Tyrone ore body. But until the right technology came along in the mid-1960s, the low-grade deposits were not worth mining. Only in 1966 did Phelps Dodge announce the reopening of Tyrone—this time as an open-pit mine. The following year it started removing the overburden, and in 1969 it commenced production. In this instance, the word *overburden* seems especially poignant because it included most of Goodhue's Tyrone.

The Company Store

APART FROM MINES and processing works, two institutions that effectively defined the urban landscapes of every Phelps Dodge town across the Southwest were the company hospital and the company store, or "Merc," as it is familiarly known today. Phelps Dodge Mercantile originated in Bisbee when the Copper Queen Consolidated Mining Company purchased Mary Crossey's general store in 1886. The widow Crossey's store dominated the local market, and like most shopkeepers of her day she sold everything from groceries to hardware, though in Bisbee hardware included picks, pans, guns, and dynamite.

In pioneer days, it was usually merchants and not miners of the American West who grew rich from the new diggings. Unfortunately, some of the shopkeepers who closely followed the mining frontier sold mainly shoddy merchandise and substandard foodstuffs at exorbitant prices. Customers could do little but grumble because there was nowhere else nearby to shop. That was true too in early Bisbee. Complaints by workers about high prices and inferior merchandise bothered Dr. Douglas, who believed that Copper Queen miners might well benefit if his company could purchase good-quality wares in volume and resell them at reasonable prices. That was his intent when Douglas purchased Crossey's emporium and renamed it the Copper Queen Store. The partners of Phelps, Dodge & Company, with more than fifty years of experience as a mercantile and trading company, had no trouble sanctioning his purchase.

On December 14, 1911, because growing trade at its several company stores required greater efficiency, the corporation reorganized and incorporated them all as Phelps Dodge Mercantile Company. This saved the new firm money by allowing it to purchase large quantities of merchandise and to expand along with Phelps Dodge's mine holdings. William H. Brophy, a charming Irishman and the original manager of the old Copper Queen Store, continued as general manager of the $5 million enterprise that hired approximately 450 people in 1912. In all, Brophy remained at Phelps Dodge for thirty years.

A large new Mercantile opened at the company's coal property in Dawson, New Mexico, in 1913. In the three-story brick store were the usual fresh meats and vegetables, hardware, furniture, drugs, clothing, and shoes; there was also a bakery, along with an ice plant, auto service station, and mortuary. With the company hospital nearby, Phelps Dodge paternalism in Dawson extended from cradle to grave. The Mercantile that opened in Goodhue's Tyrone in 1917 was not only the largest commercial emporium in New Mexico but also one of the most beautiful buildings in the West. It featured an Italian marble staircase, an elevator, a lofty bell tower (minus the bell), and an elaborate bas-relief facade.

In 1919 the six stores and general offices of Phelps Dodge Mercantile employed 595 people, some 33 percent of them women. Until 1931, the operation expanded rapidly and established a number of new stores at company mine and smelter operations. At various times the list included stores in Leopold, Tyrone, and Dawson, New Mexico; Bisbee, Lowell, Warren, Morenci, Clifton, Stargo, and Ajo, Arizona; and Pilares and Nacozari, Mexico. It added three stores in

the Clifton-Morenci area when the parent corporation purchased the Arizona Copper Company in 1921. Prior to that time the Arizona Copper and the Detroit Copper Mining Company (popularly known as A.C. and D.C.) stores competed with one another for local business.

Phelps Dodge stores also closed along with mining operations. The original Tyrone store locked its doors for a final time in the early 1920s, and the Dawson store did likewise when the company ceased coal-mining operations in northern New Mexico in the early 1950s. Expansion of the Morenci open-pit mine in the early 1960s required Phelps Dodge to raze the original company store along with the old town, but when it built an attractive new shopping mall in 1966, Phelps Dodge Mercantile was the centerpiece. A Mercantile commenced operation at new Tyrone in the late 1960s when open-pit mining and milling began nearby, and another in the newly established smelter town of Playas in the early 1970s.

Especially in the latter half of the nineteenth century, the term *company store* had a decidedly negative connotation. In coal towns of the East, company stores gained a reputation for gouging employees and holding them forever in debt: "I owe my soul to the company store," in the bitter refrain of a popular song. Contrary to common practice, Phelps Dodge neither pressured its employees to trade at the local Mercantile nor used its dominant position in a community to prevent competition, charge unfair prices, or sell shoddy merchandise. New employees of the mining company invariably received "instant credit" at Phelps Dodge Mercantile as soon as they started work.

In few places did the Phelps Dodge store have a monopoly. Dr. Douglas himself opposed the idea of captive customers. In the early days of retail trade in the Bisbee, Douglas, and Morenci areas, the company store regularly solicited business door-to-door by boys hired for that purpose. In Bisbee in 1923 the competition included thirty independent grocery stores, at least six clothing merchants, and one very large and prosperous department store.

All through the 1920s and 1930s the Mercantile repeatedly checked "keybills" to make certain its prices were similar to competitors'. When price tags in Dawson seemed out of line with those in Raton and Trinidad, P. G. Beckett, vice-president and general manager out West,

wrote to A. W. Liddell, Mercantile general manager: "Please watch out for your Dawson keybill, which looks somewhat high compared with the others."[4]

Running a company store was never easy. "Every professional labor organizer, many professional politicians, and all other persons who for any reason are against the companies are constantly hounding them for maintaining stores," noted an observer of Arizona's copper towns whose findings were published in the *Saturday Evening Post* in 1923. That year the state legislature debated a bill to make it illegal for any mining company to operate a general store and deduct from wages the cost of purchases. "One notable feature about the company stores I saw in

☙ MAINTAINING A WELL-RUN COMPANY STORE ❧

OLD LETTERS REVEAL *that Phelps Dodge general managers like P. G. Beckett took no less interest in Mercantile matters than they did in mining or smelting. Writing from his office in Douglas in 1929, Beckett complained to A. W. Liddell, general manager of Phelps Dodge Mercantile, that clerks in the men's clothing department of the Bisbee store "give a very poor impression to the public." Beckett was distressed to see them talking and laughing together and "sprawled all over the counters." If nothing else, he suggested, keep them busy through the workday rearranging and checking stock, because what he saw "wouldn't be tolerated in any well-run store." A short time later, the Mercantile sent all its employees a pamphlet that offered detailed advice on how to dress and behave on the job, along with an outline of their duties and benefits. The tiny guide closed with words full of meaning to the wise: "We hope you will enjoy your association with us and that you can stay until you can collect your retirement benefits."*

(Based on a letter from P. G. Beckett to A. W. Liddell, September 11, 1929; *Phelps Dodge Mercantile Employees' Guide*, copy in Phelps Dodge Mercantile records, Phoenix.)

the Southwest," continued the observer, "was the high quality of the goods carried. Evidently company stores do not dare carry shoddy goods." That was still true in the late 1990s. At Phelps Dodge Mercantile in Morenci the quality of the meat and produce was equal to, and often better than, that found in metropolitan supermarkets. Most prices were comparable.[5]

To diffuse criticism "so often used by labor agitators that all wages paid are returned to the same company in exorbitant prices for merchandise," Phelps Dodge in the early 1930s debated whether to turn all its Mercantiles into cooperative stores. At the time only the recently acquired New Cornelia Cooperative Mercantile at Ajo operated that way, and it continued to do so until 1975 when Phelps Dodge Mercantile fully absorbed it. Incidentally, the New Cornelia store between 1917 and the end of 1936 paid out a total of $807,000 in Christmas rebates to miners and their families. In 1936 the average refund was slightly more than 12 percent.[6]

For many years, Phelps Dodge Mercantile charged no interest on installment accounts. An employee who wished to buy major household items, such as new furniture or appliances, simply made arrangements with the store for stated deductions from each paycheck to cover the cost of the purchase. Phelps Dodge Mercantile even extended credit to employees during times of strikes, and it carried employees who could not pay their bills on time, as was often the case during the Great Depression. Many times it simply wrote off bad debts. In late 1929, at the start of the long economic downturn, the Mercantile in Dawson accumulated a considerable number of delinquent accounts

because discharged workers occasionally slipped out of town without paying their bills. The corporation's bad account problem only grew worse in the months to come.

Because of the hard times, Louis S. Cates, the new president of Phelps Dodge, urged P. G. Beckett in November 1931 to skip the usual year-end rebate at the New Cornelia Cooperative Mercantile in Ajo, which it had paid every Christmas since 1917. Phelps Dodge had acquired the store a few months earlier as part of its Calumet and Arizona deal, but Cates added, "however, if you still feel that we are not taking the proper course in this matter, I would be pleased to hear from you further." Beckett, who thought that such a drastic step would make Phelps Dodge look like Scrooge to its new Ajo employees, made a strong case for a rebate of at least 5 percent (down from the customary 10) and won Cates over. "This will be mighty good news to the people in the district," Beckett emphasized.[7]

Phelps Dodge Mercantile in early 1999 still operated a store in Morenci, with fifty-five

THE FRONT COVER OF a book of coupons redeemable at Phelps Dodge Mercantile. The mining company deducted the cost of $20 from an employee's paycheck. If a worker bought too many of these little books, it was possible to receive a paycheck with a zero balance.

thousand square feet, and a smaller one in Playas, with fourteen thousand square feet. It ran the Morenci Motel, the Morenci Lanes Bowling Alley, and a convenience store too. In Playas it ran the Copper Pins Bowling Alley and Fitness Center. The Merc operated a wholesale division that sold hardware to both "internal and external customers." Its catalog was as thick as any that Sears Roebuck ever issued.

As for the company towns, during the decade of the 1990s, Phelps Dodge itself sold off several hundred dwellings in modern Tyrone and Ajo to individual buyers, as well as Ajo's attractive central plaza. It retained both Morenci and Playas. Neither place fits the stereotype of a grim company town composed of row upon row of identical houses slapped together from the lowest-cost materials to squeeze maximum

returns from renter-employees. And neither one is a throwback to nineteenth-century corporate paternalism. Nor are Morenci and Playas urban dinosaurs.

Only consider the countless ghost towns scattered across the West. With few exceptions (notably Aspen, Colorado), when local mines played out and prosperous times ended, real estate values plummeted and never rebounded. Often all that remained of once lively settlements were a few slanted storefronts and half-collapsed houses where "empty windows stare like wistful eyes down streets where nothing moves save memories and the wind." When Phelps Dodge chose to provide housing in remote one-industry towns, it shouldered the risk inherent in the real estate markets associated with boom-and-bust industry; it protected employees from loss of

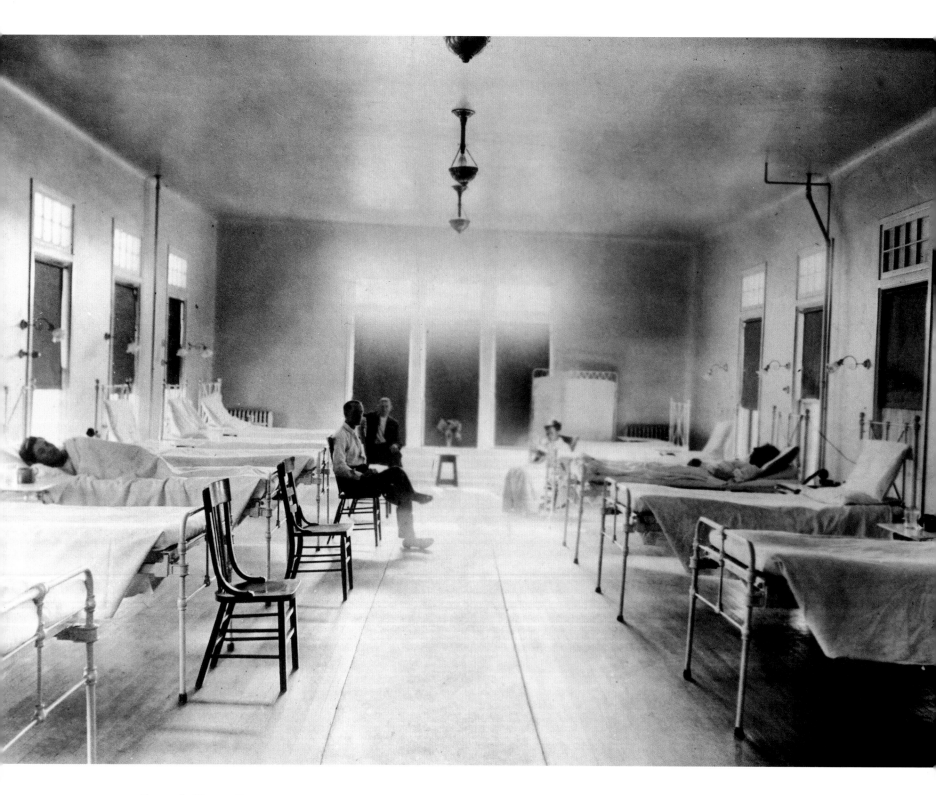

equity should the mines or smelters halt production. At Playas, so supremely isolated in a vast expanse of rangeland that extends south to the Mexican border, employees who paid low rents for company housing often purchased homes of their own in other parts of New Mexico, including several resort communities, that would serve them well after they retired from Phelps Dodge.[8]

BESIDES THE COMPANY store, every turn-of-the-century Phelps Dodge town in Arizona and New Mexico had its company hospital. Not only was Dr. Douglas the inspiration behind the first Copper Queen store in Bisbee, but he also took great personal interest in all Phelps Dodge health-care facilities, including the Thomas S. Parker Hospital that took shape in Goodhue's Tyrone (though he apparently never visited it in person because of his advanced age). Many

experts regarded the facility as one of the best-equipped clinics in the Mountain West. For this and other kinds of Phelps Dodge health care, a married employee paid two dollars per month for an entire family, and a single employee half that amount. In a Phelps Dodge town, it is true, no real separation existed between life on and off the job. Mining activity not only helped fund community amenities like good schools and hospitals but also determined the location of major buildings. In addition, it defined the ordinary rhythms of life, most notably by encouraging steady work during good times and forcing layoffs in bad. Unfortunately, this close association meant that the whole community suffered when a disaster struck. That was why everyone living in a Phelps Dodge town benefited when the corporation promoted safe production in its mines and processing facilities and why safety became the centerpiece of its landscapes of labor.

～ *Chapter Six* ～

LANDSCAPES *of* LABOR

WHEN VARICOLORED FLAMES shoot forth and sparks fly as workers periodically force compressed air into the copper matte to burn out the sulfur and iron impurities, visitors to the converter aisles at Phelps Dodge's Chino and Hidalgo smelters witness an industrial process every bit as dramatic as a Fourth of July fireworks display. The skills that overhead crane operators use to move and pour massive cauldrons filled with glowing molten copper are impressive. No less impressive is the evangelical fervor with which Phelps Dodge preaches *and practices* the gospel of safety at its various production facilities.

From a Columbian Chemicals Company carbon black plant in West Virginia to a Phelps Dodge International wire and cable plant in Thailand, there is a common emphasis on safety. It forms an integral part of the corporate culture. Workers monitor one another for safe practices, and safety signs are everywhere. A common one admonishes, "If it's not safe, don't do it." The most often repeated mantra is Safe Production. Those two words are emblazoned on company trucks, hard hats, signs, and

---

～ BISBEE'S TURN-OF-THE-CENTURY *underground miners faced many hazards that workers on the surface could scarcely imagine. If a miner expected to survive any length of time in the subterranean world, he learned to avoid a careless misstep that sent him plunging down a ladder into the abyss, and he made certain that the bucket lifting him up the shaft to the surface did not swing dangerously from side to side, catch on an unseen projection, and spill him out.*

jackets, and they emphasize that no distinction can be made between work and safety concerns. No one enters a Phelps Dodge mine, mill, or manufacturing facility without first putting on a hard hat and adding ear plugs, goggles, respirator, and special clothing where appropriate.

Since earliest times, danger attended the work of mining and processing copper. That work still requires gargantuan equipment designed to dig, crush, and burn. As molten copper pours into the anode casting wheel at the Chino Mines Company, for example, an electronic gauge registers a blistering 2,300 degrees Fahrenheit (paper, by contrast, ignites at 451 degrees). Yet in the United States, the facilities of Phelps Dodge Mining Company have recorded accident rates less than one-third the industry's national average. As one safety-conscious employee phrased it, "Our job is to take hell and make it a safe place to work."

The rise of safety consciousness, like so many attitudes and practices of today's Phelps Dodge Corporation, dates back to Dr. James Douglas, who in 1913 launched one of the metal industry's first comprehensive safety programs at the Copper Queen Consolidated Mining Company in Bisbee, Arizona. That was a memorable year for heightened safety consciousness because on October 22 at Dawson, New Mexico, 263 coal miners perished in the worst single tragedy in Phelps Dodge history (and the second-worst mine disaster in United States history). Only two days earlier, the state coal mine inspector Rees H. Beddow completed a two-week check of local conditions, and before he left Dawson he officially pronounced Mine

No. 2 to be "totally free from traces of gas, and in splendid general condition." Obviously, everyone had more to learn about safety.[1]

Safety became a key feature of the landscape of labor as Phelps Dodge evolved during the first two decades of the twentieth century, years that easily rank among the most prosperous and optimistic of times for the United States copper industry. World electrification, though still confined mainly to urban areas, continued to generate unprecedented demand for red metal. Moreover, the industry achieved astonishing decreases in production costs as companies mechanized their underground mines or embraced the lately developed open-pit technology. Such changes benefited Phelps Dodge too, for it enjoyed exceptionally good years—if measured strictly in terms of technological improvements and economic gains. However, those were also the years of the Dawson tragedy and World War I (1914–18) with its horrific bloodletting on the battlefields of Europe.

Amidst all the change there was one constant: Bisbee remained the biggest and brightest star in Phelps Dodge's universe. It was a tribute to the timeless importance of both Bisbee and the Copper Queen that when Phelps Dodge transformed itself from a holding company into an operating company in 1917—from Phelps, Dodge & Company into Phelps Dodge Corporation—it used the original charter issued to the Copper Queen Consolidated Mining Company thirty-three years earlier. In fact, even today the official seal of Phelps Dodge Corporation includes the date August 10, 1885, when formal incorporation of its Copper Queen property took place.

A GROUP OF *underground miners is hand drilling blast holes beneath Tombstone, Arizona, in the early 1920s. Phelps Dodge gained the Bunker Hill Mines Company, an old Tombstone silver property, along with the venerable but bankrupt Tombstone Consolidated Mining Company in 1914. Its Tombstone mines earned a small profit, made no major discoveries, and were in the end plagued by water rushing into deep passageways.*

On the other hand, Bisbee was also the site of the single greatest controversy in Phelps Dodge history: the mass expulsion of more than a thousand striking mine workers in 1917. Critics still invoke the Bisbee deportation (often conveniently forgetting its wartime context) whenever they want to assert that Arizona wears a "copper collar."

Alive with Ceaseless Movement

NO RESIDENT OF early Bisbee or Morenci ever doubted that mining was the community's primary reason for existing. During the era of underground work, successive shifts of men drilled, blasted, mucked, and trammed to send copper ore to the surface. There the noise of dynamite blasts and rock crushers competed with bellowing steam hoists and steam railway locomotives laboring to surmount mountainous grades to reach the tangle of tracks, trestles, and embankments that defined the local landscape. When their mines were in full production, both towns resembled beehives alive with ceaseless movement.

Bisbee was home to approximately nineteen thousand people in 1913, a year when the combined mines of the Copper Queen, the Calumet and Arizona, and the Shattuck and Arizona companies employed as many as six thousand men and shipped more than thirty thousand tons of ore to the Douglas smelters each week. The district's monthly payroll reportedly averaged $750,000, of which these three mining companies accounted for about two-thirds. Metal miners earned some of the fattest paychecks in the United States, and only rarely did a family man quit his job.

The miner shared the surface landscape with all townspeople, but the semidarkness of the stopes and drifts below was his special domain. In 1903 the Copper Queen alone contained so many levels of subterranean workings that it was hard for outsiders even to visualize the maze of passageways. Underground miners developed their own distinctive lingo, hierarchy of jobs, and work customs and superstitions. Any man who had worked regularly since the early 1880s had seen about as many changes underground as above, on the streets of Bisbee or Morenci. Massive pneumatic drills, introduced shortly after 1900 to reduce fatigue and improve output, had rapidly transformed the much slower hand drills

into museum pieces, and electric motors that powered mine trams put the last mules out to pasture. On their heads miners wore carbide lamps that replaced flickering candles for illumination, and superseding the glow of burning carbide were the powerful beams of electric storage battery lamps. For explosives, black powder gave way to nitroglycerin and then to sticks of dynamite. Nitroglycerin-based explosives were powerful, but miners complained of "nitro headaches," an unwanted side effect.

Certain rhythms of life remained unchanged. Miners in Bisbee and Morenci celebrated the Fourth of July, Memorial Day, and Washington's Birthday with parades, sporting events, dances, and often fireworks and dynamite explosions. During holidays they especially enjoyed watching or participating in contests of strength and endurance. Typically on the Fourth of July they thrilled to rock-drilling contests finished off by hotly contested games of baseball between Morenci and Clifton or between Bisbee and Douglas. They commemorated foreign holidays with gusto as well, and often at Bisbee's Pythian Castle the honored country's flag flew over the building just below the Stars and Stripes. Another day worth celebrating was the tenth of each month, payday at one time for many miners throughout Arizona.

Even as the landscapes of production evolved, so too did the composition of town populations. As onetime mining camps matured, they became cities of families. More and more newcomers arrived with their wives and children, and local stores advertised the latest in fashions for both men and women. The coming of families, in turn, intensified local efforts to clean up streets and walkways, control diseases, and eliminate vices often associated with saloons. By the era of World War I, Arizona had adopted statewide prohibition, a reform once considered wholly unthinkable in the hard-drinking mining camps.

Further altering town populations was the increased mechanization of mining in early twentieth-century America, which reduced the level of skill required for many jobs. Thus it became possible—even desirable from the perspective of management in a highly competitive industry—to reduce labor costs by hiring less-skilled workers at lower wages. That was why mines and smelters of the Southwest came to rely on immigrant labor, often from southern and eastern Europe. Giving Bisbee a cosmopolitan quality were the Serbian, Irish, Welsh, Cornish ("Cousin Jacks"), Hispanic, and other immigrant groups that worked for the Copper Queen and competing mining companies. Many new arrivals came from Italy, Germany, Sweden, and Switzerland, too. The diversity of population was much the same in Morenci, although it had a much higher percentage of Mexicans and Mexican Americans in its workforce than did Bisbee.

No doubt about it, living in a Phelps Dodge mining town at the beginning of the twentieth century was different from living elsewhere. The community's collective pulse beat at a pace determined by world demand for copper; and especially for workers and their families, the mine or smelter whistles defined distinct hours of labor and leisure. There was no doubt about who had power and who didn't. It was no less visible on the shop floor or in the mines than in the community itself, where a geography of power extended from on-the-job relationships into different neighborhoods and everyday social contacts. Bisbee residents in those days were unabashed about class distinctions, and they easily identified dwellers on each of the town's four major hills with different economic and ethnic status. In Dawson too, the different national groups tended to live in separate sections.

In mining communities of the Southwest, various forms of residential and work segregation reflected the prevailing biases of American society—at least until the 1960s. For years the Morenci Social Club closed its doors to the town's many Hispanic residents, who formed a separate Spanish American Club instead. Popular prejudice relegated Mexicans and Mexican Americans to less prestigious and lower-paying jobs, such as mucking and tramming underground, although Bisbee at one time did not permit them belowground at all. In early twentieth-century Arizona, even labor unions supported a variety of laws hostile to Hispanic workers. Nearly all mine foremen, engineers, and mechanics were native-born whites or immigrants from northern and western Europe. Women did not hold jobs in the mines, but a few worked as secretaries and stenographers in company offices.

Most residents accepted the hierarchical structuring of their lives as the natural order of things, and no doubt many workers believed that if they only showed initiative and drive they too might rise to positions of authority. Other workers knew that certain jobs were not available to them, given the prejudices of the time, but they were still glad for steady work that provided

money to feed, clothe, and house a family. Yet what many workers valued as corporate benevolence—low rents for employee housing, the hospitals, libraries, and credit freely granted by the company store—others scorned as paternalism. Sometimes workingmen displayed their resentment by joining one of the labor unions that first became significant in mining and smelter towns of the southwestern borderlands during the decade before World War I. Such was life for Phelps Dodge employees in the turn-of-the-century West.

The Search for Safe Production

IN EARLY BISBEE and Morenci few people voiced public concern for the safety and well-being of employees. In the 1880s and 1890s such indifference was common in industries across the United States, and even among workers themselves. It was all too easy for a miner to be

THE NEW ITALIAN Band of Morenci. In 1911 about seven hundred men were employed in mining in the camp, 74 percent of them Mexican and 10 percent Italian. Music was such an important part of mining town life that Bisbee miners who were also musicians often worked in a special "band stope." Because of their important civic duties, which included attending all funerals and marching in front of the hearse from the undertaking parlor to the cemetery, good bandsmen were not expected to labor too hard underground.

injured while working with a drill, explosives, or high-voltage electricity or to fall down dark shafts, get kicked by mules, step on nails, be hit by a falling tool dropped by a careless worker, sprain his back, or hurt his eyes. Yet as late as 1910, the year the federal Bureau of Mines was established, most observers accepted industrial accidents as part of the inevitable hazards of work.

Attitudes changed and equipment improved, but underground mining in the early twentieth century remained an exceedingly dangerous and difficult occupation. That truth haunted the consciousness of everyone, not just the working miners, who lived in an active copper or coal town. No one knew when accident or death on the job might unexpectedly disrupt the pulse of everyday life for the whole community and rob it of one or more of its inhabitants.

Safety devices by today's standards were crude, but the mines of Bisbee were actually safer than most. In fact, the main threat to a miner's health came from diseases contracted in town. The Copper Queen Consolidated Mining Company noted in 1904 that it was typhoid fever, not accidents, that accounted for more than ten thousand days that employees lost from production that year. As towns cleaned themselves up, so too did the mines. Companies practically eliminated the danger of typhoid contracted in the mines by providing sanitary latrines underground and piping city water to drinking fountains located in various parts of the workings. These protective measures together with modern change houses, where clothes worn for subterranean work were thoroughly dried between shifts and where each man had a locker

and access to hot showers, helped maintain the comfort and health of miners. So too did adding artificial ventilation underground.

Even as worker health improved as a result of turn-of-the-century modernization, the conviction grew throughout the industry that mining jobs themselves could be made safer. That was especially true for coal mining, which easily ranked among the most dangerous occupations in the United States. The same high probability of accident and death applied to Phelps Dodge coal operations in northern New Mexico, and the shocking disasters that occurred there in 1913 and again in 1923 only spurred the search for safety, not just in Dawson but in Bisbee and other properties as well.

Phelps Dodge purchased the Dawson Fuel Company in 1905 to secure coke to fuel its copper smelters in Arizona as well as coal for steam locomotives on its growing network of railroad lines. During its fifty-year history as an active Phelps Dodge property, Dawson Fuel dug a total of ten underground mines into a continuous seam of low-sulfur coal that measured more than ten miles long, five miles wide, and often as much as eleven feet thick. At one time, several hundred coke ovens located at the edge of the prosperous coal camp vented fire and smoke day and night.

The mines of Dawson operated under the name Stag Cañon Fuel Company, a wholly owned subsidiary, until its name changed in 1917 to Phelps Dodge Corporation, Stag Cañon Branch. The town of Dawson itself was located at 6,393 feet in the New Mexico highlands about halfway between Cimarron and Raton and about eight miles south of the Colorado–New Mexico border. Isolated at the end of a railroad branch, it was thirty-five miles from any town of similar or larger population. Dawson once claimed a population of almost nine thousand residents, and like Bisbee and Morenci, it ranked among the most culturally diverse of the two hundred or so company towns that at one time dotted the American West. In fact, more than 80 percent of its coal miners were foreign born.

With tree-shaded sidewalks, sanitary sewers, and a large and prosperous business district, Dawson in appearance was a far cry from the stereotypical coal camp clustered about a mine mouth and surrounded by unsightly piles of rock and ashes, all blanketed with smoke and soot. Phelps Dodge intended that Dawson be a model among the several coal-mining communities of the Southwest; soon after it acquired the property, the company arranged to build hundreds of modern homes for employees and their families, pave town streets, and landscape both the business and the residential districts. In time, Dawson could boast of a hotel, dispensary, movie theater with daily shows, bowling lanes, billiard parlor, swimming pool, and one of New Mexico's first golf courses.

Two churches provided regular religious services. Dawson's opera house was one of the most popular theatrical centers of the Southwest, and many celebrities appeared on its stage. An imposing three-story red brick Mercantile opened in 1913. Dawson schools, including the James Douglas High School that Phelps Dodge built in 1920 for $100,000, ranked among the best in the nation. Back in 1906, shortly after it acquired the property, the company opened a thirty-two-bed hospital. Both it and the town's

dispensary were designed by Henry C. Trost, the architect who did so much to redefine the urban landscapes of early twentieth-century Morenci and Bisbee.

Phelps Dodge was justifiably proud of its Stag Cañon coal mines and of Dawson. From its model ranch and dairy to its modern swimming pool, golf course, and tennis court, Dawson was obviously "different and distinct from coal camps in general on account of [the Stag Cañon Fuel Company's] progressive policies and its care of the workmen and their families." The company proudly described its coal operations as representing the "highest achievement in modern equipment and safety appliance[s] that exists in the world." No resident of the close-knit coal camp thus anticipated the tragic events of October 22, 1913.[2]

Apart from a fire that killed three miners in 1905, the safety record compiled by the Dawson mines was among the most laudable in the United States. That included Mine No. 2 where the 1913 tragedy originated when a worker attempted to set off an explosive charge by momentarily touching a shooting wire to a live trolley feed. Company rules forbade the risky procedure. The ensuing explosion blasted air and flames down the tunnel and, by simultaneously lifting and igniting fine particles of coal dust, triggered a second, even more horrific detonation and fire.

To eyewitnesses nearby, flames that shot a hundred feet out of the mine mouth looked like a "volcano in eruption." Accompanying the dense smoke that belched from the portal was a blast of rocks and debris that almost buried two miners standing just outside. Reaching up from beneath

Cars on new main line — at Unfinished ovens

A photographer in *the early twentieth century aimed his camera down the tops of a string of railroad boxcars. They divided his image of workmen building new coke ovens at Dawson (left) from company housing (right). The Stag Cañon Fuel Company was once New Mexico's largest coal-producing property. The biggest year was 1916 when sixteen hundred Phelps Dodge employees mined and processed nearly 1.5 million tons of black diamonds.*

the earth came a deep rumble that rocked homes of townspeople located a mile and a half away. When a siren sounded only moments later, everyone in Dawson sensed a disaster. Dropping everything, they raced into the streets and up the dirt road leading to Mine No. 2.

Soon joining them was the advance guard of an army of volunteers who headed to Dawson from coal camps across northern New Mexico and southern Colorado. Two special government trains, one from Pittsburg, Kansas, and another from Rock Springs, Wyoming, brought more rescuers. Dashing along the tracks from El Paso, a special Phelps Dodge train arrived with nurses and doctors. Directing the recovery attempt

was J. C. Roberts, an expert on safety and rescue from the United States Bureau of Mines. In total, 263 workers died. Phelps Dodge recorded the tragedy in its annual report but emphasized that, when it came to safety, "apart from the terrible loss of life in the explosion of October 22 last, the number of fatal accidents was the lowest in the history of the property."[3]

Safety consciousness became more important than ever before. Dr. Douglas authorized creation of the first formal safety department at the Copper Queen Consolidated Mining Company. During the dozen years after 1913, various branches of Phelps Dodge adopted safety programs of their own, although all of them

Phelps Dodge's Stag Cañon Fuel Company instituted numerous safety improvements at its Dawson mines after 1913. For example, it restricted to specially trained workers the hazardous job of loading holes with dynamite and connecting blasting wires, it installed more facilities to sprinkle water on coal dust to keep it from becoming explosive, and it increased the number of Edison electric safety lamps that miners used underground. But illustrating how difficult it was to implement truly safe production was yet another disaster that struck Dawson, this time on February 8, 1923, when a railcar jumped the track underground in Mine No. 1 and in a split second knocked down timbers "to which the trolley feed line was attached, raising a quantity of [coal] dust which was ignited by an electric arc [stemming] from the feed wire coming in contact with one of the iron pit cars." The 1923 explosion claimed 120 lives.[4]

The "deplorable loss of life" appalled P. G. Beckett, the man who served as general manager of all Phelps Dodge western mining properties from January 1920 until 1936. He reaffirmed that it was and had "always been the aim of the Phelps Dodge Corporation to make its mines as safe as it is humanly possible to make them. The safety of its workmen comes before everything else in our operations." Increasing its vigilance, the company went so far as to search all miners for matches before allowing them underground to work. No more major disasters occurred at Dawson. In fact, Stag Cañon was among five honorees in the 1928 National Safety Competition held to recognize coal mines compiling the best safety records.[5]

The 1923 casualties at Dawson encouraged Beckett and his colleagues to launch a detailed

Before the rise *of safety consciousness: a bit of horseplay with an electromagnet at the United Verde Copper Company in Jerome.*

mainly emphasized the need for better safety devices. As for raising the safety consciousness of workers themselves, the dominant belief was that no book of rules or special safety procedures could overcome the careless or bad work habits that inclined some miners to take risks. That defeatist attitude changed only gradually and after various forms of prodding, such as occurred in 1918 when the Copper Queen offered its first Practical Mining Course to assist new employees. Fully one-third of the training program was devoted to safety and accident prevention.

As part of its successful Safety First campaign in 1920s Mexico, Phelps Dodge's Moctezuma Copper Company used staged photographs to warn Nacozari mine workers of the dangers of high-voltage trolley lines.

Safety First, a 1920s admonition at a headframe in Nacozari, which begins with "I am careful and thoughtful in all my actions."

study of accidents at all Phelps Dodge metal mine properties. The resulting statistics highlighted the need to find still more ways to ensure safe production. For 1923 they revealed that the Copper Queen, Morenci, and Moctezuma branches and Old Dominion (the latter mine always included as part of operations where matters of safety were concerned) recorded a total of 21 fatalities, 516 serious accidents, and 1,461 minor mishaps.

Safety concerns at Phelps Dodge remained largely decentralized, each branch implementing its own measures as needed, until early in 1925 when the corporation launched a wide-ranging program of safety committees and plant safety engineers modeled after one at United States Steel. Phelps Dodge instituted basic safety measures that every person, employee or visitor, entering one of its mine properties today must accept. By the late 1920s the list of company-required safety attire included hard hats (then known as hard-boiled hats), goggles, steel-toed shoes, and safety belts, where appropriate.

At all Phelps Dodge branches first-aid teams competed for cash prizes and trophies, and by the early 1930s, basic knowledge of first aid was fast becoming a requirement for employment. During that decade the company compiled a nine-hundred-page textbook dedicated to training workers in safe production practices. Not all safety innovations, alas, worked as well as intended—for example, lightweight aluminum hard hats introduced in the 1950s tended to blow off in the stiff winds of Morenci. Nonetheless, Phelps Dodge was clearly headed in the right direction to achieve safe production. Statistics based on the frequency of accidents suggest that heightened safety consciousness prevented more than thirty-eight thousand disabling injuries during the years from 1914 to 1930 alone.

Toil and Trouble

AMONG MANY LEISURE-TIME activities available during the years before World War I, fraternal organizations were extremely popular across the United States, and no less so in Phelps Dodge mining and smelter towns. Every lodge and brotherhood maintained secret rituals and costumes; and their members marched in parades, organized memorial services, and held public dances and fairs. Collectively, the lodges created an atmosphere of extended family, generated business contacts, and provided mutual aid in the form of sickness and death benefits. The appeal of such benefits cannot be emphasized enough because in those years an individual had difficulty purchasing life insurance except as a member of an organization such as a fraternal lodge or trade union—and insurance was a matter of no small concern to miners, especially the married ones.

One distinct type of fraternal organization, the trade union, did not loom large at first in the several urban landscapes of Phelps Dodge. In early days, a familial spirit bound the Copper Queen Consolidated Mining Company to its employees, and much the same was true for Arizona Copper and Detroit Copper in the Clifton-Morenci area. But during the opening decade of the twentieth century, as copper production boomed and mining companies grew substantially larger, some workers no longer felt the old loyalty to managers. In fact, over time all

PROUD OF THEIR accident-free record, day-shift employees of the Southwest Mine, Copper Queen Branch, pose for a photographer in May 1926. In 1925, the first year of the new corporationwide safety campaign, every department at Morenci competed to achieve the lowest number of accidents. When the machine shop won, workers went home at noon, got cleaned and dressed up, and returned four hours later for food and awards. First came a drawing in which each person received something, then a movie, and finally the dinner. At a long table set up in the machine shop everyone enjoyed a huge pot of beans and ham.

of the large and prosperous western copper communities comprised both homebody and tramp miners, the latter typically being young, single, and restless men who drifted from job to job without developing any attachment to employer or place. Rates of annual labor turnover averaging one-quarter of the total workforce grew common in western copper mines. That disturbing trend caused the United States secretary of labor to worry in 1918 that so large a pool of migratory labor might develop a sense of injustice and serve as "inflammable material for beguiling agitators to work upon."[6]

Bisbee mining companies had long maintained the principle of the open shop. Management at Phelps Dodge firmly believed that if it took good care of workers they had no need to turn either to labor unions or to radical agitators; and at the turn of the century, many miners regarded themselves as employed by Phelps Dodge for life. After all, by that date many of them had worked for the Copper Queen Consolidated Mining Company for nearly two decades. Often if an old-time miner had a grievance he sought out Dr. Douglas, who was known to sit down on the curb to hear his complaint.

Bisbee was not unfamiliar with unions. In 1916 the community had nine labor organizations that encompassed a variety of trades, from bootblacks, cooks, and bakers to meatcutters, barbers, and carpenters. There was also a local of the Western Federation of Miners, though that union had a difficult time maintaining a presence in the local mines. In the early twentieth century, all of Bisbee's major mining companies adhered to the principle of the open shop. The Copper

Queen, moreover, paid better than union scale wages and carefully observed the eight-hour day mandated by an Arizona law that took effect on June 1, 1903.

A major strike occurred in 1907 when workers grew disgruntled after the introduction of modern equipment reduced the number of skilled workers underground at the same time that a recession caused the Copper Queen and the Calumet and Arizona to reduce wages. About three thousand workers walked out. Both companies fired the strikers and hired replacement workers, mostly from Mexico. Bisbee mines remained nonunion for the next decade, until World War I gave rise to conditions favorable to renewed efforts to organize copper miners there and elsewhere.

Conflict of Heritage

FOR NEARLY THREE years after the Great War erupted in Europe in August 1914, the United States tried to steer a neutral course between the Allies (Great Britain, France, and Russia) and the Central Powers (Germany, Austria-Hungary, and Turkey). Failing that, in early April 1917, Congress formally declared war on the Central Powers. After the nations of Europe marched into battle two and a half years earlier, rising demand had caused the price of copper to climb from 26½ cents a pound during the summer of 1916 to 37 cents a pound by the following March. The rising price in turn animated copper centers in Arizona and New Mexico. Copper production for Arizona alone in 1916 totaled 675 million pounds, an amount

PRELUDE TO *World War I: rising tensions in the Balkans sparked patriotic demonstrations in Bisbee. Each year various national and religious festivals and holy days highlighted the community's rich ethnic heritage. By 1900 more than a third of Bisbee's seventy-five hundred residents were citizens of England, Ireland, Mexico, Finland, Austria, and Serbia.*

equal to that produced by the entire United States just ten years earlier.

Mines worked around the clock in successive shifts to supply red metal that modern warfare demanded for both munitions and telephone and telegraph communications networks that extended into frontline trenches. Every rifle cartridge contained about half an ounce of pure copper, and stepped-up production of warships, tanks, trucks, and other forms of transport required still more copper. In all, from 1914 to 1918 about 80 percent of the world's copper supply went into the manufacture of war-related materials.

Rising along with copper prices, however, was the cost of living. Wages simply did not keep pace. Although pay envelopes grew fatter, every dollar bought less and less as the cost of living skyrocketed 42 percent between 1913 and 1917. That added to growing stresses in the workplace. Some miners left for better-paying jobs elsewhere, while those who stayed behind produced at a hectic pace. Many of them grew weary and discontented. Some turned to labor unions that promised quick relief.

One such organization was the Western Federation of Miners and its successor union after 1916, the International Union of Mine, Mill and Smelter Workers. During the decade before the United States joined the Great War, the union and Phelps Dodge had seldom, if ever, met face to face across a bargaining table; and whenever the two sides attempted to size up one another, they always did so from widely divergent points of view. Theirs was truly a conflict of heritages.

Phelps Dodge's heritage included a tradition of paternalism, and in the eyes of management and probably a good number of longtime employees, that was a positive attribute. For more than eighty years, from formation of the original partnership until 1917, Phelps Dodge had seldom been involved in any labor dispute, large or small. From Anson Phelps to James Douglas, the men who headed the company acted according to their conviction that they had a moral obligation to provide well for their workers. It greatly worried Douglas that the growth of large-scale corporations tended to loosen the close relations between managers and workers common in the old days.

That may have been one reason he introduced medical care for employees and their families for a nominal fee at the company's several hospitals and clinics, and he never failed to inspect those facilities during his frequent trips to the Southwest. Phelps Dodge also provided old-age pensions, carried on a vigorous safety campaign by the standards of the time, and maintained a voluntary insurance plan in which over 80 percent of its employees participated.

Both Dr. Douglas and his son Walter strongly opposed the existing system of labor unions. Union critics in response derided various forms of Phelps Dodge's generosity as nothing more than an underhanded attempt to control workers both on and off the job, and no labor organization was more outspoken at the beginning of the twentieth century than the Western Federation of Miners, which gained a reputation for militancy and confrontation. Its heritage of conflict derived from two violent disputes in the silver mines of northern Idaho in the 1890s that gave rise to the union and shaped its thinking. Scholars and journalists who wrote about those conflicts, as well as about the several vicious episodes of labor violence in Colorado and elsewhere, characterized them as a Rocky Mountain Revolution. *Revolution* is far too strong a word, yet it conveys some of the intensity of the discord.

The Western Federation of Miners always denied responsibility for any acts of violence, but it helped foster the popular impression that it was a bloodthirsty organization. At its annual convention in 1897, and using radical language that would haunt the union for years to come, President Edward Boyce advised members that "every union should have a rifle club. I strongly advise you to provide every member with the latest improved rifle. . . . I entreat you to take action on this important question, so that in two years we can hear the inspiring music of the martial tread of 25,000 armed men in the ranks of labor." This and numerous other examples of fiery rhetoric were important if only because publications hostile to the Western Federation of Miners frequently repeated them.[7]

Perhaps the stark differences between the two heritages would never have mattered much had not America's entry into the Great War set in motion forces that brought the two contrasting ways of looking at the world into conflict. Even before then, labor agitation in the mining West had begun to assume what many people regarded as genuinely frightening dimensions, and after April 1917 it only got worse. Menacing rhetoric and strike activity that people tolerated during times of peace seemed seditious during war.

Not only did tension between labor and management increase at Bisbee, but also complicating matters, two different unions competed for the loyalty of copper workers. The more moderate organization was the Western Federation of Miners' successor, the Mine, Mill and Smelter Workers, an affiliate of the American Federation of Labor. Ironically, the once militant Mine-Mill union could be described as moderate only because its competitor was an affiliate of the Industrial Workers of the World (the IWW, or Wobblies, as its members were popularly known), a union that the Western Federation of Miners itself had helped to organize back in 1905. Apparently not enough room remained on labor's Left for both of them, and the two had a bitter falling out.

Wobblies had originally conducted their most visible organizing efforts among the most alienated of western laborers—migratory harvest hands, timberworkers, and similar pools of unskilled labor—a fact clearly stated in the preamble to the IWW constitution: "The working class and the employing class have nothing in common." Instead of the conservative unionist's stated quest for "a fair day's wages for a fair day's work," the red banner of the IWW raised the revolutionary cry "Abolition of the Wage System." Wobblies believed that all workers in each major industry should belong to the same union and that the industrial unions should run society.[8]

What worried Americans, aside from militant labor's fiery left-wing rhetoric, was that growing industrial violence punctuated the opening years of the twentieth century—most notably in the West where words sometimes gave way to deeds. The bombing of the *Los Angeles Times* building in 1910 cost twenty-one lives and resulted in the subsequent conviction of most officers of the International Association of Bridge and Structural Iron Workers for illegal transportation of explosives in interstate commerce.

The tide of strikes and other forms of labor unrest rose dramatically after the nations of Europe went to war. In Arizona's copper centers, a wave of strikes in late 1914 and early 1915 disrupted production at twenty-four major mines and processing plants and threw some seventeen hundred men out of work. In all, the stoppages resulted in the loss of almost a million workdays. On September 11, 1915, when a general strike was called in the Clifton-Morenci area, approximately five thousand miners joined it. Until then the district had been comparatively free of union-inspired work stoppages. After strikers ran replacement miners out of town, Governor George W. P. Hunt (himself once a mucker in the Old Dominion Mine in Globe) dispatched some 450 members of the Arizona National Guard to keep the peace. The Clifton-Morenci strike was finally settled in January 1916. Workers won wage hikes, but they agreed to forswear allegiance to the miners union, which was anathema to copper management because of its long-standing legacy of violence and radicalism. Peace proved short-lived because another strike erupted the following July.[9]

WITH THE *"little red songbook,"* the Industrial Workers of the World intended to *"fan the flames of discontent."* The union's militant radicalism won widespread public attention by means of colorful protest songs like *"Dump the Bosses off Your Back"* and unorthodox organizing tactics that favored street-corner demonstrations.

Because Arizona supplied 28 percent of the nation's essential copper, labor turmoil in that state threatened serious shortages of a vital material—and in April 1917 the nation was in no mood to tolerate anything that interfered with the war effort. An idealistic President Woodrow Wilson promoted United States participation as if it were a crusade, asking Americans to fight not for materialistic or territorial gain but "to end all wars" and "to make the world safe for democracy." Wartime zealotry and a special spirit of conformity spread across the land. Self-appointed patriots applied the derogatory term *slacker* to anyone who failed to contribute to the war effort. Members of Congress enacted several punitive measures that further whipped up public emotions by constricting the boundaries of permissible speech.

Even in the best of times, the Industrial Workers of the World frightened many Americans with their radical agenda, wild words, and reputation for confrontation. Clearly the Great War years were not the best of times for any form of dissent, including labor militancy. But Wobbly agitators ignored the changing public mood when they continued to foment labor unrest.

In mid-May 1917, Wobblies captured the Bisbee local of the Mine-Mill union. Scarcely a month passed before they compiled a lengthy list of grievances and presented it to Gerald Sherman, superintendent of the Copper Queen Mine. Sherman, a well-educated son of the frontier, was a Columbia University graduate in civil engineering who earlier had gained experience building irrigation systems in the wilds of Idaho. He was a man of few words who

rather than waste time arguing with Wobblies chose to shred their ultimatum in public. Sherman intended his flamboyant gesture to impress upon everyone that he was not afraid of a showdown with radical labor, but he so infuriated local Wobbly leaders that they responded by calling a strike without benefit of any vote by their Bisbee members. So much for labor democracy.

Probably no more than four hundred out of Bisbee's nearly five thousand mine workers actually paid dues to the Industrial Workers of the World, and many of those were also members of the Mine-Mill union. Nonetheless, approximately three thousand miners and virtually the entire surface crews of both the Copper Queen and the Calumet and Arizona mines walked off the job on June 27 and formed picket lines. For the next two weeks the strike remained peaceable, although tensions rose visibly with each passing day.

Using Bisbee's city park as a forum, Wobbly organizers distributed literature and sought to spread dissension among the most receptive miners by encouraging them to nurse old grievances against the copper companies. At one point rumors circulated that both the Copper Queen band and baseball team had joined the walkout. That proved untrue, but a Wobbly strike leader did once interrupt a public dance to deliver a quarter-hour harangue to the band leader, and militants threatened workers who failed to support their cause. Wobbly posters advocating sabotage further stoked the rumor mill. Stories circulated that Wobblies had cached weapons and dynamite in the hills above Bisbee and would turn their words into deeds by blowing up the mines. Taking a longer range view, the *New York Times* blamed German agents for fomenting the trouble. Whoever was responsible, it was no longer business as usual.

Wobbly tactics left little, if any, room for negotiation. With a war on, patriotic Americans regarded work stoppages of any kind in the copper industry as a threat to military victory overseas. "Patriotism strongly demands the opening of new mines," emphasized the *Arizona Copper Miner* for July 1, 1917, adding that miners show a "patriotic desire to serve their country." The many disruptive tactics of the Wobblies, on the other hand, branded them agents of Germany in the eyes of many people. Wobblies, after all, had publicly blasted "Wall Street's War" and openly sought to encourage public opposition to the draft.

Among their several songs of protest was one from the Wobbly songbook that contained these highly inflammatory words:

> *Onward Christian Soldier! Rip and tear*
> *   and smite!*
> *Let the gentle Jesus bless your dynamite.*
> *Splinter skulls with shrapnel, fertilize the sod,*
> *Folks who do not speak your tongue deserve*
> *   the curse of God.*
> *Smash the doors of every home, pretty maiden*
> *   seize;*
> *Use your might and sacred right to treat her*
> *   as you please.*[10]

In the midst of war, such talk seemed traitorous, and in the popular mind the words *Wobbly* and *traitor* became increasingly synony-mous. Not surprisingly, leaders of other unions desperately sought to distance themselves from the Industrial Workers of the World. On July 3, Charles Moyer, head of the Mine-Mill union, wired Arizona's governor to castigate the Bisbee local as a maverick and denounce its strike as unauthorized. Three days later Moyer revoked the charter of the Bisbee local and called for all workers to ignore Wobbly picket lines.

More Blood on the Border

CLOSELY LINKED TO public fears of Wobblies, sabotage, and German intrigue and subversion in the southwestern borderlands were growing public concerns about Mexican revolutionaries. Bisbee was a border community, and ever since the Mexican Revolution erupted in 1910 troubles south of the border had been part of its consciousness. As various factions battled for advantage, Arizona hardware and sporting goods stores did a land-office business in the arms trade, both legal and illegal. Rebel armies fought pitched battles on Bisbee's doorstep in Naco, Sonora, both in 1913 and 1915, and again in Agua Prieta in October 1915, almost within shouting distance of the Copper Queen's Douglas smelter.

About six months after the battle of Agua Prieta, in the dark morning hours of March 9, 1916, a ragtag band of nearly five hundred rebels led by Pancho Villa slipped into southern New Mexico. There they raided Columbus, a small desert outpost of two hundred people located about three miles north of the border and astride

they shoved into the street and gunned them down there. The Villistas then torched the hotel.

In casualties, the Columbus raid did not rank high. Seventeen Americans and about one hundred of the rampaging rebels died. But because of the deliberate violation of national sovereignty, the affair infuriated Americans, and frightened dwellers along the border demanded retribution. With the blessing of President Wilson, General John J. Pershing led sixty-six hundred American troops across the border and three hundred miles into Mexico in futile pursuit of Villa, who seemed easily to dance away from danger and thus gain heroic stature in the eyes of many of his countrymen. The prolonged and muddled intervention by the United States succeeded only in inflaming anti-American passions south of the border.

the tracks of the El Paso & Southwestern Railroad. Plunder was their main objective, but Villistas seldom hesitated to burn and kill. With lusty yells of "Viva Villa" and "Mata los Gringos" (Kill the Americans!), the wild men smashed storefront windows with the butts of their rifles as they hurried along.[11]

Shouting and shooting, one band galloped along the El Paso & Southwestern tracks to the Commercial Hotel, a small two-story building located near the railroad station. As they went from room to room hunting down each person hiding there, they bayoneted each bed for good measure. In all, they murdered four of the nine registered guests, all of them men. One they dragged away from his bride of a few weeks and shot him on the stairway; two others

Determined to make the most of the volatile situation were German spies who lurked along both sides of the border. They sought to create trouble for the United States wherever possible, although in March 1916 the two nations were still a year away from declared war. If, for example, German agents somehow succeeded in halting the flow of copper from Bisbee mines, even temporarily, they benefited their fatherland. They also intrigued with various Mexican factions. The German government's famous Zimmermann telegram of January 16, 1917, suggested to its ambassador in Mexico City that "we make Mexico a proposal of alliance: . . . make war together, make peace together. Generous financial support and understanding on our part that Mexico is to reconquer the lost territory in

Texas, New Mexico and Arizona." Reconquer Arizona with German help! For Americans resisting war with Germany, the Zimmermann telegram provided the proverbial straw that broke the camel's back.[12]

When the United States and Germany did formally declare war on one another, it became all the more imperative that German spies stop the flow of copper from America's mines, and one way to do that was to use the Industrial Workers of the World to foment trouble among Arizona's miners. Perhaps some kind of conspiracy had been hatched among Mexican revolutionaries, Wobblies, and German spies—at least that was how it appeared to many frightened and angry citizens in Bisbee in mid-1917, and they nervously pondered how best to protect themselves and their families from this unholy alliance. In truth, no one thing caused the town's unease, but everyone seemed on edge.

Always rumors and more rumors swept up and down the canyons of Bisbee. How many of them originated as cynical attempts to discredit the Wobblies and how many were sincerely believed, it is hard to know today, but popular fears of impending trouble made sense in Bisbee within the context of those troubled times.

A portent of what could happen next occurred on July 10 when vigilantes in Jerome deported sixty-seven "troublemakers," all alleged Wobblies, aboard a cattle car to Kingman, Arizona. Only two days later, the winding streets of Bisbee echoed with the footsteps of a posse comitatus two thousand men strong. Armed with a previously prepared list of names, they carried out a deportation of a size and scope that was unprecedented in American history. The headline

in the *Bisbee Daily Review* for July 12, 1917, warned "All Women and Children Keep Off Street Today." To ensure success, the vigilantes allowed no messages to leave Bisbee.

Their leader was the Cochise County sheriff Harry C. Wheeler, and backing him were the Workman's Loyalty League and the Bisbee Citizens Protective League. Two bands of armed men—a posse identified only by white armbands and reminiscent of the wild frontier West— raided homes of known strikers and Wobbly sympathizers, including some lawyers, tradesmen, and businessmen, and took them into custody. One Wobbly killed a deputy who sought to arrest him and was slain in turn by a fellow deputy. Those were the only deaths to occur during an otherwise efficient roundup.

The armed citizens marched first to the downtown plaza and then to the ball park in Warren, where they offered each prisoner a last chance to return to work. Some did. They then marched the remaining men into twenty-four waiting box- and cattle cars. They provided deportees with water and food, contrary to labor's claims, and dispatched them aboard a freight train that rattled across the desert to a makeshift camp near Columbus, New Mexico. Most of the thousand or so deportees quietly slipped away during the next several weeks and never returned to Bisbee.

Walter Douglas, who was often accused of being the mastermind behind the deportation, considered the event within the context of the war effort and his patriotic duty to see that Phelps Dodge produced needed copper. Only a week before the Bisbee deportation, he wrote to Nicholas Murray Butler, president of Columbia

University (Douglas's alma mater) to complain that "the strike here is purely anti-American and with the avowed object of cutting off copper supplies for the Allies." Douglas feared sabotage, perhaps as a result of German spies lurking just south of the border, and he was afraid that Wobblies had collected dynamite to that end. To the logical and tidy mind of a well-trained engineer, deportation of "undesirables" must have seemed a simple and straightforward solution to a messy and dangerous problem.[13]

What most distinguished the Bisbee deportation was the number of people involved and the fact that it had the tacit approval of both business leaders and law enforcement officials. With the exception of the labor press and left-of-center journals like the *Nation* and *New Republic*, public reaction to the Bisbee deportation was generally supportive, and especially so in Arizona. In truth, since early days on the mining frontier in California, banishment had been a common way to deal with individuals publicly perceived as troublemakers. On several occasions in the early twentieth century, the state of Colorado had banished striking union miners to Kansas and New Mexico. Miners even practiced banishment themselves, as happened in Telluride, Colorado, and Goldfield, Nevada, during the first decade of the new century.

It should be noted too that Bisbee was not alone in giving vent to anti-Wobbly hostility. The Industrial Workers of the World's unorthodox forms of protest had previously landed hundreds of them in jail in Spokane, Washington, and resulted in violence in Fresno and San Diego, California. In the sawmill town of Everett, Washington, several people died during a blazing gun battle between Wobblies and lawmen in November 1916. Wobblies claimed for years that they were unarmed and innocent victims of the "Everett massacre," but newly discovered documents reveal that many indeed bore firearms.[14]

After the Bisbee deportation, Woodrow Wilson dispatched a group of five federal investigators known as the President's Mediation Commission to study labor unrest in Arizona. The group's secretary and general counsel was the Harvard law professor Felix Frankfurter. Members of the commission spent nearly two months investigating conditions in Bisbee, Tucson, Miami-Globe, and Clifton-Morenci. The report they issued on November 6, 1917, was written by Frankfurter, who branded the deportation "wholly illegal and without authority in law, either State or Federal." Soon a federal grand jury in Tucson indicted Sheriff Wheeler and twenty Bisbee mining, business, and professional men—including Walter Douglas, the new president of Phelps Dodge—for violating the rights of the deportees. A federal district court later invalidated the indictment, a decision upheld on appeal to the United States Supreme Count, which said the matter should be decided by Arizona courts.[15]

The state of Arizona arraigned 210 Bisbee citizens, including Sheriff Wheeler. After weighing the evidence presented during a three-month trial held in the historic two-story courthouse in Tombstone, jurors agreed in February 1920 that the deportation was justified by the "law of necessity," which in Arizona permitted a person to take the life of another in self-defense. They rendered a verdict of not guilty. Acquittal on the first ballot led the court

to dismiss criminal charges against all other defendants. As a result, almost all plaintiffs in the civil damage suits that totaled more than $6 million also dropped their claims, though defendants settled a few of them out of court for small sums.

With encouragement from the copper companies, laborers from Mexico permanently replaced many former strikers, and joining them was a new category of worker—young white men from the agricultural and mining districts of the Midwest. Miners unions virtually disappeared from the copper communities of Arizona, at least in the 1920s and early 1930s.

As for the Bisbee deportation, whatever its justification as a wartime necessity, it would not be forgotten. More than two decades later the young law professor who wrote the report of the President's Mediation Commission on the Bisbee deportation was appointed to the United States Supreme Court, and there Justice Felix Frankfurter took the side of labor in a famous case called *Phelps Dodge Corporation v. National Labor Relations Board* (discussed in chapter 9). More recently, during the 1980s and 1990s, memories of Bisbee's long-ago deportation were used to hammer Phelps Dodge during times of tense labor negotiations. The episode will likely remain part of labor's folklore in the Southwest for many years to come, although few people will recall its whole context. Bloodshed by Villistas at Columbus, shrill wartime patriotism, and the conflicting heritages of business and organized labor—all of these contributed to rising tensions in Bisbee and permanently altered the landscape of labor.

Chapter Seven

THE MODERN CORPORATION
from MINE *to* MARKET

S YMBOLIZING HOW DRAMATICALLY Phelps Dodge changed during and immediately after the Great War was the blast that took the top off Bisbee's Sacramento Hill in early 1917 and launched the corporation's first open-pit mine. Daniel Jackling had pioneered the revolutionary technology back in 1907 in Utah's Bingham Canyon: as the first person to apply mass-production methods to copper mining, he transformed the industry, just as Henry Ford transformed automobile manufacturing. Phelps Dodge used Jackling's techniques to recover metal-bearing ores that, "like raisins in a cake," were disseminated throughout a rocky mass beneath Sacramento Hill. Copper men had originally scorned porphyritic rock, which nearly always contained less than 2 percent ore, as only so much waste—but Jackling's innovative technology changed their minds.

Development work at Bisbee continued around the clock as seven steam shovels and fifteen brawny locomotives coordinated their labors to strip away millions of tons of rocky overburden before actual mining of copper ore began in 1923. By the time the Sacramento Pit halted all production in mid-1929,

---

~ IN ADDITION TO *its numerous underground mines, Phelps Dodge's new Sacramento Pit, its first open-pit mine property, contributed to the river of copper that flowed from Bisbee during the 1920s. An image of the pit on April 1921 shows a portion of the red mountain landmark still standing. At the time, familiar icons of open-pit mining included steam shovels, oil-burning locomotives, standard-gauge railroad tracks, and twenty-ton-capacity dump cars.*

the mountain landmark had vanished and a huge crater streaked with the colors of the rainbow had emerged in its place. That was not the end of open-pit mining at Bisbee or at Phelps Dodge, nor was it a technological false start. During the years ahead as it acquired or developed new mines, Phelps Dodge used open-pit methods to recover an ever greater percentage of its copper until in 1998 the triumph of the technology seemed complete and it closed its final two underground copper mines—at least temporarily.

The dawn of the open-pit era was just one of the many significant changes that occurred between 1917 and 1937, a time perhaps most noted for several dizzying boom-and-bust cycles in the copper market. A crippling downturn during the early 1920s was followed by a far worse one during the Great Depression of the 1930s, but during each bout of hard times Phelps Dodge turned adversity into advantage. It was during those two difficult decades that the company transformed itself into an industrial giant by selling its controlling interest in the El Paso & Southwestern Railroad, acquiring a trio of major mining firms, extending its reach from mine to market, and relocating its headquarters to Wall Street after it first was listed on the New York Stock Exchange in 1929. The years 1917 through 1937 witnessed the birth and remarkable growth, despite the hard times, of the modern Phelps Dodge Corporation.

Milestones in a Corporate Landscape

LEADING OFF THE transformation were several major events that occurred during the war years and early 1920s. Besides launching its first open-pit mine in early 1917, effective April 1 Phelps, Dodge & Company was reborn legally and operationally as Phelps Dodge Corporation. At that time it increased the sum of its capital stock from $2 million to $50 million. Restructuring promised to enhance its total efficiency by eliminating duplicate accounts, offices, executive meetings, and federal tax payments (for Uncle Sam had levied claims on the incomes of both holding companies and their subsidiary operations). Thus nearly all formerly separate mine and mill enterprises became branches of the new parent corporation. That is, the Copper Queen Branch combined mining properties at Bisbee with the big smelter at Douglas, and the Detroit Copper Mining Company at Morenci became the Morenci Branch, soon to be enlarged by acquisition of the Arizona Copper Company. In addition there were the Stag Cañon Branch at Dawson, the Burro Mountain Branch at Tyrone, and the Copper Basin Branch near Prescott. Moctezuma, because of its location in Mexico, remained a subsidiary company, as did Phelps Dodge Mercantile Company and Bunker Hill Mines Company, a marginal operation in Tombstone.

Even after Phelps Dodge Corporation transformed itself, it maintained a rigid division of labor between the suit-and-tie management in New York and the denim-and-heavy-boots

A PHOTOGRAPH OF Bisbee's Sacramento Pit in 1928 shows a dramatically altered landscape. Railroad mines like this one had a beautiful oval shape defined by the curvature of the track, but the technology created all sorts of problems by tying up valuable ore beneath the tracks.

workforce in the Southwest. New York handled all general administrative duties, while from offices located in Douglas, Arizona, a general manager dealt with matters related to western mine, mill, and smelting operations. The man who formed the vital personal connection during the 1920s and 1930s was P. G. Beckett, a Canadian-born (1882), British-educated mining engineer. Beckett's own career, which spanned more than forty years at Phelps Dodge and included service on the company's board of directors after 1940, illustrated the tensions that might arise between the two distinct geographic personalities within the corporation.

A large, ruddy-faced, fair-haired man, Beckett was a confirmed westerner in 1917 when Phelps Dodge recruited him from Globe's Old Dominion Mining Company to New York to serve as assistant to its new president, Walter Douglas. Beckett, who never adapted to the pace of big-city life, returned to Arizona a few years later. As general manager of Western Operations from 1920 to 1936, he embodied the link between top executives and the managers of Phelps Dodge's branch and subsidiary operations, who in turn were responsible, with the aid of an assistant manager or general superintendent, for the many departments within their own jurisdictions. The size of departments varied from a tiny staff of four members at Bunker Hill Mines, who engaged mainly in overseeing leases in the Tombstone area, to the mighty Copper Queen Branch, which counted more than fifteen hundred employees in Bisbee and another eight hundred at the Douglas Reduction Works. The corporation's organizational structure was clearly

military in nature, its various properties corresponding to bases of varying size and importance.

Despite Beckett's numerous day-to-day responsibilities, the one thing longtime Phelps Dodge employees remember most about their usually courteous general manager was his fanatical insistence on promptness. It became a ritual after his return to Arizona in 1920 that, whenever he entertained at home, his guests gathered on the sidewalk at least fifteen minutes prior to the announced hour of dinner and occupied themselves by walking back and forth to make certain they arrived on time. On one occasion Beckett invited George R. Drysdale to dinner, but when the corporate secretary arrived fifteen minutes late, he found the door locked for the evening. There were no second chances. No wonder the punctilious general manager seemed such a formidable figure to a generation of Phelps Dodge employees. A confirmed bachelor, P. G. Beckett (no one ever called him by his given name, Percy) seemed to devote every waking hour to company business.

In terms of profit and loss, the mining properties that Beckett oversaw were all considered integral parts of the parent corporation. Only the El Paso & Southwestern and its allied railroads remained legally separate, though individual Phelps Dodge executives continued to oversee the El Paso & Southwestern system until they sold it along with its land development and timber subsidiaries to the Southern Pacific in 1924 for nearly $64 million in stock, bonds, and cash. After the sale, Phelps Dodge and Southern Pacific officials worked

PORTRAIT OF
*Walter Douglas. The younger
Douglas (1870–1946) was an
energetic and imaginative
executive whose formal education
and hands-on experience provided
him a good grasp of the copper
industry during troubled times.
Fundamentally a mining engineer,
he reveled in the outdoor life his
profession entailed. To the public,
however, he seemed a shy and
retiring man—some people
thought him cold and arrogant—
and he was forever not only
consigned to the lengthy shadow
cast by his remarkable father but
also unfairly blamed for the Bisbee
deportation.*

together, and for many years a Phelps Dodge representative sat on the railroad's board of directors.

During the war years, Phelps Dodge passed another milestone: on June 25, 1918, Chairman James Douglas died at his New York estate at Spuyten Duyvil. Flags flew at half staff throughout the corporation, and for one hour as he was laid to rest all Phelps Dodge plants paused to show respect—all, that is, except for mines still required by war regulations to maintain their hectic production schedules. Well before his death, Douglas had become a legendary figure in North America's mining industry. For almost four decades, his judgment, knowledge, and energy had guided Phelps Dodge as its business activity contributed to the general economic betterment of Arizona and New Mexico.

Walter Douglas had succeeded his father as president of Phelps Dodge in 1917. The family tie was important, but the son brought valuable knowledge and skill of his own. A quarter century earlier the younger Douglas had completed his postgraduate training at Columbia University's School of Mines (then located on the same Park Avenue site where many years later Phelps Dodge maintained its headquarters). Walter first worked as a metallurgist at some of Phelps Dodge's minor properties and then for a smelting and refining company in Kansas City. It was in 1894 that he returned to Phelps Dodge to begin his long, steady climb to the top. Douglas needed his wealth of experience, and considerable good luck besides, during the thirteen rough-and-tumble years he guided the corporation.

Copper's Not-so-roaring Twenties

EVEN AFTER THE guns of war fell silent on November 11, 1918, high-ranking federal officials refused to hope the conflict was truly over. They wrongly encouraged American copper companies to keep production well above normal levels until weeks after postwar consumption had plummeted. In all, the producers added seven hundred million more pounds before Uncle Sam permitted them to reduce production two months later. The surplus of red metal in the United States grew so immense that every copper mine in the nation could have closed for a year without causing any shortage. That was only one example of the topsy-turvy world Walter Douglas inherited as president of Phelps Dodge.

As a result of the postwar glut, copper prices tumbled from a high of twenty-seven cents a pound early in 1917 to a mere thirteen cents four years later. The nation's copper companies had no choice but to curtail production if they hoped to survive. At Phelps Dodge that meant cutting output from a high of 186 million pounds in 1918 to about half that amount in 1920, and finally to a comparative trickle of less than 22 million pounds the following year. Having greatly expanded its production capacity during the war, the company suspended underground mining at Tyrone and Morenci and scaled back and then finally halted output at Bisbee and Nacozari after prices plummeted in 1920 and 1921. The Douglas smelter shut its gates for six months.

For a time production of copper virtually ceased throughout Arizona and New Mexico.

"The casual visitor would have concluded, with reason, that the industry was virtually dead." It proved to be merely in "a state of coma." Many mines did reopen a year later, but Tyrone, despite its gem of a company town, remained dormant for nearly half a century. As profits plummeted, so too did wages and dividends. In March 1919, Phelps Dodge eliminated extra dividends, and after suffering two more years of falling demand for copper, it reduced even its regular quarterly payout from $2.50 to $1.00.[1]

The price dip between 1919 and 1924 reached its nadir in 1921, the worst year of the so-called forgotten depression of the early 1920s—forgotten, that is, only because the Great Depression of the 1930s was so much worse. Copper prices likely would have sunk still lower in 1922 had not boom times in the nation's electric utility industry boosted copper consumption for wires, cables, and conduits. The world economic climate gradually improved, but by no means did copper producers enjoy the robust earnings of some other industries, notably the automobile manufacturers, who caused the decade to be popularly remembered in America as the Roaring Twenties.

After copper prices rebounded somewhat in 1922, Phelps Dodge resumed production at most of its mines and mills; however, by the following year it was clear that America's copper industry had better prepare itself for a prolonged period of dispirited prices. For several more years copper only limped along at between thirteen and fourteen cents a pound, a disheartening level that allowed only the lowest-cost producers to earn even modest profits.

THE FRONT COVER *of the May 1944* Bulletin of the Copper and Brass Research Association *highlighted ways that copper and its alloys played a vital role in transportation between the 1920s and 1940s. The industry formed the association in 1926 to promote use of copper and its most popular alloys in sanitary and long-lasting plumbing, window screening, roofing, and numerous other products.*

Promoting Copper by "Keeping Out the Enemy"

Phelps Dodge and *major American copper producers during the 1920s struggled to deal with the glut of red metal. First, with Uncle Sam's blessing, they formed a temporary export association that ultimately disposed of two hundred thousand tons of surplus copper outside the United States at a price five or six cents a pound higher than prevailed at home. Phelps Dodge next helped to organize the Copper and Brass Research Association, a trade group that America's copper giants launched in 1926 to discover and promote new uses for copper and brass. That same year they also established Copper Exporters, Inc., a cartel that included both United States and foreign firms. Its members accounted for approximately 85 percent of world output in the late 1920s, but the group broke apart in the early 1930s when prices plunged disastrously during the Great Depression.*

In the mid-1920s, Phelps Dodge conducted a vigorous campaign of its own to promote the use of copper. During the 1920s and 1930s its advertisements urged consumers to buy more copper products. One, headlined "Electricity lightens Mother's work," showed a smartly dressed woman plugging in an appliance, and it recommended that the modern home contain at least two electrical outlets in every room. Another advertisement called attention to the durability of copper by noting that George Washington worshiped under a copper roof at Christ Church, Philadelphia. Yet another urged the use of bronze screens to keep out insect pests. That one, which showed a giant fly prevented by a bronze screen from reaching a sleeping child, had a caption that read, "Keeping Out the Enemy." Whether this or any other industry effort had any real impact on depressed copper prices around the world was never clear.

(Based on a bound collection of advertisements in the
Phelps Dodge Corporation archives, Phoenix.)

Consolidation in the Clifton-Morenci District

Postwar troubles presented Phelps Dodge a buying opportunity of great significance in 1921 when the Arizona Copper Company, the venerable enterprise with mining operations in the Clifton-Morenci district that dated back to the early 1880s, foundered. Like all other copper producers, the Arizona Copper Company enjoyed soaring profits during the war years, but dulling the luster of the impressive financial achievements of those times were heavy war taxes imposed by both the United States and Great Britain and the strikes of 1915 and 1916, when labor strife convulsed almost all major copper-mining camps in the Mountain West. Its directors made the mistake in 1919 of buying the holdings of the nearby Shannon Copper Company just as the copper price tumbled. The deal put Arizona Copper in such a dangerously over-extended position that it could not survive the impending financial tempest. During 1920, its closing year of operation, Arizona Copper properties produced nearly thirty-six million pounds of red metal and lost money on every pound, or a sum of about a million dollars. The firm ceased production.

By the autumn of 1921 the price of Arizona Copper Company common stock had dropped like a stone. Because it was "weakened by luxurious living," in the words of its retired president James Colquhoun, the firm ran out of maneuvering room and sold out to Phelps Dodge, which gained control of nearly all copper-producing properties in the Clifton-Morenci district. The deal was the biggest merger in Arizona History to that time.[2]

Acquisition of the Arizona Copper Company cost Phelps Dodge fifty thousand shares of capital stock (10 percent of the company's total authorized amount), but its Morenci Branch gained a large mercantile business, valuable railway properties, a total of eight mines, including the Longfellow, the Coronado, and the old Metcalf several miles up Chase Creek, and mill and smelter facilities far better than any Phelps Dodge had in the area. Undoubtedly of greatest significance for the future was a huge low-grade deposit known locally as the Clay ore body that adjoined Arizona Copper's other

Morenci claims and extended over into Phelps Dodge's holdings. At the time of the 1921 merger no one realized that the Clay ore body actually contained copper reserves amounting to 450 million tons, a staggering quantity that ranked it among the largest of all copper deposits in North America. During the years to come, acquisition of the Arizona Copper Company proved a far richer prize than anyone dreamed in 1921: the Clay ore body would form the foundation for Phelps Dodge's open-pit mine operation in the area.

To make the deal all the sweeter from Phelps Dodge's perspective, it had purchased Arizona Copper for only $18.2 million; two years later, in March 1923, the Anaconda Copper Mining Company paid the Guggenheims $77 million for the South American company Chile Copper, the largest cash transaction known on Wall Street to that time. For that bundle of money, Phelps Dodge's longtime business rival gained the fabulous Chuquicamata deposit, the largest known copper deposit in the world. That mine was one of the few operations that ranked ahead of Morenci in production in the late 1990s, but well before then, Anaconda had lost Chuquicamata to Chilean expropriation during the early 1970s.

In Prosperity and Adversity

DURING THE MID-1920s, even when much of America seemed to be prospering, Walter Douglas continued to fret over what he termed "discouraging" conditions in the copper industry. "In apparent contradiction to the generally accepted economic law that demand exceeding output results in increased prices" of a commodity, he lamented to stockholders, copper still sold at too low a figure to bring the industry "a reasonable return." Only toward the close of 1928 was there a dramatic improvement in copper prices, and that was because, finally and for the first time since the Great War, there was no surplus metal on hand when demand for copper surged during the winter of 1928–29.[3]

"It is gratifying," began Douglas in his letter to stockholders covering activity in 1928, "after a period of six years of intense depression" in the copper market to report a "greatly improved" condition. Copper prices during the year had advanced steadily from fourteen to sixteen cents a pound. The following year they climbed to twenty-four cents a pound, copper's highest price since 1919 and a figure well above what it cost Phelps Dodge to produce it.[4]

The industry considered the near doubling of the price of copper during 1928 and 1929 miraculous. After all, just a short time earlier a leading newspaper had called the commodity "one of the most downtrodden necessary staples of life." Now some analysts worried that the American public might feel it was being "soaked" by the high price. Others asked, "Will substitutes be used successfully and extensively?" That was a good question, although apart from overhead high-voltage power lines fashioned from aluminum on a steel core, the substitution menace became a major headache for copper producers only in the 1950s and 1960s. What virtually no one foresaw as the experts debated what twenty-four-cent copper portended was the catastrophic depression lurking just over the economic horizon.[5]

On October 29, 1929, the stock market crashed. If the Wall Street collapse did not cause the Great Depression that followed, it was at least a symptom of an economy in deep trouble. The prosperity that copper producers enjoyed so briefly now vanished once again. In fact, with the coming of the worst period of hard times in United States history, the price of copper went into free fall, plunging first to thirteen cents, then to nine, and finally to less than five cents a pound in 1932. That was one of the lowest prices ever recorded in the long history of Phelps Dodge.

Walter Douglas took the collapse of copper prices personally. For more than a dozen years he had guided the company through very trying economic times, when every success was hard won. His sudden retirement in April 1930, ending forty-five years of Douglas leadership at the Copper Queen Consolidated Mining Company and then at Phelps Dodge itself, "excited considerable gossip and surprise." Exactly why members of the board of directors decided to replace him is not known. They were not angry with him, although his growing health problems must have caused them some concern. Most likely, they simply wanted to acquire the special talents of Louis S. Cates, who because of his earlier successes developing low-grade porphyry copper deposits was uniquely qualified to lead the corporation's anticipated development of the Clay ore body at Morenci. The vast size of the proposed operation, involving major feats of railroad engineering and the need to push the limits of open-pit technology, would surely appeal to him.

In any case, board members gave Cleveland E. Dodge, a vice-president and the heir of two of the corporation's founders, the delicate and awkward task of conveying their decision to the president. Douglas apparently stepped aside without animosity toward Phelps Dodge. Certainly he was no invalid, for he remained active as a member of the Southern Pacific board of directors until his death sixteen years later.[6]

Cates Takes Command

W HEN THE DIRECTORS of Phelps Dodge turned to Louis Shattuck Cates,

PORTRAIT OF LOUIS SHATTUCK CATES. *Though not a tall man, his large head, solid build, and deep voice gave him an intimidating presence. Cates left behind a larger-than-life personal legacy at Phelps Dodge, much like Dr. Douglas before him.*

it was the first time they had entrusted top leadership to a stranger having no previous background, training, or experience at the company. The new president, *Fortune* magazine informed readers in its July 1932 issue, "is not an impressive man until he speaks. His face is strikingly big-jowled and red. He seems to be smaller than he is, and rounder. But when he speaks the impression is gone. The mouth in his round face is decisive. His eyes are both wise and merry. He is considerably younger than his white hair makes him seem, but he has traveled a long way in his fifty years."[7]

Back in 1910, only eight years after he graduated from the Massachusetts Institute of Technology, Cates became general manager of Utah Copper Company, the Bingham Canyon pioneer fast becoming an industry giant that later took the name Kennecott. It was at Bingham that he formed a lifelong friendship with Daniel Jackling and became an understudy in the application of open-pit mining methods. At Utah Copper and Jackling's other properties, Cates added to his impressive record of technological and business achievements. He was, it soon became clear, a wise choice to head Phelps Dodge, for during the next thirty years he piloted the corporation through good times and bad— but always away from financial dangers that sank many larger enterprises.

Cates traveled east from Salt Lake City and assumed the presidency of Phelps Dodge on May 1, 1930. The enterprise he headed rested on a solid financial foundation and was well equipped with mines, machines, and talented personnel. Yet, ominously, only a month after he took over, copper prices began their long and

THE RECLUSIVE *Arthur Curtiss James (1867–1941). The great-grandson and richest living descendent of Anson G. Phelps in the 1920s, he was a director of Phelps Dodge Corporation. He and Cleveland H. Dodge, chairman of the board for a time during the 1920s, were instrumental in hiring Louis S. Cates as president in 1930.*

The Normal Flow of American Copper

LEGEND

- ■ Flow of Copper to American Centers
- ▨ Export Copper
- □ Centers of American Consumption

Width of Band is Proportional to Quantity of Copper

The refining of copper is largely done at tidewater localities adjacent to large domestic markets and facilities of export to foreign markets.

Los Angeles Harbor is the nearest port to the bulk of North American copper as it comes from the interior smelters and combines most economically the West's largest metropolitan and industrial market with low-cost electric power, for refining, and cheap transportation outlets to world markets.

⟦Industrial LOS ANGELES COUNTY⟧
Manufacturing Executives Are Invited to Write to the Industrial Department, Los Angeles Chamber of Commerce, for Detailed Information Regarding this Great Western Market

〜 THIS MAP ILLUSTRATES *why El Paso was the logical place to refine red metal coming from the mines and smelters of the Southwest in the late 1920s*

steady slide. Cates had a good grip on the copper industry, but nothing could have prepared him fully for the Great Depression that lay just ahead.

Acquisition of the Nichols Copper Company

SEEMINGLY WITHIN DAYS after he took the helm in early 1930, Cates proceeded to galvanize Phelps Dodge into action. As soon became clear, his goal was nothing less than complete vertical integration, that is, extending the corporation's reach from mine to market. Phelps Dodge took a first giant step in that direction in September 1930 when it acquired the Nichols Copper Company based in New York, a refining enterprise that reflected the technological genius of William Henry Nichols. A chemist who during the 1890s discovered an inexpensive and profitable method of producing sulfuric acid from copper pyrites, Nichols built one of the first large-scale electrolytic copper refineries in the United States. Phelps Dodge became his most valuable customer because 90 percent of all copper that Nichols refined at Laurel Hill during the next thirty years came from its mines.

Increased copper production during the mid-1920s had taxed the limits of the venerable New York plant and necessitated expansion of its refining capacity, preferably with an all-new facility located out west closer to the copper mines. El Paso was the logical site. The industry had long discussed the need for such a plant, because nearly all of the copper mined and smelted in the Southwest passed through El

SEP 30 1947

Paso's rail yards on its way east. But because Nichols Copper itself lacked the necessary capital, both Phelps Dodge and the Calumet and Arizona Mining Company helped finance the Texas refinery project. Between them they invested $3.5 million in exchange for stock in the Nichols Company.

In the desert eight miles east of downtown El Paso and near tracks of the Southern Pacific and Texas & Pacific railroads, groundbreaking for the Nichols refinery took place in February 1929. The site, which was little more than a large expanse of sand and sage, was located so far out of town that employees without cars of their own had to walk nearly three miles from the end of the nearest streetcar line—although the modern city of El Paso now completely surrounds it. In all, the landscape of production amounted to an aggregation of twenty buildings on 580 acres.

Workmen cast the first refined copper on March 8, 1930, and a shipment of wirebars left the El Paso plant six days later. After the Texas refinery commenced production, the Nichols plant at Laurel Hill mainly processed shipments of newly mined copper arriving from Africa and South America and substantial quantities of scrap copper collected in the large cities of the Atlantic coast.

From the beginning, the El Paso plant was one of the most modern copper refineries in the world. Its original annual capacity of 100,000 tons of cathode more than doubled by 1944 to 240,000 tons, which made it also the largest copper refinery in the world. Today annual production totals nearly half a million tons. In addition to a modern continuous-cast rod mill that opened in 1981 and a precious metals

refinery in 1991, Phelps Dodge's El Paso refinery complex embraces a tank house the size of a football field, powerhouse, furnaces, casting wheels, narrow-gauge railroad, battery-powered locomotives, water storage tank, laboratory, and administrative offices. Towering above the landscape of production stands a large brick smokestack, a landmark more than four hundred feet tall that still carries the name Nichols Copper Company in enormous white letters.

The stack has a story. A popular tale is that during the depression decade, when Nichols was still involved in the operations, Cates asked how much it would cost to remove the Nichols name. He and C. Walter Nichols, who succeeded his father as company president in 1917, had

A DRAWING OF *the attractive Spanish colonial–style*
entrance to the El Paso refinery. The towering stack with the words
Nichols Copper Company stands in the background. The image was
much the same in 1929 as in 1999.

apparently experienced a bitter falling-out. When word came back that the job required $5,000, big money during hard times, Cates only growled, "I hate that son of a bitch, but I don't hate him that much."[8]

Vertical Integration: Wires and Pipes

FOR THE LARGEST American copper producers, vertical integration became a fashionable concept during the 1920s. Unlike Standard Oil or United States Steel, for example, "the copper industry has been slow to follow the lead of other industries in supervising the utilization of its raw materials and their transformation into finished marketable form," observed a major industry publication in late 1921. Only a short time later, Anaconda, the world's largest copper producer, took up the challenge by pursuing a merger with the American Brass Company, the world's largest independent fabricator of nonferrous metals. Convinced that slumping copper prices could best be stabilized by vertical integration, Anaconda paid $45 million for American Brass in 1922. It was one of the biggest deals in American industrial history up to that time. Ironically, among the Naugatuck Valley plants that Anaconda purchased was one in Ansonia that Anson G. Phelps pioneered back in the 1840s.[9]

The rest of the industry followed Anaconda's lead. Kennecott purchased Chase Brass and Copper Company of Waterbury, Connecticut, and Cleveland, Ohio, in 1929. Phelps Dodge was the laggard among America's Big Three copper

producers in integrating vertically; yet within four months after Cates arrived in New York in May 1930, it too took giant steps toward fully extending its reach from mine to market. First came the merger with Nichols Copper in September, followed only one month later by acquisition of the National Electric Products Corporation through an exchange of shares, the method Phelps Dodge had used earlier to acquire the Nichols Copper Company. Its latest deal gained it a large fabricating complex capable of processing more than two hundred million pounds of refined copper annually.

Only months after he came out of the West, President Cates, by vigorously pursuing earlier Phelps Dodge negotiations, had transformed the company from an ancient and honorable family-based firm with an assortment of aging mines and a substantial bank account into one of the great copper mining and fabricating companies of the world. As if to herald its rebirth as a modern corporation, Phelps Dodge sold its original headquarters building at the corner of John and Cliff streets in New York early in 1930 and moved to rented offices in a modern skyscraper at 40 Wall Street (location of the Trump Center today). "The address 99 John Street, New York, has had a significance in western mining for over half a century," observed the *Mining Journal*, "and for over a century has been the address of the Phelps Dodge Corporation, its original home and headquarters for fifty years before that company was engaged in the mining business." The *Mining Journal* added that "it is unusual in the country of rapid progress when any firm rounds out a century with but one address." The symbolism of the move was important

because Wall Street formed the economic heart of America—for good or ill.[10]

As recently as 1925 Phelps Dodge recorded only seven hundred stockholders, but that number grew rapidly as the company extended its reach. During the first nine months of 1929, Phelps Dodge enjoyed greater prosperity than during any other time in the 1920s. Curiously, the canny Scotsmen of the former Arizona Copper Company held tight to their 50,000 shares of Phelps Dodge stock through a four-for-one split in February and then sold the entire 200,000-share block for $80 each. In order to provide the Scotsmen a convenient way to dispose of their shares as well as enable the corporation to sell two million additional shares of its common stock, Phelps Dodge was listed on the New York Stock Exchange for the first time in mid-May 1929. "The great investing public is to be congratulated on this opportunity to participate in the business of America's third most important copper-producing company," noted the *Engineering and Mining Journal.* "Operations are intelligently conducted and no one can visit its plants in the Southwest without sensing the admirable spirit that pervades its personnel."[11]

It's a Wired World

THE NEW MANUFACTURING landscape that Phelps Dodge assembled during the early 1930s had little in common with its traditional mining landscape except for the copper connection—that plus the technological geniuses who made each operation profitable. When the electrical age dawned in the 1880s, the mechanics of drawing wire through a tapered hole in a die to reduce its diameter to a desired size was well-known technology that dated back to ancient times. Artisans as early as the seventeenth century learned to draw wire for the strings of musical instruments, often using urine or stale beer to lubricate their dies. By 1861 and the start of the American Civil War, steam- and water-driven machinery drew tons of wire for suspension bridges, telegraph lines, and many other industrial and domestic uses, even for hoop skirts so fashionable in the mid-nineteenth century. If John Roebling, the genius behind the Brooklyn Bridge, could return to tour one of Phelps Dodge's modern wire and cable plants he would easily recognize the basic processes used more than a century ago to make individual strands of huge cables that still suspend his famous bridge high above the East River.

Despite civilization's lengthy history of wire drawing, insulating wire to carry electric power was something surprisingly new-fashioned. There was simply no precedent for the would-be fabricators to draw upon. At the time Edison addressed the problem of how best to insulate the underground conduits that radiated from his Pearl Street station in New York in 1882, it was already obvious that electric cables needed to be installed in a variety of harsh environments, not just beneath city streets, and hence they were subject to the ravages of weather, sun, ice, salt, oil and gasoline, brackish waters, acids and alkalis in the soil, and to attack by animals, birds, insects, and fungi. But unlike Edison, who became a household name because of his incandescent light bulbs, William Martin Habirshaw and George A. Jacobs, pioneers in the largely unsung field of insulating wire to make light bulbs and a host of other electrical applications possible (and founders of two of the companies that Phelps Dodge acquired in the early 1930s), remain largely unknown.

The challenge of figuring out how to make durable insulated conductors to transmit electric power safely intrigued William Habirshaw, an analytical chemist and one of a group of eight men (including the copper refiner William Henry Nichols) who founded the prestigious American Chemical Society. Just four short years after Edison lit up New York's financial district, Habirshaw organized the Triplex Company in 1886, a name he soon changed to the India Rubber and Gutta Percha Insulating Company (after those popular natural insulating materials). In the late nineteenth century the various plastics we take for granted were unknown, but gutta-percha trees growing in the jungles of southeast Asia yielded a plasticlike material that was used to make everything from chess pieces to cable insulation. With the latter item in mind, Habirshaw opened a plant in lower Manhattan, an early mecca of the electrical industry.

The young chemist quickly gained fame as one of the nation's leading experts on power cables. His newly formed business grew so rapidly that in 1890 he moved its fabricating facilities farther up the Hudson River to a large plant in Yonkers (an industrial suburb that remained its home until Phelps Dodge sold its North American wire and cable businesses in the early 1980s). In 1908 the enterprise became the Habirshaw Wire Company, which by the time its founder died the following year had already emerged as a leading fabricator of insulated wire

and lead-sheathed cables for underground and marine applications.

During the 1920s the Habirshaw Company made more rubber-covered cable than any other company in the world. It was all part of a mind-boggling array of specialty products that included "Brewery Cord," "Packinghouse Cord," "Fire Alarm Cable," and "Floating Dry Dock Cable." Because rubber by itself quickly deteriorated in a wet environment, Habirshaw had to engineer each product to survive in destructive environments created by excess dampness or vibration or both. The company also manufactured durable cable needed to power electric motors in damp underground mines. Design innovation continued after Habirshaw became part of Phelps Dodge in 1930, the research facilities remaining up-to-date for work with metals and insulating materials for both power and communication cables.

Apart from various wires and cables used to carry messages or electric power, special challenges of another sort confronted makers of insulated wire used in generators and motors that produced electrical currents or employed them. Originally, insulation of that type used hand labor to apply a variety of thick coating materials, such as coarse silk or cotton threads and tar, to individual wires. The process was still slow and labor intensive when George Amidon Jacobs, a chemist working for Sherwin-Williams, a paint company based in Cleveland, Ohio, came up with a better idea.

An electrical engineering graduate of Worcester Polytechnic Institute in 1900 and a man blessed with an inventive mind, Jacobs proposed to make an insulated coating for wire

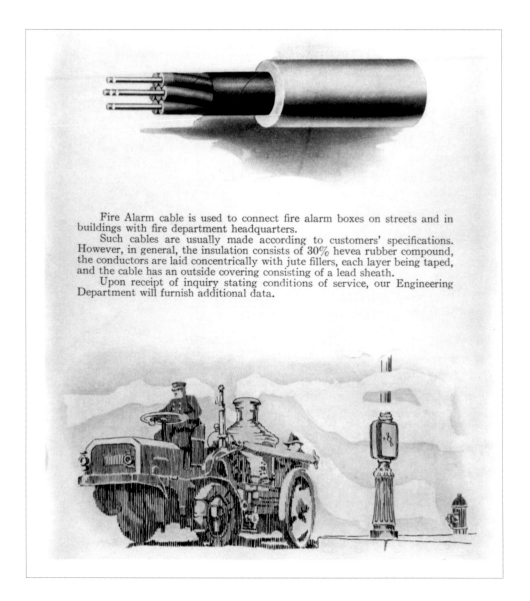

Fire Alarm cable is used to connect fire alarm boxes on streets and in buildings with fire department headquarters.

Such cables are usually made according to customers' specifications. However, in general, the insulation consists of 30% hevea rubber compound, the conductors are laid concentrically with jute fillers, each layer being taped, and the cable has an outside covering consisting of a lead sheath.

Upon receipt of inquiry stating conditions of service, our Engineering Department will furnish additional data.

"FIRE ALARM CABLE" was one of many related products that the Habirshaw Electric Cable Company featured in its products catalog in the 1920s.

that was thin, durable, and easily applied by machine. He cooked up a variety of smelly and volatile compounds at home on his wife's cookstove and baked them on wires in her oven. His coatings cracked and peeled with frustrating

regularity, but during numerous trial-and-error experiments Jacobs discovered a thin enamel compound that was easily applied and did not crack or break off the wire when flexed.

Joined by his father-in-law, a wealthy Indiana hardware merchant, Jacobs incorporated a new firm, the Dudlo Manufacturing Company, in Fort Wayne in 1914. Building a series of ever-larger plants, Dudlo soon became a giant of the young industry and a leading supplier of ignition coils—one for each cylinder, with frequent replacements—for Henry Ford's popular automobiles made in nearby Detroit. By 1917 it provided jobs for six thousand workers and claimed important customers ranging from Ford and Dodge to Magnavox and Westinghouse. By 1924, when Dudlo ranked as the world's largest producer of magnet wire, it consumed millions of pounds of copper rod annually.

Jacobs sold Dudlo in 1927 to an aggressive combination of manufacturers that became General Cable. The new owners relocated magnet wire production from Fort Wayne to Rome, New York, three years later. During the months that followed the multimillion-dollar buyout, it became increasingly clear that Jacobs had become too wealthy and too restless to work as a vice-president for General Cable—or for anyone else. He resigned, and in Fort Wayne on July 30, 1929, he incorporated the Inca Manufacturing Company, a new magnet wire enterprise capitalized for a million dollars.

Phelps Dodge gained Inca as part of the National Electric Products Corporation in 1930. Two years later it added Inca to the cluster of manufacturing enterprises that were consolidated under the name Phelps Dodge Copper Products

Corporation. For another decade Jacobs continued as special consultant to Inca (having no formal title but maintaining an office in the Fort Wayne plant). For two more decades, from 1941 until 1961, his nephew S. Allan Jacobs headed Phelps Dodge's magnet wire business. Inca was a profitable investment for Phelps Dodge because even during the depths of the Great Depression, when nearly all its mines halted production, the big Indiana plant continued to produce a steady supply of magnet wire.[12]

In March 1932 Phelps Dodge divided its metal-fabricating business between the National Electric Products Corporation and Phelps Dodge Copper Products Corporation, which included the Inca Division as well as several other units. Under the leadership of Wylie Brown, the president of Phelps Dodge Copper Products, the newly formed subsidiary easily ranked among

the major tube, wire, and cable manufacturers in the United States. Cates allowed Brown so much autonomy that for two decades the company's manufacturing sector, while part of Phelps Dodge Corporation, remained fundamentally an extension of the personality of that hard-driving, but sometimes quirky, industrialist.

Back in the late 1890s Brown launched his remarkable business career in the castings shop of the Bridgeport Brass Company. During the next few years he literally worked his way through the plant by holding a variety of jobs that enabled him to learn the brass trade in intimate detail— and what a trade it was. The Connecticut firm, which pioneered the development and sale of much of the nation's trolley wire, remained a booming business until the triumph of the automobile during the 1920s. Moreover, just as its brass trolley wire was indispensable to electric

A STYLIZED *Inca chief, symbol of the old Inca Manufacturing Company that George Jacobs formed in 1929 and named after the early people of Peru, who were skilled metal workers. Phelps Dodge acquired the company and retained the Inca name for three decades, changing to Phelps Dodge Magnet Wire in the mid-1960s. Several bas-relief portraits of a stylized Inca chief still decorate the facade of the large manufacturing plant on Fort Wayne's New Haven Avenue.*

∾ "STINKY INCA" ∾

At one time, *Fort Wayne residents knew when a neighbor worked at the Inca Manufacturing Division, "Stinky Inca," simply from the smell of his clothing. The aromatic secret of magnet wire was its enamel coating, which was mixed in a great underground room and pumped from there up to the production line. At no time did more than one person besides George Jacobs know the formula, and the pair always worked alone in the underground room. To this day, formulations and engineering specifications of wire drawing and coating machinery remain highly guarded secrets, and security is tight at the big Phelps Dodge Magnet Wire plant in Fort Wayne. The Indiana city can rightfully claim to be the magnet wire capital of the world because so many competitors are based there, nearly all of them able to trace their roots back to the Dudlo Manufacturing Company.*

transportation, so too were its brass condenser tubes to ocean commerce when great and luxurious steamships ruled the waves.

Brown, a Cornell-trained engineer, soon became unhappy because he thought Bridgeport Brass paid him too little; so he transferred to the sales side of the business. During the opening decade of the twentieth century he traveled to every corner of the United States, observing the nation's industrial growth and selling service as well as brass condenser tubes. In this way he came to see that the market for tubes and pipes—unglamorous products, to be sure, but a vital part of the modern infrastructure of the country—offered rich rewards to a man with his knowledge and vast network of business contacts.

In 1914 Brown resigned from Bridgeport Brass, where he was secretary and general manager, and struck out on his own. He was certain he would never attain his goal of becoming a millionaire if he worked for someone else. Not only did he become a supersalesman of the condenser tubing required by many ships built during World War I, but he also formed a new company to supply it. In fact, even before Uncle Sam placed his first orders for the product, Brown organized the British American Tube Company in 1914 and secured machinery and a suitable plant in Plainfield, New Jersey. All during the Great War the factory ran at top speed day and night, and for a time it was the largest producer of condenser tubes in the world.

When the conflict ended, Brown's bank account overflowed. The newfound wealth made him expansion minded. That was the main reason he joined with investors in 1920 to form the American Copper Products Corporation. Next he used the firm's financial clout to buy the Bayway plant of the Waclark Wire Company from the legendary Montanan William Andrews Clark (of United Verde fame), who back in 1897 added the fabricating works to his extensive holdings. Located in an industrial waterfront area called Bayway (later part of Elizabeth, New Jersey), the plant helped to make American Copper Products a major manufacturer of copper wire during the 1920s. Brown served as its vice-president and general manager.

American Copper Products sold copper rod to many customers including the Habirshaw Cable and Wire Company in Yonkers, William C. Robinson's National Metal Molding Company in Pittsburgh, and the Inca Manufacturing Company in Fort Wayne. That was how all three enterprises fell under Brown's spell. The blunt manner in which he approached George Jacobs of Inca was vintage Brown. He asked, "Why the hell don't you sell out. You certainly have enough money." Jacobs, who was contemplating taking more time away from business to spend some of his fast-growing fortune, responded, "Sure thing, if you pay me enough." Obviously he did. Further expanding his manufacturing empire, Brown added a tube mill in Los Angeles to settle a bad debt.[13]

Perhaps Brown alone could sort out the innumerable technical and financial details of his enterprise, which under the inclusive name National Electric Products Corporation (formed in 1928) ranked among the largest consumers of copper in the world. The multimillion-dollar combination of metal fabricators, together with such subsidiary firms as Habirshaw Cable and Wire, produced some two hundred million

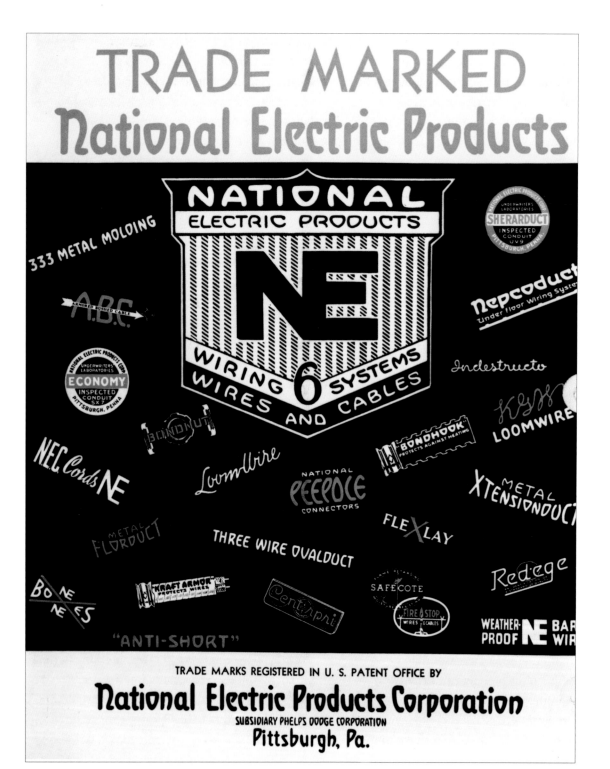

Various trademarks of the *National Electric Products Corporation*, a onetime Phelps Dodge subsidiary based in Pittsburgh. On January 1, 1932, the parent corporation divided *National Electric Products* into two distinct subsidiaries. One was *Phelps Dodge Copper Products Corporation*. The other, the world's largest maker of steel conduits, fittings, outlet boxes, and copper building wire, kept the *National Electric Products* name and was subsequently sold back to a group of investors headed by William C. Robinson, one of the original owners, in 1935. The public reason was that a subsidiary specializing in steel had not meshed well with Phelps Dodge's copper business, but in reality the attempt to marry firms headed by two such strong personalities as Robinson and Wylie Brown proved impossible.

pounds of copper products a year. Any business utilizing so much copper could not escape the attention of Louis Cates and Phelps Dodge. Only months earlier Walter Douglas had been "too Scotch" to pay Brown his asking price, but Cates was different.

At his request, the new Phelps Dodge president met with Brown for an escorted tour through National Electric's big fabricating plants in Fort Wayne and elsewhere. It was during the tour that the two executives casually discussed how they might combine their assets for their mutual benefit. After further negotiations, Phelps Dodge struck a deal with Brown later in 1930 that resulted in a merger of the two enterprises through the exchange of shares of common stock. The deal turned out to be made in heaven. Neither firm had any funded debt, and its new fabricating operations, which accounted for approximately 55 percent of the total value of its market copper in 1930, allowed Phelps Dodge to end the depression year of 1931 with practically no surplus copper on hand. With the addition of the National Electric Products Corporation and the Nichols Copper Company, declared Cates, Phelps Dodge formed "the third largest copper unit in the world, and one which will be second to none in the diversification and completeness of its line." Among its customers were General Electric, Western Electric, and Westinghouse.[14]

This, in summary, was the manufacturing enterprise that Louis S. Cates assembled: of Phelps Dodge's reformulated landscape of production in the early 1930s, it could be said that the corporation dug ore at Bisbee, Morenci, and Nacozari, smelted copper at Morenci and Douglas, refined it at El Paso and Laurel Hill,

and drew it into wires at Bayway, New Jersey, and at several fabricating plants it operated in the greater New York area. It also manufactured magnet wire in Indiana. All parts of Phelps Dodge Copper Products Corporation had in common that they made metal products such as copper tubes, pipes, wires, and cables. In 1937 the five plants of Phelps Dodge Copper Products Corporation produced about ninety-six thousand tons of finished products a year.

The Unsinkable Wylie Brown and the Copper King

LIKE EVERY LARGE corporation, Phelps Dodge has employed some offbeat geniuses who nonetheless made very positive contributions to the enterprise. By all accounts, Wylie Brown was one such character: a brilliant businessman but an irascible executive, he tyrannized close associates within his fiefdom. When Phelps Dodge was headquartered at 40 Wall Street, the offices of its Copper Products subsidiary were located on a floor separate from the rest of the corporation. On one occasion, Brown chastised a manager for eating lunch with a friend from another part of Phelps Dodge, warning him in no uncertain terms to have nothing to do with "those people" on the other floors. As president, he alone wanted to be the point of contact between Phelps Dodge Copper Products and its parent corporation.

Most people in the corporation knew nothing about Brown's fiefdom, "nor did he want them to know," and it should not be surprising that Brown gave a new definition to the command-and-

control style of leadership. He once publicly humiliated his assistant secretary and treasurer by calling in other officers to warn, "This man made a mistake, and it cost us money. I've told him to stand in the corner for ten minutes."[15]

For good reason, Phelps Dodge executives left Brown pretty much alone. Perhaps it was out of fear, but probably it was because he enjoyed the confidence of Louis Cates as few other colleagues did. Those two, who lived in the same exclusive residential club in Manhattan, often met for dinner and informal conversations that more than likely dwelt on company matters. Cates didn't suffer fools lightly, but if he liked someone he was very loyal, and Brown was his man.

Cates himself could be quite an intimidating,

even scary, presence. When he presided over Phelps Dodge's annual meeting in the corporation's New York boardroom, he usually began, "Any questions?" But before anyone had a chance to respond, he already had adjourned the meeting. His idea of a vacation was to put his Rolls Royce on the *Queen Mary* bound for France, where he would remain out of touch with the company for six weeks. During Cates's regular tours of Phelps Dodge mining properties out West, his preferred way to relax was to join with a few close associates for drinks and a few games of poker. During poker nights in Morenci, the onetime general superintendent of the complex (and future Phelps Dodge president) Warren Fenzi felt lucky to lose only ten or fifteen dollars to his boss.[16]

With most people, Cates had little interest in small talk. Once at a dinner party given by the general manager P. G. Beckett, Cates was seated next to Walter Douglas's wife, Margaret. A bright and well-traveled woman, she attempted to converse with him about all sorts of things but without success. His attention seemed totally fixed on the contents of several liquor glasses he carefully arranged around the edge of his plate. Finally, Cates placed his hand on her arm and made this solemn pronouncement: "Margaret, the art of conversation is dead." Naomi Kitchel, the Douglases' daughter, recalled that he was "a terribly rough diamond." Regardless of any personal shortcomings, Cates possessed the vision Phelps Dodge needed to meet the challenges of the Great Depression—and that vision included buying and developing copper mines to maintain the corporation's recently forged mine-to-market commitment.[17]

Chapter Eight

MEETING *the* CHALLENGE

of the GREAT DEPRESSION

A FTER BLACK THURSDAY, October 24, 1929, the price of copper retreated rapidly from twenty-three cents to eighteen cents per pound, and mining stocks plummeted. Further, even as Phelps Dodge multiplied its business activities by acquiring several fabricating operations, the dollars flowing into its corporate treasury dwindled ominously. As the clouds of the Great Depression gathered, the world business climate grew colder and harsher with each passing day.

Fortunately, Phelps Dodge's financial position in the early 1930s was rock solid, its organization and leadership were top-notch, its smelting and refining capacity was great enough to handle far more copper than its current mines supplied, and its new fabricating subsidiaries easily ranked among the best in the industry. But the corporation still required one more asset to insure long-term success: low-cost copper. That is, it needed mines able to produce copper profitably even at the dismal prices that prevailed during much of the 1930s.

Ironically perhaps, at the very time Phelps Dodge emerged as a mine-to-market giant, its producing mines were rapidly running out of low-cost metal. Until 1923, ore from Bisbee's

---

A STEAM SHOVEL *loads ore at the New Cornelia Copper Company in Ajo, one of the major mine properties Phelps Dodge acquired during the depression decade. As this picture illustrates, open-pit mining is essentially an earth-moving operation.*

A Perfect Fit

Cates saw a simple solution close at hand in the company's old rival, the Calumet and Arizona Mining Company. Buoyed by profits from its twenty-acre Irish Mag claim, which contributed to a rich and steady dividend, it had expanded its Bisbee holdings in the early twentieth century to include hundreds more claims there, and then it reached south to acquire lead mines in northern Sonora, east to gain the Eighty-five Mining Company near Lordsburg, New Mexico, north to the Verde Central near Jerome, and west for John Greenway's New Cornelia Copper Company and the beautiful town of Ajo. Mainly it was the mine and mill holdings at Ajo—a desert oasis located in an isolated, mountain-ringed valley 120 miles west of Tucson—that made Calumet and Arizona so attractive to Cates. The New Cornelia Mine held plenty of copper ore to supply Phelps Dodge's smelting and refining facilities for years to come and to provide copper for its fabricating plants. Cates regarded the Ajo property as just what his vertical integration of Phelps Dodge required.

Auspiciously, as Cates reminded his stockholders, their company had cash and securities it accumulated in periods of favorable earnings "available for possible future expansion." Other copper companies were not so fortunate. Calumet and Arizona paid such liberal cash dividends during the 1920s that the disastrous downturn after 1929 left it short of money and on the verge of bankruptcy. Here, as in the case of the struggling Arizona Copper Company in 1921, was a buying opportunity for Phelps

STOCK CERTIFICATES issued by the Calumet and Arizona Mining Company and its subsidiary, the New Cornelia Copper Company.

underground mines was so rich in copper that it could be shipped directly to the Douglas smelter; but as its quality dwindled, Phelps Dodge depended more than ever before on modern mining and milling wizardry to extract profitable ore from the huge porphyry deposits in the area. But after a few years those sources too appeared exhausted, and despite a $2 million search the company was unable to locate any new deposits of substantial size on its Bisbee properties. It closed the Sacramento Pit in 1929, partly for lack of ore but also for lack of space needed to expand its landscape of production. That happened even as shafts of some of its underground mines drove through the bottom of their ore bodies. All that was grim news. Cates needed to develop or buy some new mines—and soon.

Dodge. Cates fully realized that gaining Calumet and Arizona would greatly strengthen his corporation where it was most vulnerable—in long-range production. Between them the two companies produced almost 360 million pounds of copper in 1930. That year Phelps Dodge's assets were $285 million, and the Calumet and Arizona's were approximately $85 million. Cates dreamed of the powerhouse the combined companies could become.[1]

Calumet and Arizona's president, Gordon R. Campbell, had joined the company shortly after it was organized in 1901 in Michigan's Copper Range and served as its chief executive since 1921. The relationship between Calumet and Arizona and Phelps Dodge had always been friendly enough, but after spending thirty years successfully mining copper in Arizona, Campbell had no desire to end his career by taking orders from Cates. The Calumet and Arizona president fought the merger openly, but sensing that the battle was lost, he resigned in April 1931. All during the spring and summer months when members of the Calumet and Arizona board hammered out merger terms, their office of president went unfilled. In the end they drove a hard bargain: their stockholders exchanged thirteen Calumet and Arizona shares for four of Phelps Dodge (receiving 2.6 million shares in all), and they profited from a special premerger dividend of $2.50 a share. All this meant that the wealth and power of Phelps Dodge were distributed into a far greater number of hands.

To carry out the consolidation in September 1931, Phelps Dodge doubled its capitalization to almost six million shares and added nine new members to its board of directors, for a total of twenty-one. From the former Calumet and Arizona board it gained Louis D. Ricketts and James C. Rea, an eminent Pittsburgh businessman who effectively headed Calumet and Arizona after Campbell quit. Rea was also a member of the Phelps Dodge family because he had married Cleveland Hoadley Dodge's daughter Julia in 1911.

The Calumet and Arizona merger gained Phelps Dodge extensive holdings in Bisbee, which included the Cole, Junction, and Campbell mines, the once great Irish Mag Mine, and several smaller properties. From 1931 on, with the exception of the Shattuck companies, the Warren mining district was wholly a Phelps Dodge operation. Now there was plenty of room to reactivate and expand its open-pit operation. A further payoff was the New Cornelia Copper Company at Ajo.

When Phelps Dodge acquired Calumet and Arizona, besides the well-constructed mining and milling complex at Ajo and a virtually untouched sulfide ore body to meet future needs, it gained another railroad (the Tucson, Cornelia, & Gila Bend) and one more model company town. Ajo was truly an attractive oasis. Back in 1916, having finally solved the riddle of how to remove New Cornelia's overburden profitably, Calumet and Arizona tackled the water problem. Because the Gila River ran fifty miles north of town, residents of old Ajo had to rely instead on whatever drinkable water they could accumulate in tanks and barrels during the summer cloudbursts. For the company the obvious solution was to drill, and after several failed attempts, a crew using an oil-drilling rig at a site some six miles north of town broke into an ancient lava flow and

～ NEW CORNELIA ～

From Rogues to Riches

AJO AND THE *New Cornelia Branch became the latest addition to the Phelps Dodge landscape of production in 1931, although mining in that corner of the Sonoran Desert actually dated back many decades to Spanish and Mexican prospectors. The first Americans arrived there in late 1854, shortly after the United States acquired that portion of Arizona south of the Gila River known as the Gadsden Purchase, but lack of water and good transportation facilities discouraged all but the most persistent prospectors.*

Perhaps the single most frustrating thing about Ajo was that its sulfide ore lay buried deep beneath a surface deposit of granite and some twelve million tons of low-grade oxide copper not readily amenable to concentration. Just to remove the rocky overburden that capped the valuable copper would cost millions. Over the years the challenge of solving this mineralogical puzzle had attracted a veritable "Who's Who" of mining to Ajo, along with mine promoters, harmless cranks, and a few clever swindlers. Moreover, the challenge at Ajo showcased some of the most imaginative technology in mining history, though much of it was just as bogus as the fabled philosophers' stone supposedly able to transmute base metals into gold.

Among the several hucksters who blackened Ajo's reputation, none was more flamboyant than Professor Fred L. McGahan. Promising that his vacuum smelting method would unlock the area's copper riches, he fashioned a large furnace consisting of a brick-lined steel cylinder twenty-five feet high and six feet in diameter, and to this he attached several smaller horizontal cylinders containing the oxygen and hydrogen gases that entered into the reaction. A bewildering array of pipes, gauges, and spigots stuck out in all directions. McGahan glibly claimed that, after he pumped regular air out of the furnace cylinder and carefully fed in copper ore mixed with a little fuel oil and just enough oxygen to burn it, he could regulate the temperature so precisely that he could melt every element and separate each out one by one.

Molten gold would sink to the bottom first, where it could be drawn off through the lowest spigot, then came pure silver, copper, and other metals from spigots located progressively higher, until finally the ones at the top yielded oxygen and hydrogen gases. Once it was fired up, or so McGahan claimed, his remarkable furnace could recycle oxygen and hydrogen to maintain heat without needing any outside fuel. Nothing was lost and every element came out chemically pure. Even more impressive, he promised to achieve all this without first having to crush the ore.

It was all pseudoscientific nonsense, the technological equivalent of a perpetual motion machine, but before the professor put his wondrous vacuum smelter to the test, he hastily departed Ajo, leaving behind a note that demanded another $50,000 before he would reveal the secret of its operation. Whether the professor was crazy, or whether he was a rogue who deliberately took advantage of the childlike ignorance and greed of mine investors is no clearer today than in 1907. For years the rusted hulk of McGahan's "vacuumizer" served to stimulate local conversation, and if nothing else, it provided an inventive work of industrial art.

Other would-be entrepreneurs used the more conventional technologies of stamp mills and concentrate tables, but the ores of Ajo

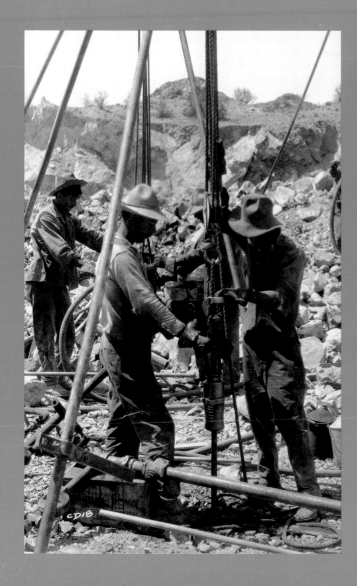

thwarted each one. Out of this harsh land came the New Cornelia Copper Company, named for one speculator's wife. At first it was no more successful than any other outfit. But in 1911 along came John Campbell Greenway, who had headed west a year earlier to run the underground mines at Bisbee for Calumet and Arizona, and with him was Louis D. Ricketts, the genius who had earlier helped Phelps Dodge launch its operations at Nacozari. Both radiated the influence of Daniel C. Jackling and his open-pit mining technology, although Jackling himself was never actively involved at Ajo.

On the advice of Greenway and Ricketts, Calumet and Arizona immediately secured a controlling interest in the New Cornelia Copper Company. The two men began five years of patient experiment and research. During most of the time their quest seemed hopeless. They would ruin their company and their reputation, renowned engineers warned, if they kept on. What drove Ricketts and Greenway, a Rough Rider in 1898 with Teddy Roosevelt in the Spanish-American War and a man of heroic qualities who allowed nothing to stand in his way, was that further test drilling revealed an immense sulfide deposit estimated to contain at least fifty million tons of copper ore beneath the unforgiving Ajo landscape.

Their persistence won the day. In early 1916, Greenway and Ricketts appeared before the Calumet and Arizona directors to secure their approval to build a leaching plant at Ajo capable of processing five thousand tons of ore per day. When workmen completed the facility in May 1917, it was a monument to engineering efficiency and a rebuke to the naysayers. Once the big leaching plant reached full operation, open-pit mining of the valuable ore body started at Ajo. Yielding a steady stream of copper that sold at the high wartime price of twenty-five cents a pound, the Ajo mine and mill complex more than covered its operating expenses and the cost of removing the oxide overburden, developing the sulfide ore below, and paying regular dividends to shareholders.

(Adapted from Ira B. Joralemon, *Copper* [Berkeley: Howell–North Books, 1973]) at a depth of 1,250 feet tapped into an almost inexhaustible underground river.

With an abundance of life-giving water close at hand, Calumet and Arizona proceeded to construct a remarkable company town. Inspired perhaps by Bertram Goodhue's Tyrone, planners employed the Spanish colonial style and determined to make Ajo as beautiful as it was functional. Residential areas were carefully located away from the mine and mill complex, yet near enough so that employees could walk to work. Many young shade trees soon lined wide, paved avenues that fanned out from a spacious plaza where the well-watered common and the oleander bushes contrasted dramatically with the desert that surrounded Ajo. The company constructed good-quality homes for its workers along the newly landscaped roadways.

THE NEW CORNELIA's boss and Ajo's primary creator, John Greenway, built a bungalow of his own on a ledge overlooking the village, though he did not live there long enough to see both the town and the company become Phelps Dodge property. While in New York late in 1925, the Colonel, as Greenway became affectionately known, suffered severe stomach pains. The problem was gallstones, and although the

--

PHELPS DODGE SALUTED *Clarkdale and Ajo, two of its most attractive towns in the mid-1940s. Ajo's predominately mission-style architecture was intended to honor the tradition of Spanish padres, who were some of the first European residents of the Sonoran Desert. In a community where every drop of water was precious, the company intended the lush plaza to serve as everybody's front yard. Ajo's visually arresting layout remains largely unchanged today.*

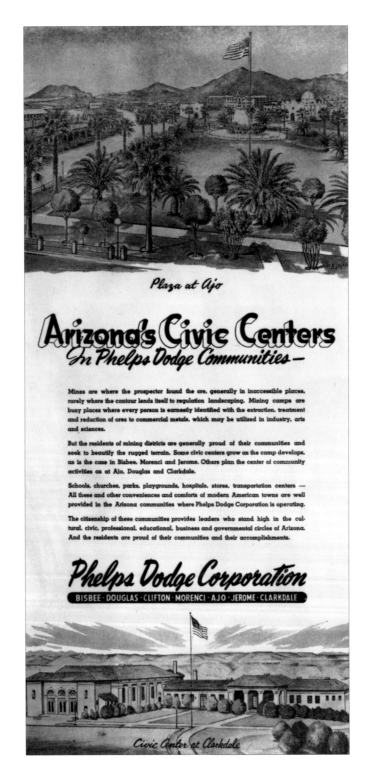

Plaza at Ajo

Arizona's Civic Centers
In Phelps Dodge Communities—

Mines are where the prospector found the ore, generally in inaccessible places, rarely where the contour lends itself to regulation landscaping. Mining camps are busy places where every person is earnestly identified with the extraction, treatment and reduction of ores to commercial metals, which may be utilized in industry, arts and sciences.

But the residents of mining districts are generally proud of their communities and seek to beautify the rugged terrain. Some civic centers grow as the camp develops, as is the case in Bisbee, Morenci and Jerome. Others plan the center of community activities as at Ajo, Douglas and Clarkdale.

Schools, churches, parks, playgrounds, hospitals, stores, transportation centers — All these and other conveniences and comforts of modern American towns are well provided in the Arizona communities where Phelps Dodge Corporation is operating.

The citizenship of these communities provides leaders who stand high in the cultural, civic, professional, educational, business and governmental circles of Arizona. And the residents are proud of their communities and their accomplishments.

Phelps Dodge Corporation
BISBEE · DOUGLAS · CLIFTON · MORENCI · AJO · JEROME · CLARKDALE

Civic Center at Clarkdale

operation to remove them was a success, the patient died four days later on January 19, 1926, from a blood clot. Fifty-three at the time of his death, Greenway was buried in an imposing private graveyard overlooking the open-pit mine he did so much to develop.

On another hilltop was the New Cornelia Hospital, a well-staffed, thirty-five-bed facility with the most modern equipment. Employees in the 1920s paid two dollars or less per month for medical and hospital services for themselves and their families. Greenway spent another $85,000 to establish the New Cornelia Cooperative Mercantile in early 1917, a store overseen by a committee of workers recruited from various departments of the New Cornelia Copper Company. It regularly returned any profits to its customers. When the store first opened its doors in 1917, many employees were skeptical. But before Christmas that first year, 472 workers received a 15 percent rebate, an amount that totaled some $12,000. Year-end rebates became a regular feature of the rhythm of life in Ajo and were one reason that employee turnover was very low.

Dark Days

THE MERGER IN the fall of 1931 joined Phelps Dodge's well-filled treasury and management skills with Calumet and Arizona's huge copper reserves. But it did little to brighten the troubled business climate. The early 1930s were not kind to the American copper industry, or to any other large industry for that matter. With the United States mired deep in the Great Depression, diminished demand for copper, falling prices, and foreign competition combined to force Phelps Dodge to cut production by almost 75 percent. "Conditions in the copper industry in 1931," wrote a dispirited Louis Cates, "grew steadily worse."[2]

Cost cutting was the order of the day. Following the Calumet and Arizona merger, Phelps Dodge sold several boxcars of surplus machinery from shops in Bisbee and Douglas to the companies constructing Hoover Dam on the Colorado River. To effect even larger savings, Phelps Dodge abandoned the old Copper Queen smelter at Douglas and enlarged the nearby Calumet and Arizona works to process ore from the consolidated companies. Prior to this both smelters operated at only 50 percent capacity.

For producers generally, hard times required down-the-line belt tightening and overall cuts in production. Even so, with copper selling for less than five cents a pound at a time when Phelps Dodge could scarcely deliver metal to the refinery for less than eight cents, the corporation lost money on mining. Its trifling profits in refining and fabricating, where production slowed to about one-third the normal rate, could not stem the flow of red ink. After trimming its payout at the start of the decade, Phelps Dodge dropped dividends altogether in 1932 and 1933 until it finally granted a "special distribution" in February 1934. The corporation, which had paid out $130,574,000 in dividends from 1909 through 1931, had an open market value of only $21,512,704 in 1932. Cates's push for vertical integration seemingly counted for little in the face of hard times.

Day wages, which customarily fluctuated

according to a sliding scale that corresponded to copper's selling price, reflected the downturn almost immediately. First, production workers received thinner pay envelopes, and then none at all as layoffs spread. Phelps Dodge slashed office and managerial pay but tried in every way to lessen the hardship, such as by favoring employees who had families, spreading hours of work, and making generous gifts to local charities. It contributed, for example, $20,000 to unemployment relief funds at Bisbee and Douglas before Christmas 1931.

At Ajo, Phelps Dodge quickly organized the New Cornelia Branch and spent a considerable sum of money to upgrade the mine, processing plant, and town. But because copper prices continued their sickening slide and the hard times lasted far longer than Cates anticipated, in 1932 Phelps Dodge had to suspend production at all its mines except for the richly endowed Campbell Shaft of the former Calumet and Arizona operation in Bisbee. For a time, nearly all the labyrinthine passageways beneath the Warren district fell silent. There and elsewhere, idle machinery rusted in the desert sun. Of the ten thousand miners that Phelps Dodge employed during good times, fewer than a thousand retained their jobs. In Bisbee the one miner in ten who still reported for work was likely a longtime company employee.

The ongoing adversity Phelps Dodge faced during the Great Depression can be summarized by a series of bleak news items that reported "Moctezuma Copper Closed," "Morenci Branch Shut," and "Operations at Bisbee [Reduced from] 15-Day Schedule to 12 Days a Month." Everything "depends on the copper situation,"

reflected a pensive Beckett. "How long we can continue even on the present greatly curtailed basis is impossible to say."[3]

As the mines closed, nearby towns suffered both widespread unemployment and an exodus of residents in search of jobs elsewhere. After the Morenci Branch ceased operations on June 23, 1932, so many people left the mining district that the population of Clifton dropped from almost five thousand residents to only fifteen hundred in a matter of months. Phelps Dodge furnished a special passenger train to take hundreds of stranded Mexicans back to Juárez. Many found it difficult to adjust to life in Mexico after living in the United States several years, and so they drifted back, some walking the entire four hundred miles. A few who returned to Clifton earned a precarious living panning for gold in the San Francisco River.

As hard times grew worse, once busy streets in Bisbee, Douglas, Morenci, Ajo, and other depressed mining centers became almost deserted. As for Nacozari, you could leave out the qualifier *almost*. Effective September 21, 1931, the Moctezuma Copper Company was forced to lay off indefinitely virtually all its employees. The Mexican copper center virtually ceased to exist as a town after Phelps Dodge bought a railroad ticket for each of its sixty-eight hundred residents.

People wondered when, and if, the mines of Nacozari would ever reopen. Obviously, it would not be any time soon. Stacks of unsold wirebars crowded the storage yards at Nichols Copper refineries in El Paso and New York, and still more surplus copper existed at other Phelps Dodge properties. Domestic producers, usually content with a six-week supply (approximately

The Mercantile Has a Heart

ON DECEMBER 10, 1931, *Andrew W. Liddell, general manager of Phelps Dodge Mercantile, wrote to his boss, P. G. Beckett, vice-president and general manager, to explain that the "credit problem in Bisbee is a very difficult one and the tendency is for it to become more acute." Liddell went on to note that, "rightly or wrongly, the employee losing his job or suffering a curtailment in income and having his credit stopped, feels he is hit from both sides—by the Mining Branch when he loses his job and by the Mercantile Company when his credit is stopped." He continued, making a case for leniency: "Accounts which were paid promptly prior to the curtailment, these customers, due to the reasons indicated, find it difficult to pay their bills now." The "credit problem" in the Bisbee area, a subject of several letters exchanged by Liddell and Beckett in late 1931, offers a reminder that in Phelps Dodge communities the Mercantile provided a needed social service as well as groceries and other merchandise.*

(Based on a letter from A. W. Liddell to P. G. Beckett, December 10, 1931, Phelps Dodge Mercantile records, Phoenix)

A NIGHTTIME PHOTOGRAPH *of Phelps Dodge Mercantile in Bisbee. After a fire destroyed the original store and all its additions except for the warehouse, the Mercantile constructed this new facility on the site in 1939. It was the first major building project undertaken by a young contractor named Del Webb, who later became famous for his "Sun City" retirement communities and ownership of the New York Yankees. This store closed in July 1976, not long after Phelps Dodge shut down its last Bisbee mines.*

120 million pounds of blister or refined copper or both), by the spring of 1932 confronted a mind-boggling one billion pounds of above-ground stocks.

Seeking New Markets: The Dymaxion Bathroom

WITH THE WORLD once again awash in an ocean of low-priced copper and with the collapse of markets in the electric, railroad, building, and automotive industries, the best thing for Louis Cates and Phelps Dodge to do, apart from further cutting costs, was to create new markets for red metal. One unusual result of that effort was the Dymaxion Bathroom. In what was one of the corporation's stranger business ventures, Phelps Dodge hired a controversial young industrial designer named Richard Buckminster Fuller, Jr., to fashion a one-piece bathroom entirely from copper. The man known to history as Bucky Fuller joined the technical staff at the Laurel Hill Refinery in 1935 to develop a unit called the Integrated Bath or Five by Five (for the number of square feet it occupied).

Using Fuller's basic design, Phelps Dodge produced twelve prototypes of the Dymaxion Bathroom. Technicians handcrafted each one from a single large sheet of copper that they hammered into shape over dies, hardened, and then sprayed with an alloy of tin and antimony intended to resist soap stains and keep the copper shiny and bright. A complete bathroom, which included an integral lavatory, toilet, and tub, weighed 404 pounds, about as much as the usual porcelain bathtub alone. Two men were supposed to be able to handle a home installation in three hours because of the minimum number of plumbing and wiring connections. Best of all, or so Phelps Dodge scientists claimed, renters could purchase and easily install a Dymaxion Bathroom themselves. When they moved on, they could just as easily remove the unit and haul it to their next home. The latter feature was considered important to the many mobile Americans during the uncertain times of the 1930s.

Fuller, incidentally, coined the term *Dymaxion* from the words *dynamic* and *maximum*. In 1933 he invented a Dymaxion car, a three-wheel rear-engine model that apparently inspired the researchers at Phelps Dodge to pursue making bathrooms that could be easily stamped out like so many automobile bodies. By late 1937, Phelps Dodge had invested some $20,000 in the quixotic project, no small amount during hard times when a new Arrow dress shirt cost less than $1. Cates took such personal interest in the project that he even had a Dymaxion Bathroom displayed outside the corporation's new head-quarters on the twenty-ninth floor at 40 Wall Street. Up in midtown Manhattan in 1937, the Museum of Modern Art offered the curious public another look at the Dymaxion Bathroom. In truth, no one knew what to make of the strange object. Was it art or science or sheer lunacy? Press observers were no less certain of Fuller himself, whom they variously described as a genius, a crank, or a madman.

Both the quirky bathroom and its no-less-quirky inventor were studies in contradiction. Fuller, the son of an aristocratic New England family, was expelled from Harvard for skipping too many tests and later arrested at the Waldorf-

Astoria Hotel for failing to pay his bills. There was talk of sending him to a reformatory, but a friend saved Fuller by finding him a job as a meat handler with Armour and Company in New York. One popular profile described him as a stubborn but good-natured man who seemed to stare out at the world with disconcerting intensity.

For Phelps Dodge the Dymaxion Bathroom was purely a business proposition. It offered an innovative way to respond to the copper market during times of abnormally low demand, energize the company's struggling mines, and put employees back to work. Hoping to sell each unit for less than $400, Phelps Dodge anticipated a tremendous market for the Dymaxion Bathroom in government housing projects. But because its first units sold for more than $700 apiece, not one ended up in low-cost housing. Quite the contrary. Two were installed in the Jasper Morgan residence in New York City and two more in the John Nicholas Brown residence on exclusive Fishers Island in Long Island Sound.

Fuller received patent no. 2,220,482 for his Dymaxion Bathroom in 1940, which he promptly assigned to Phelps Dodge. He may have intended his invention to become part of a complete Dymaxion Dwelling Machine, a futuristic house he had evidently conceived back in the 1920s. He had two prototype dwellings built by Beech Aircraft in Wichita, Kansas, during the months immediately following World War II. Both of the homes were circular structures fabricated from a combination of lightweight plastics, aluminum, and stainless steel and hung from a central mast instead of resting on a conventional foundation. In truth, they were visually arresting—looking much like flying saucers, which, incidentally,

people first spotted in western skies in 1947. But neither the modular bathroom nor the full-blown dwelling machine entered production. Well before 1947, Cates had grown disillusioned and halted Phelps Dodge participation in Fuller's projects.

During World War II the company sold one Dymaxion Bathroom to a secretary at its Laurel Hill plant for the junk value of the unit, about $50. That was pocket change compared to approximately $25,000 Phelps Dodge invested in its odd venture in industrial design. Apparently most of the rest ended up as copper scrap. Only two units survive today: one is at the Montreal Museum of Decorative Arts and another is at the Henry Ford Museum and Greenfield Village in Dearborn, Michigan.[4]

One cannot fault Cates and Phelps Dodge's research personnel for their initial interest in the Dymaxion Bathroom. Those were desperate days. Shares of Phelps Dodge common stock that sold for three hundred dollars apiece in 1929 had dropped to seven by 1932, the year that copper prices bottomed out at slightly less than five cents a pound. Earlier, after a two-week trip to the East in mid-1931, P. G. Beckett had reported back to his colleagues in Douglas, Arizona, that "in the copper business everything is just about as sad and unhappy as it could be." And, he added, the situation "is getting worse rather than better."[5]

A New Day at Old Jerome

THE YEARS OF the Great Depression presented Phelps Dodge with yet another great opportunity at Jerome, Arizona, where in

1935 it bought the old United Verde Copper Company, a purchase that positioned it ahead of Anaconda and second only to Kennecott among America's leading producers. In Jerome, as in Ajo and elsewhere, Phelps Dodge picked up the strands of several decades of history along with its newly acquired property. Among Jerome's claims to fame was that it was home to the world's richest individually owned copper mine, the United Verde. Another was that Jerome's mines yielded enough wealth to make three separate fortunes—one each for William Andrews Clark and Phelps Dodge at the United Verde Copper Company and another for "Rawhide Jimmy" Douglas and his associates from the United Verde Extension Mine, which they discovered almost literally under Senator Clark's nose.

United Verde first began to produce a significant amount of copper matte and precious metal bullion back in 1883, though not necessarily at a profit. In 1885, as Montana commissioner to the New Orleans Cotton Centennial Exposition, William Andrews Clark examined an exhibit of United Verde ore. Legendary for his attention to detail, Clark noticed gold and silver in addition to copper. Three years later, in 1888, shortly after his first visit to the isolated site, he purchased the United Verde property and developed it into an impressive mine, smelter, and railroad complex.

By 1900, when the United Verde Copper Company ranked ahead of the Copper Queen as Arizona's most productive copper mine (and the sixth most productive in the world), it yielded

GROUND WAS BROKEN *in 1912, and the first furnace of the United Verde Copper Company's Clarkdale smelter was "blown in" on May 16, 1915. The following year it processed three-quarters of a million tons of copper ore. Phelps Dodge acquired the property in 1935. Jerome and the United Verde's original smelter site are on the slopes of the distant mountains at the far left.*

Inside the United Verde's *drill-sharpening shop built in 1919–20. This was a small but vital cog in a major mining enterprise, which itself formed only one portion of the vast and complex industrial machine assembled by William Andrews Clark. Among his holdings were metal mines, wire-fabricating plants, electric utilities, sugar, rubber, and coffee plantations, and America's largest bronze factory. Phelps Dodge gained valuable pieces of the Clark empire in the form of the Bayway plant in New Jersey and the United Verde Copper Company in Arizona.*

about forty million pounds of red metal a year. The torrid pace of production held steady for at least twenty years. By the second decade of the century, however, Clark's smelter was unable to keep up with the flood of ore. And because the smelter, a network of railroad tracks, and various shop buildings sat atop the underground mine workings, maintaining the complex had become both difficult and expensive.

In fact, the whole works started to slip downhill because of the miles of tunnels that honeycombed the property, giving rise to the quip that Jerome "is a town on the move." That was one reason Clark bought a number of ranches along with their valuable water rights in the Verde Valley in 1910. There, about six miles downhill from Jerome, he built a new reduction works and the model town of Clarkdale to attract and hold stable, family-oriented employees. His utility

company provided water, light, and power for Jerome and administered Clarkdale. Its services ranged from police protection to garbage collection.

United Verde production temporarily peaked at the end of World War I in 1918, when the company mined about 870,000 tons of ore, most of it coming from an open-pit that commenced operation on the old mine site a year earlier. The postwar slump in copper prices temporarily slowed production, but in 1923 after the copper market recovered somewhat from its doldrums, the company's output broke all records at more than a million tons. The United Verde was a very profitable property, and from the late 1880s until the mid-1920s, most dividends flowed straight into Clark's own deep pockets. Unlike the usual business tycoon of his day, he avoided mergers and partnerships, and he preferred to take on only family members as associates. At the time of his death in March 1925, Clark and his family owned 99.7 percent of all shares issued by United Verde, and a few trusted colleagues owned the remaining handful.

Clark's heirs were not as interested in mining as he was, and when copper prices plummeted during the Great Depression, they halted all production at United Verde in May 1931 and a short time later opted to sell the property. Mining experts had only recently advised them that its ore reserves were nearly exhausted. Further, a massive slide of six million cubic yards of rocky overburden occurred along a steep wall of the open-pit mine on March 23, 1931, enveloping Jerome in dust and creating a thunderous roar heard several miles away. The slide required five years and considerable expense to remove, and during that time Clark's sons and a grandson died.

Meanwhile, Phelps Dodge's own geologists and engineers, upon investigation, came to believe that the United Verde Copper Company still possessed enough potential to make it a worthwhile investment, which during the next quarter century it proved to be. After waging a brief contest with the American Smelting and Refining Company, which probably had its eye on the smelter at Clarkdale, Phelps Dodge acquired the property on February 18, 1935, for slightly less than $21 million.

For that sum it gained the large Clarkdale smelter as well as the nearby company town, connecting railroads, and local utility company, in addition to the mines and mining claims in the Jerome area and the wholly owned Compañía Minera de San Carlos, S.A., a lead-silver property located ten miles south of the Rio Grande River in Chihuahua, Mexico. From its United Verde holdings Phelps Dodge earned more than $40 million in profits from copper, zinc, silver, and gold ores before it ceased large-scale mining at Jerome in 1953. It subsequently sold off the utility company, railroad, smelter, and Clarkdale itself. Today the town is an active community where wide streets and several Spanish-style municipal buildings grouped around a plaza recall its origin in 1914 as Senator Clark's ideal company town.

Slow Return to Prosperity

BY THE MIDDLE years of the 1930s, Phelps Dodge's finances seemed secure. The price of copper still hovered around eight to nine cents a pound, a figure that did nothing to lift the prevailing gloom but, because of shrinking inventories and a four-cent protective tariff levied by Congress in 1932, would not likely drop down to rock-bottom levels seen during the worst of the depression. Cates reported to stockholders in early 1935 that the mines of Moctezuma and Morenci remained closed, Bisbee was producing copper only on a limited basis, but the New Cornelia at Ajo had reopened the previous July 1 and "operated steadily" for the last half of 1934. Moreover, to strengthen its balance sheet Phelps Dodge sold the Tombstone mine properties that it had leased out for years and the steel-oriented National Electric Products Corporation, which it had never successfully integrated with other components of Phelps Dodge Copper Products Corporation. In general, the business climate was improving.

The Copper Products subsidiary traveled down the same road to moderate recovery as did its mining counterpart in the late 1930s. By that time, plant additions and a corresponding enlargement of its workforce were accelerating; and some new orders, such as for hundreds of miles of heavy cable for the new Bonneville Power Administration in the Pacific Northwest, were sizable enough for the business press to take notice. Additional encouragement came from Arizona, where deposits in Bisbee continued to yield copper ore as workmen deepened old shafts and sank new ones. In anticipation of a bright future for copper, Phelps Dodge expanded the power plant and enlarged the Douglas smelter by adding three modern furnaces to increase its capacity beyond current needs.

ELECTRIC HAULAGE IN United Verde's Hopewell Tunnel. Although steam shovels generally loaded ore into railroad cars for the trip up and out of an open-pit mine, in the 1920s at Jerome's United Verde Mine they loaded it into some of the first haulage trucks used in the industry. These took the ore to an old vertical shaft hundreds of feet deep and dumped it. At the bottom, railcars carried it through a long horizontal drift called the Hopewell Tunnel to the outside, where more men loaded it aboard the cars of the Verde Tunnel & Smelter Railroad for a seven-mile trip down the mountainside to the Clarkdale smelter.

By the spring of 1937 something approximating prosperity once again energized America's copper producers. The improving business climate meant revival of activity at almost all Phelps Dodge branches and opportunity to launch a number of new building projects. At Ajo's New Cornelia operation alone, the company spent more than $2 million on a host of improvement programs that ranged from increasing the capacity of the open-pit mine to building fifty more workers homes. Among P. G. Beckett's important contributions to mine operations was a switch from old-fashioned steam shovels that operated along railroad tracks to more flexible and productive electrical shovels that traveled on their own moving treads.

Back in the dark days of 1932 who could have predicted that only four years later the Ajo open-pit mine would emerge as Arizona's leading copper producer, yielding an impressive ninety-six million pounds of red metal in 1936? The mine remained at the head of the pack until 1942. It earned a profit at a time when the price of copper averaged less than ten cents a pound, and Ajo ore contained only ten pounds of red metal per ton.

On the whole, Phelps Dodge emerged from the Great Depression in much better shape than it had entered it. By permanently closing its underground mines at Morenci, temporarily halting production at its open-pit mine at Ajo, and focusing its activity at Bisbee, Phelps Dodge minimized production costs during the worst of the hard times. And occasionally a ray of sunshine pierced the gloom, as it did in 1934 when underground crews at Bisbee discovered a massive sulfide ore body averaging 9.5 percent copper. By the late 1930s, corporate earnings had climbed to the highest level since the boom times of World War I, and the number of miners Phelps Dodge employed rebounded from a thousand or so during the worst months of the depression to ten times that number.

By 1937 the future looked brighter than at any time since Cates became president seven years earlier. For Phelps Dodge management it was time to start to develop the colossal low-grade copper deposit at Morenci, where its plans for an open-pit mine had been put on hold. It was fortunate that the man who stood at the corporate helm during that challenging undertaking was someone who had made his reputation developing large porphyry ore bodies. Assuredly the technological challenges did not worry him. In fact, after steering Phelps Dodge safely through the murky waters of hard times, probably very little worried him, not even rising international tensions in Europe.

In 1938, as men and machines battled to prepare the Clay ore body for mining, the Nazi legions goose-stepped their way into Austria and then Czechoslovakia. They launched an invasion of Poland in September 1939. Phelps Dodge had nearly completed its Morenci project when wave after wave of enemy warplanes launched a surprise attack on Pearl Harbor on December 7, 1941. After that "day of infamy," what a few years earlier had seemed like an enormous and risky undertaking was too modest to satisfy the insatiable wartime demands for copper.

WORKMEN INSTALL Habirshaw's special "Park Cable" in a bucolic area of Saint Louis, Missouri, in this undated image (apparently from the 1920s). The product was used "where it is preferable to bury the cables directly in the ground rather than put them in ducts."

Chapter Nine

EVOLVING ENTERPRISE
in WAR *and* PEACE

O NE THING THAT greatly irritated Louis S. Cates in 1941 was the ghost of the old Bisbee
deportation as invoked in a landmark United States Supreme Court case called *Phelps Dodge
Corporation v. National Labor Relations Board.* Therein lies an episode that illustrates how
much more Uncle Sam intervened in business dealings after 1937 and why the New Deal and World
War II years brought dramatic and enduring alterations to Phelps Dodge—and not just in the area of
government relations.

A Ghost of Old Bisbee

T HE STRENGTH OF organized workers in the United States declined after World War I when the
nation entered an era of applied scientific management and labor cooperation that prevailed until
the early 1930s. Locals of the International Union of Mine, Mill and Smelter Workers often identified
their interests with those of management, such as when a Miami local publicly urged Congress in 1926

~ THE OPEN-PIT MINE AT *Morenci, seen here in 1940 before workers spiked railroad haulage lines into place along the benches. At
one time the mine contained more than seventy-five miles of standard-gauge track.*

to impose a heavy tax on copper imports. The union firmly believed that a protective tariff would lift domestic copper prices enough to permit American companies to increase wages.

The era of labor-management harmony ended abruptly when mines throughout Arizona and New Mexico either closed or greatly scaled back production in response to the Great Depression. Encouraged by the National Industrial Recovery Act of 1933, which seemed to commit the federal government to support labor, organizers stepped up the pace of union activity among copper workers. At Phelps Dodge, after more than a decade of relative calm, labor disputes increasingly beset its mine and mill operations—even after the economic picture brightened somewhat in the later 1930s.

Exemplifying the unprecedented support labor received during the Franklin D. Roosevelt administration was congressional passage of the National Labor Relations Act, or Wagner Act, which according to one historian of the New Deal era, "threw the weight of government behind the right of labor to bargain collectively, and compelled employers to accede peacefully to the unionization of their plants," yet "imposed no reciprocal obligations of any kind on unions." No one, concludes William E. Leuchtenburg, "fully understood why Congress passed so radical a law with so little opposition and by such overwhelming margins." Not even Roosevelt embraced the measure until he saw that Congress would likely approve it.[1]

The third branch of government, the Supreme Court, likewise stepped up its involvement in labor relations during the late

Courtesy Crane Company

He's aiming at a Nazi from 3,000 feet underground

1930s. Among its milestone decisions was one that resulted from a strike that began at the Copper Queen Branch on June 10, 1935, or some three weeks *before* the National Labor Relations Act went into effect. The timing was important. When the strike ended a few weeks later, on August 24, the Mine-Mill union charged Phelps Dodge with engaging in unfair labor practices

⌐ THIS PHELPS DODGE *advertisement from World War II emphasized that production for all-out war began in mines deep beneath Arizona. Phelps Dodge was ready when war came because of the new landscape of production it built at Morenci after 1937.*

as defined by the Wagner Act. Specifically, it accused the company of firing several strikers simply because they belonged to the union.

After the recently established National Labor Relations Board found those charges valid, it mandated that Phelps Dodge reinstate and pay any former employees who had not found jobs elsewhere the wages they would have earned had the company not discharged them. One snag was that each worker's case was different, and especially so for two men the company discharged before either the Wagner act took effect or the strike began. Phelps Dodge appealed the ruling, and the Second Circuit Court upheld its right to fire the employees. It then remanded the case to the National Labor Relations Board, which ruled yet once more in favor of the discharged men. Phelps Dodge appealed to the United States Supreme Court in early 1941, six years now having elapsed since the dispute began.

The company selected a young Arizona attorney, Denison Kitchel, to argue its case. As a student at Harvard Kitchel had studied law under Felix Frankfurter, one of the nine "old men" who now had to decide the complex matter. A day or two before oral arguments, Kitchel enjoyed a pleasant visit with his former mentor, but just as he prepared to leave, Frankfurter asked him, "Did you ever see this document?" In his hand he held a printed copy of the government's report critical of the Bisbee deportation of 1917, a document Frankfurter himself had written as counsel for the President's Mediation Commission established by Woodrow Wilson to investigate labor unrest in Arizona.[2]

For Kitchel the report resurrected a frightening specter from old Bisbee. "I immediately went to the telephone and called Louis Cates in New York and said, 'We've had it. Felix is sitting up there and he's going to bury the hatchet right between my shoulders.'" And he was right. When Kitchel presented his oral arguments for Phelps Dodge (with his mother, father, and fiancée watching), Frankfurter showed his former student no mercy. After every statement he would respond, "and also Mr. Kitchel," as if to undercut all his arguments. Chief Justice Charles Evans Hughes appeared willing to support Kitchel, who recalled feeling like a ball being swatted back and forth between the two eminent justices.

When the high court later announced its decision in *Phelps Dodge Corporation v. National Labor Relations Board*, the outcome surprised few observers. Kitchel was on his honeymoon in Mexico with his wife, Naomi (daughter of Walter and Margaret Douglas), when he learned of the vote from an angry Cates. His phone message was exactly two words long: "We lost!" No wonder Kitchel always recalled Cates as a very tough and brusque man.

In a split decision, a majority of the justices—all five of Roosevelt's New Deal appointees—voted to uphold the National Labor Relations Board. Frankfurter, their spokesman, stirred memories of the Bisbee deportation of 1917, as Kitchel had feared, when he wrote a rather slanted "history" of union organization that formed a major part of his opinion in the case. Frankfurter probably should have excused himself because he was hardly impartial. That would not be the last time that the ghost of actions taken in wartime Bisbee in 1917 returned to haunt Phelps Dodge.

Mountains of Copper for Democracy's Arsenal

IN THE LATE 1930s, at a time when the American economy slowly improved and the nation's businesses entered a new era of labor and government relations, Phelps Dodge formally launched development of its open-pit mine at Morenci. Underground production from its consolidated properties at the branch had remained relatively high during most of the 1920s, but sagging copper prices after the Wall Street crash forced the company to suspend production there in mid-1932. Even had prices not plunged, Phelps Dodge was running out of ore that it could mine profitably at Morenci using existing underground technology, but because of hard times the company wisely postponed the Herculean task of converting to an open-pit mine.

In fact, three things combined to delay the Morenci project: questions about the best mining methods, concerns about how to finance so large an enterprise, and fears that low copper prices at that time made low-grade ore not worth mining. Phelps Dodge, however, dared not hesitate too long. Louis Ricketts warned his fellow directors in 1935 that the declining output from the mines of Ajo, Jerome, and Bisbee made developing major new sources of ore a priority. Fortunately, the price of copper climbed to ten cents per pound in 1936, and that was encouraging news for Phelps Dodge.

The mine proposed for Morenci was not to be operated by tiny ore cars, hand-operated windlasses, and shovels and picks. It demanded an immense, complicated, thoroughly mechanized, and very costly industrial operation. That meant standard-gauge railway tracks spiraling down long horizontal benches to where steam shovels loaded trains with ore for the long climb back to the rim and across to a concentrator of unprecedented size. Early-day mining companies had recovered copper metal from ores by smelting alone, but smelting any ore composed of 98 percent rock and merely 2 percent copper would require so many furnaces and consume such staggering amounts of fuel that no company could earn a profit from it. To make smelting of low-grade ore feasible, a producer needed a large and efficient mill to eliminate as much waste rock as possible *before* it sent the resulting concentrate to the smelter, the succeeding step in a reduction process that yielded copper of ever-increasing purity. The mine itself was thus only one part of a complex and costly enterprise.

The question for Phelps Dodge was whether it should merely refurbish and add to its existing processing facilities—crushers, concentrator, smelter, and the like—or spend considerably more money to fashion a new and thoroughly modern landscape of production. The existing Morenci concentrator and its many ancillary structures were not large enough to handle huge tonnages of ore that the new mine was expected to produce, nor were they advantageously located. Corporate executives correctly reasoned that any attempt to enlarge and modernize them would in the end prove costly and disappointing.

A BOLD UNDERTAKING AT MORENCI

NOTHING COMPARABLE IN *size to the proposed open-pit mine and processing complex at Morenci had been attempted before in Arizona, and consequently its cost would be staggering— totaling perhaps as much as $42 million, an unprecedented figure. In addition, unlike all mines previously developed or acquired by Phelps Dodge, where rich surface deposits essentially underwrote the cost of digging deep to reach large and profitable ore bodies, Morenci had an enormous unproductive overburden that would have to be removed before the low-grade ore body could be worked. In other words, even before the nation's first major new copper mine in at least twenty years yielded a single pound of copper, Phelps Dodge must strip away fifty million tons of overburden (most of which the mining experts then considered waste) and transform a ridge and mountain crest into something like a vast Roman amphitheater—all at enormous cost. The company might have to wait as long as five years to recover its initial investment.*

Fortunately, the size of the Morenci challenge appealed to Louis Cates and gave him a chance to apply what he had learned earlier working with Daniel Jackling at Bingham Canyon, Utah. As president of Phelps Dodge Cates did not fear spending millions of dollars to develop a mammoth mine that over a thirty-year span was predicted to yield some five billion pounds of copper.

That was why Phelps Dodge directors decided to build everything new and locate each structure to maximize production.

Preparatory work at Morenci required construction of roads, railroads, and buildings for offices, shops, change rooms, and garage and purchases of electric shovels, trucks, bulldozers, tractors, and graders by the score, jackhammers, churn drills, blasting explosives—and so much more. Engineers estimated that the company needed to lay at least forty miles of standard-gauge railroad track to reach down into the pit for ore and haul it back up to the mill and

smelter, and this track work alone would cost $1.5 million. To that sum Phelps Dodge must add another $2.5 million for diesel and electric locomotives and dump cars, $800,000 for twelve power shovels, $600,000 for electric power installations, and finally $250,000 for cranes, air compressors, hand drills, and other necessary equipment.

The shopping list grew longer and longer until all the projected expenses totaled approximately $42 million. To finance so enormous an undertaking, Phelps Dodge sold 3.5 percent convertible debentures worth more than

$20 million to the public in June 1937 and paid the rest of the expense from current earnings as construction progressed.

Directing work on the Morenci project was Harrison M. (Harry) Lavender, who succeeded P. G. Beckett in 1937, the year the longtime general manager of western branch operations relocated to New York, where he continued as a vice-president. Lavender, who conformed to the popular stereotype of the rugged, tough, and gruff mining man, was demanding and at times even tyrannical. Because he didn't sleep well, he frequently phoned subordinates any time of the day or night. Often when he returned to his Douglas office following a whirlwind inspection of the various branches, Lavender rapidly dictated twenty pages of notes to his secretary, who frantically wrote everything in shorthand, all the while knowing that he might add, "I want that in the 10:30 mail." Julia Mazanek, who joined Phelps Dodge as a secretary in the General Manager's Office in 1947 and was still working for the corporation more than fifty years later (at its Phoenix headquarters), recalled that Lavender "was a very approachable person despite his aura of authority."[3]

Still, in his drinking days he could be "pretty wild," recalled the former president Warren Fenzi, and he was an incorrigible practical joker, yet on the job Lavender always possessed energy needed to get a prodigious amount of work done. Further, he was very bright man who could hold many facts and figures in his head without difficulty. Lavender came to Phelps Dodge from Calumet and Arizona, where his knowledge of operations had greatly impressed Cates during merger negotiations.

Working closely with Lavender was Walter C. Lawson, a brilliant mining engineer and an authority on open-pit mining (and who was as much a quiet introvert as Lavender was a boisterous extrovert). As chief engineer, Lawson was largely responsible for the success of the original Morenci project, and his careful planning permitted its further expansion, by 80 percent, only a short while later to meet stepped-up production required by World War II. Like Lavender, Lawson came to Phelps Dodge as a result of the Calumet and Arizona merger.

Besides those two men, numerous other people deserve credit for successful development of Morenci's new landscape of production, most notably Wilbur Jurden, head of engineering at the Anaconda Copper Mining Company. Two greatly revered figures in Phelps Dodge history, the elder James Douglas and Louis Ricketts, had long insisted that, although copper companies competed vigorously with one another, their scientists should nonetheless work together and exchange their findings. That kind of liberality paid big dividends when Cornelius F. Kelley, the president of Anaconda, responded to a call from Cates by lending Jurden to Phelps Dodge to help it design a new reduction works. In fact, it was Jurden and his staff of engineers borrowed from Anaconda who wisely located most of the treatment facilities approximately halfway between Morenci and Clifton, where they received ore by a downgrade haul from the mine.

On August 24, 1937, Phelps Dodge Shovel No. 1 dug deep into the Arizona topsoil at the 5,350-foot elevation and brought up a dipper full of overburden, the first step in the process of stripping away a layer of earth and rock that averaged slightly more than two hundred feet thick above the Clay ore body. The area's old-timers soon accustomed themselves to strange new sights and new mining lingo. Lines of transportation were horizontal rather than vertical, and instead of cages dropping into dark shafts, locomotives crisscrossed a sunlit landscape. Underground miners jokingly referred to the new pit as the "sunshine stope."

Above the Clay ore body, dynamite blasts raised whole banks of earth skyward. Bulldozers then pushed the shattered pieces into large piles along the benches that started to spiral down along the walls of the emerging pit. In this amphitheater setting, power shovels buried their buckets deep into the rocky piles, swung around, and dropped their loads into one of the waiting haul trucks that ranged in capacity from five to twenty-two and a half tons. The chief purpose of the fleet of haul trucks—utilized at Morenci more extensively than in any other open-pit operation in the United States at the time—was to prepare the way for the general use of rails.

During 1939, after workmen spiked down several miles of track in the pit and along a main line between the mine and the reduction works, the first trains went into service. During July of the next year, traffic by rail fully augmented truck haulage. Nine big electric locomotives supplied by Westinghouse, and "said to be the largest trolley-battery units of their kind for use in open-pit mining," started operating in 1941. Phelps Dodge purchased five diesel-powered switch engines to shuttle cars among its surface facilities.

Phelps Dodge in 1939 broke ground for a concentrator five time larger than its existing mill

at Morenci, to process twenty-five thousand tons of sulfide-rich ore per day. That same year it added a new smelter to the expanding landscape of production. Located next to the massive concentrator on the highway below old Morenci, it was completed in 1941. The smelter stack from its base to cap stood 612 feet tall, a remarkable height that underscored the immensity of the new facility. Twelve copper points at the top and heavy copper grounding cables served to protect from lightning strikes one of the largest reinforced concrete chimneys ever built.

In some respects the new mine, mill, and smelter complex at Morenci represented the ultimate in open-pit porphyry mining as Daniel Jackling originally envisioned it almost forty years earlier. It was conceived and built as a unit, on the basis of the latest technology, and without attempting to mix old facilities with new ones. By contrast, the layout at Utah's Bingham Canyon, where Cates had worked with Jackling, evolved mainly by adding new processing facilities to the old. Jackling himself described the Phelps Dodge venture at Morenci as "probably the boldest undertaking of its kind in history" given the physical obstacles and the "financial requirements of almost forbidding magnitude prior to the incidence of any production of consequence."[4]

Wartime Morenci and the Replumbing of Arizona

On January 30, 1942, wheels began to turn in the big new concentrator at Morenci, thereby launching production from Phelps Dodge's latest and largest open-pit mine.

The smelter poured its first copper anode on April 26, 1942. Weighing seven hundred pounds, it was the first of a steady stream of heavy anodes that traveled in boxcars to the El Paso refinery.

Cates and Phelps Dodge were justifiably pleased by what they had accomplished. A demanding five-year construction schedule was met on time, and by mid-1942 the Morenci complex operated at its intended capacity. Ironically, achieving that goal was no longer good enough because by then World War II had engulfed the United States. The Allies did not have enough copper to meet rising wartime needs, which dwarfed even the insatiable demands of World War I. Spurred by Uncle Sam, Phelps Dodge nearly doubled its productive capacity at Morenci by 1944.

Technically, Uncle Sam in the form of the United States Defense Plant Corporation paid for the additional Morenci expansion. The government built the facilities and rented them back to Phelps Dodge. By February 1, 1944, after an expenditure of about $68 million by the corporation and the federal government, the enlarged project was in full operation. Every bit as challenging as the recently completed complex, it increased the size and capacity of the mine, mill, smelter, and power plants and added new housing for a growing army of employees.

Ever since 1937, both Clifton and Morenci had looked like boom towns. Anticipating a burgeoning of personnel, Phelps Dodge had laid out a large-scale residential district about a mile southwest of old Morenci's town center. By the end of 1941, it built 271 low-rent houses, nearly all of them copper-roofed stucco dwellings. It housed single workers in the fully renovated Longfellow Hotel and in the former hospital remodeled into a comfortable dormitory. Phelps Dodge greatly expanded recreation and school facilities at Morenci and erected a new 34-bed hospital complete with the latest equipment—including an iron lung purchased with funds raised by the women's club to treat polio victims.

By 1942, CLIFTON had a population of five thousand and Morenci had seven thousand, but because of wartime expansion of production, housing remained in short supply. Empty lots became trailer camps; some people lived in the back seats of cars or took turns sleeping in eight-hour shifts around the clock in "hot beds." The rapid growth of Clifton and Morenci and Phelps Dodge's new landscape of production was not the total story, however.

The sheer size of the mine and processing complex, especially as a result of wartime

EMPLOYEE HOUSING AT *Morenci, May 1, 1944. Not all money Phelps Dodge spent during World War II to increase production was invested in industrial landscapes. At Morenci it constructed 770 additional homes in 1944 and increased the capacity of the company-community hospital from thirty-four to fifty-two beds. At the beginning of World War II the corporation operated a total of seven hospitals: Douglas with thirty-five beds, Bisbee with forty-three, Morenci with thirty-four, Ajo with thirty-three, Nacozari with thirteen, Jerome with fifty-two, and Dawson with thirty—all for employees and their families.*

MODERN PHELPS DODGE CORPORATION

～ THE COMMITMENT TO *safety is a top priority at Phelps Dodge Corporation, and its annual Chairman's Safety-Awards program provides a way to emphasize that fact. Douglas C. Yearley holds one of the award medallions.*

COMPANY PHOTOGRAPHER KEN MALLOQUE *took this prizewinning photograph of the converter aisle at the now-demolished Morenci smelter. He is a member of a multigenerational Phelps Dodge family. His grandfather Joseph Malloque appears in the black-and-white image on page 138 showing Morenci's New Italian Band (he is the second person to the right of the drum in the front row).*

〰 THIS DRAMATIC TWILIGHT PHOTOGRAPH *of the Morenci mine in the days of railroad operation appeared on the cover image of Phelps Dodge's 1974* Annual Report. *The long exposure captured the shovel in motion. Railroads are no longer used within any of Phelps Dodge's active copper mines.*

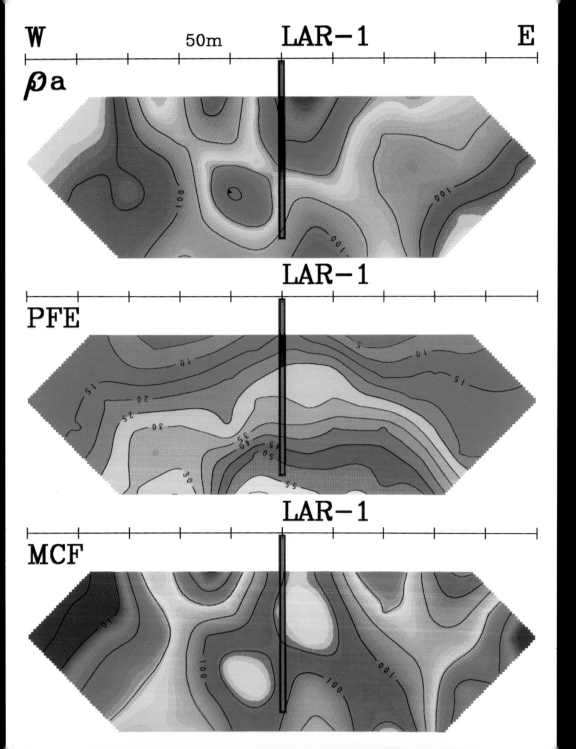

PHELPS DODGE'S *Punta Padrones mechanized clean port facility, where beginning in the 1990s ships loaded copper concentrate from Candelaria in Chile that was destined for smelters in Japan and other distant locations.*

LEADERS OF *Phelps Dodge Corporation who have been among the shapers of its recent landscapes of production. Left to right in this image of senior management that appeared in the 1979* Annual Report *are Richard T. Moolick, Arthur H. Kinneberg, Edward H. Michaelsen, L. William Seidman, George B. Munroe, and Warren E. Fenzi.*

TEN YEARS LATER: *at his retirement party in 1989, G. Robert Durham, Phelps Dodge Corporation chairman, is flanked (left to right) by Patrick J. Ryan, J. Steven Whisler, Douglas C. Yearley, and Leonard R. Judd.*

ONE MORE GENERATION: *Phelps Dodge's Senior Management Team in 1997. From left to right are Thomas M. St. Clair, Manuel J. Iraola, Douglas C. Yearley, Ramiro G. Peru, and J. Steven Whisler.*

THE FACES OF *three of more than five hundred workers at Phelps Dodge Hidalgo, Inc., who earned the Chairman's Safety Award for Best Safety Performance in 1997.*

EMPLOYEES AT *Columbian Chemicals Company's Santander, Spain, plant earned a 1997 Chairman's Safety Award for Best Performance, Small Operations.*

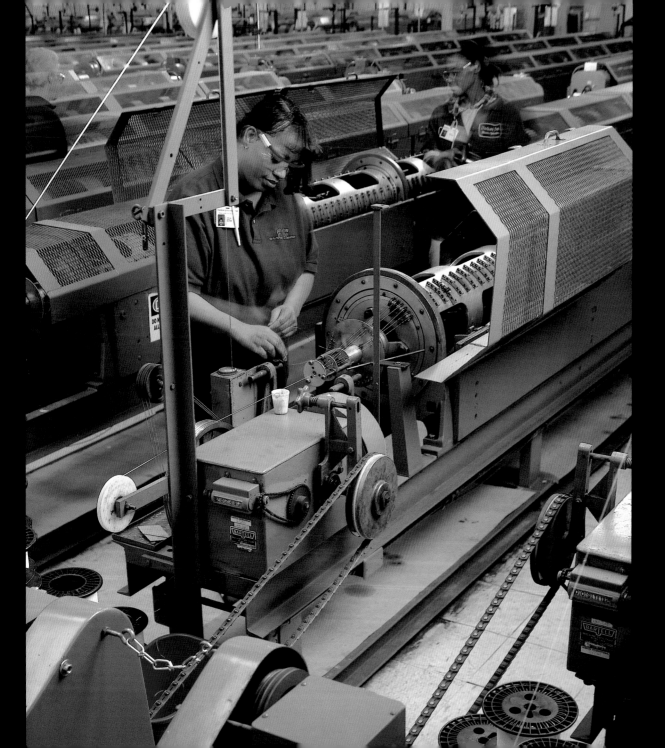

INSIDE THE *Inman, South Carolina, facility of Phelps Dodge High Performance Conductors. Arrayed across a production floor so squeaky clean that it shines like a highly polished dinner table are several dozen wire mills. In other large rooms are more machines of various sizes and shapes, all of them running around the clock to draw and spool a variety of wires and cables. Employees at High Performance Conductors plants in Inman and Trenton, Georgia, earned the 1996 Chairman's Safety Award for the Most Improved Safety Performance, U.S. Operations.*

PIONEERING THE CREATION OF VALUE"

⎯ THE MANY FACES *of Phelps Dodge, a composite portrait of some of the corporation's modern pioneers in the creation of value. The original images were used to illustrate safety brochures.*

"PD AND ME!" A collective portrait of some of today's employees in Phoenix who either grew up in a Phelps Dodge community or belong to a Phelps Dodge family spanning several generations or both. Seated left to right are Deni Kief, Julia Mazanek, Tom Gjurgevich, Raney Peru, Paula Garcia, and Ross Bacho. Standing left to right are Steve Tanner, Kelly Baker, Steve Balich, Paty Enriquez, Teresa Eberle, Jim Smith, Maurice Sandoval, Barbara Myers, and Jamie Ivey. On the wall of the corporation's boardroom hang portraits of Daniel James, Anson Phelps, and William E. Dodge, the three founders of the enterprise.

expansion, made it imperative that Phelps Dodge locate an adequate amount of water for its new mill facilities. Prior to 1937, when only underground mines operated in the Clifton-Morenci area, water rights the company had earlier acquired on Eagle Creek, Chase Creek, and the San Francisco River seemed adequate to meet all local needs. Those streams, all tributaries of the Gila River, flowed through the district and were located in general proximity to the mining areas.

When development began in 1937, Phelps Dodge anticipated that the usual sources could supply the extra water needed. That is, water would be pumped uphill to the mill from the river and creeks. However, when the federal government called for nearly twice the copper-producing capacity at Morenci to meet wartime demands, the company needed to locate still more water or trade for it. Phelps Dodge did not own any sizable water rights on the Salt or Verde river systems, two of the most obvious sources of new water. Moreover, the broken and uplifted terrain around Morenci meant that transporting additional water from where it was available to where it was needed would involve a major feat of engineering.

Phelps Dodge carefully examined each piece of a puzzle that when fully assembled illustrated the exceedingly complex nature of Arizona's water resources. It noted that the Salt River Valley Water Users Association had constructed Bartlett Dam in the late 1930s to hold Verde River floodwaters that could later be used for irrigation and to protect downstream Phoenix from damaging overflows. Unfortunately, the capacity of Bartlett Reservoir was not always

adequate to handle the deluge. As a result, every few years a major flood topped Bartlett Dam, inundated parts of Phoenix, and spilled precious water down the Gila River.

With that problem in mind, Phelps Dodge entered into an innovative agreement with the Salt River Valley Water Users Association, which also controlled other sources of water for the many irrigated fields and orchards around Phoenix. Phelps Dodge proposed to build Horseshoe Dam farther upstream from Bartlett Dam, about forty miles northeast of Phoenix, to reduce flood danger significantly and provide additional water storage for farms and cities of the Salt River valley. In return, Phelps Dodge received credits to 250,000 acre-feet of water that it could take during the coming years from the Black River in eastern Arizona, some forty miles from Morenci. Lifted by pumps seven hundred feet to the rim of Black Canyon, the water then flowed by gravity through nearly six and a half miles of buried pipeline to Willow Creek from where it continued twenty-one miles to Eagle Creek and then another thirty miles to the Eagle Creek Pump Station near Morenci, where Phelps Dodge recovered it along with the creek's normal supply of water, as in the past. It required roughly two days for Black River water to complete its fifty-one-mile journey. Its delivery to Morenci under the Horseshoe Dam agreement became Arizona's first successful experiment in transmontane diversion of water.

Phelps Dodge completed construction of $2.5 million Horseshoe Dam in the mid-1940s, the cornerstone of a complicated system that collected, stored, distributed, and recycled water needed for both domestic and industrial uses at

Morenci. Show Low Dam was next, its storage lake completed in 1953. It pumped surplus unappropriated water from a tributary of the Little Colorado River over uplands located about a hundred miles north of Morenci and into a tributary of the Salt River. That flow in turn enabled the company to pump still more water from the Black River. But Show Low proved an erratic source that some years yielded no water at all.

Phelps Dodge located additional water on East Clear Creek where in late 1965 it completed construction of Blue Ridge Dam, a thin inverted concrete arch 170 feet high, 600 feet wide, but only 6 feet wide at the top. Pumps lifted water from the reservoir formed by the dam to the top of the Mogollon Rim, from where it traveled down to the Verde River through seven miles of penstock and a hydroelectric plant that in turn generated power for the pumps (in effect coming close to creating a perpetual motion machine). By building Horseshoe, Show Low, and Blue Ridge dams, Phelps Dodge also provided Arizona with new lakes for public fishing, boating, camping, and recreation.

In all, the corporation's replumbing of Arizona meant building three large dams, reservoirs, pumping plants, pipelines, and other support facilities located in six different counties and at points as far distant from the Morenci mine as 180 miles. That was possible because of agreements negotiated with some thirteen different federal, state, and local agencies. Few, if any, American industries can claim to have developed a more extensive system for moving water than Phelps Dodge.[5]

PART OF THE *replumbing of Arizona was the graceful Blue Ridge Dam, finished in 1965 but nonetheless part of the complex water delivery system that dated back to World War II. It was Bill Evans, a Phoenix attorney and avid outdoorsman, who drew upon his extensive knowledge of the Mogollon Rim backcountry to conceptualize how Phelps Dodge could best deliver water to Morenci. He headed the firm of Evans, Kitchel, and Jenckes, which for years provided outside legal counsel to Phelps Dodge.*

Soldiers of Production

STARTING IN 1940, the copper companies of the United States produced in four years about as much copper as they had during the entire nineteenth century. From 1931, when Phelps Dodge acquired the Calumet and Arizona Mining Company, until the end of 1946, it alone invested in excess of $83 million in capital improvements, or an average of more than $5 million a year. More than half was spent at Morenci, which by the end of World War II emerged as the company's largest mine and processing complex.

Apart from its open-pit mine at Morenci—many would call it the flagship—Phelps Dodge possessed a valuable and varied fleet of other properties. Not the least of these was the venerable Copper Queen Branch. During the Great Depression, rich ore bodies and careful management had allowed some of Bisbee's underground mines to continue operating, albeit at much reduced levels. World War II, by contrast, found the Copper Queen Branch in full production. Bisbee copper operations had long yielded lead and zinc as by-products, but now those two metals were essential to winning the war and mined along with copper as never before. At Jerome and the United Verde Branch, even certain marginal or low-grade ores became worth mining, although the property overall experienced such a decline in ore production during the war years that the nearby Clarkdale smelter had enough extra capacity to relieve some of the load carried by eastern smelters. For a time it handled ore from Bisbee too.

The mines of Ajo, Arizona, worked around the clock, seven days a week, and the town's population, having declined to around four hundred permanent residents during the shutdown of the Great Depression, swelled to some six thousand people during the war. Company trains of covered steel dump cars pulled out for the big smelter plant in Douglas loaded with dark green half-dried mud, which was a concentrate of copper, silver, and gold. At one time the cars were not covered—until the company discovered that the desert heat dried out the usually moist concentrate and allowed considerable quantities of copper, silver and gold to blow off as dust in the hot winds.

As a result of the federally sponsored expansion of production at Morenci, the El Paso refinery would have formed a production bottleneck for Phelps Dodge had it not been enlarged to boost its annual capacity from 158,000 tons to 220,000 tons of copper. The cost, more than $2.5 million, was financed by the Defense Plant Corporation. When peace returned, employees of the company's two refineries could be proud that they handled all copper reaching the El Paso and Laurel Hill plants without once failing to meet their scheduled delivery dates.

Output from Phelps Dodge's mines for the first time exceeded four hundred million pounds of copper annually in 1942, but wartime production records were meant to be broken. With all properties synchronized like the parts of a fine watch, 1943 was another banner year. The corporation's annual report observed that copper production was the highest in Phelps Dodge history. Within the limits imposed at times by

SHIPBOARD NERVOUS SYSTEMS

PHELPS DODGE'S HABIRSHAW *Cable and Wire Division made special shipboard cable that functioned like the nerves to transform lifeless hulls into responsive combat and merchant vessels. Every warship required a vast network of wires and cables to control propulsion and to direct gunfire and locate enemy aircraft and ships by means of radar. A single aircraft carrier utilized more than two hundred miles of carefully engineered shipboard cable, and Uncle Sam required that every inch of it be both fireproof and waterproof.*

Shipboard cable was a surprisingly difficult item to fabricate. Although wartime shortages often forced Phelps Dodge Copper Products Corporation to use substitute materials, it still had to make certain that its shipboard cable was waterproof, fire resistant, and fully able to endure high and continuous operating temperatures. As warships grew more complex, and hence more crowded with wires and cables, manufacturers reduced the diameter of shipboard cables yet still kept them functioning flawlessly. To meet all such demands, Phelps Dodge Copper Products rushed to completion in 1943 a new plant at Yonkers devoted exclusively to fabricating shipboard cable. By war's end it was the world's largest maker of that largely uncelebrated but vital product.

manpower shortages, "all of the branches operated at capacity." That was true too for 1944 and 1945. Interestingly, Phelps Dodge was required to sell every pound of copper it produced to Uncle Sam. All buyers, including Phelps Dodge's own fabricating operations, had to apply to Washington to obtain copper.

By war's end the various Phelps Dodge metal-producing properties had increased copper production by more than 50 percent—and that was despite occasional shortages of labor. The story was much the same at its fabricating plants, where to meet military and civilian needs Phelps Dodge Copper Products Corporation carried out a major expansion program of its own. It built

new plants, enlarged and reorganized those already in operation, increased the number of machines, and added technical experts to its staff. During the war years Phelps Dodge factories manufactured billions of pounds of war materials.

Phelps Dodge Copper Products first felt the impact of war in a flood of orders from the British government for degaussing cable. That ingenious item, the result of months of research by British scientists, helped protect ships from deadly magnetic mines. After company engineers translated laboratory theory into production-line practice, Phelps Dodge Copper Products Corporation emerged as one of the world's largest manufacturers of degaussing cable.

BRAIDING MACHINES *inside the Habirshaw Cable and Wire plant manufactured navy shipboard cable during World War II.*

At plants in the Yonkers area, Phelps Dodge's Habirshaw Cable and Wire Division also fabricated endless miles of field wire for the Army Signal Corps. Soldiers unspooled small reels of field wire to extend lines of communication across the battlefields.

Phelps Dodge Copper Products made special thirty-mile lengths of submarine cable for the Russian government, which used it to supply electric power from central generating plants located in safe areas across swampy land to cities besieged by invading armies. Military police accompanied a special trainload of the cable as it traveled from the Yonkers plant to Seattle, from where it continued by boat to Russia.

Among the multitude of other special wire and cable products that Phelps Dodge made to help win the war were coaxial cables for a recently invented technology called radar. It fabricated countless miles of magnet wire for everything from ship generators and tank controls to aircraft gyrostabilizers and portable X-ray machines. In fact, by the time the war ended in August 1945, Phelps Dodge's manufactured products could be found aboard virtually all types of Allied military vehicles operating in all types of environments.

In addition to wires and cables, Phelps Dodge made rotating bands for projectiles and copper condenser tubes for steam-powered ships in the Allied war machine. Cruisers and battleships were as helpless without condenser or admiralty tubes as without their big guns, and a single aircraft carrier might utilize as much as two hundred tons of copper in piping alone. It needed millions more pounds of copper for sea-resisting castings, huge bronze propellers, and miles of onboard wires and cables.

In the months between September 1939 and December 7, 1941, when the war had not yet come to North America, Wylie Brown of Phelps Dodge Copper Products perceived a parallel between the tense times and those that preceded American entry into World War I. He grew increasingly concerned that likely shortages of condenser tubes imperiled the United States shipbuilding effort. Thus well before Pearl Harbor he launched a new division to avert a production crisis. Working with officials of the Defense Plant Corporation, which provided the needed financing, Phelps Dodge erected a facility in California to fabricate immense tonnages of condenser tubes required by West Coast shipbuilders The $21.2 million Los Angeles Tube Mill began operation in 1942 and by the end of the year was working at capacity to make seamless brass and copper tubes.

Besides mass-producing war materials, Phelps Dodge's various research and development departments cooperated with federal officials during World War II to fabricate new products, such as components necessary to build atomic bombs. For the latter assignment, Phelps Dodge Copper Products Corporation plant at Bayway rolled several million pounds of Uncle Sam's stockpile of precious silver into strips forty feet long, three inches wide, and five-eighths inch thick, which a manufacturer in Milwaukee, Wisconsin, coiled into magnets shipped to Berkeley, California, for installation in an atom-smashing cyclotron. It was all part of the famous Manhattan Project.

Perhaps no assignment for Phelps Dodge was more bizarre yet vital to winning World War II than a secret project called Operation Pluto. "Pipe Lines Under The Ocean," or Pluto for short, entailed an ingenious plan to extend a large hollow conduit beneath the English Channel to supply fuel to Anglo-American troops once they established a continental beachhead at Normandy. Success of the Allied thrust across western Europe depended in large measure upon a steady supply of petroleum products pumped from fuel depots on the west coast of England to frontline troops in France and later Germany. Military officials feared that if they used a large cable-laying ship to extend the pipeline under the English Channel, the Nazis might easily spot and destroy the suspicious vessel with torpedoes or bombs. Instead, the Allies relied on an inconspicuous tuglike vessel to pull 23-mile-long sections of pipeline into place under water and out of sight.

With that plan in mind, early in April 1944 the British government asked the United States Army Corps of Engineers if American industry could produce the unusual pipelines. Because Habirshaw Cable and Wire was a recognized leader in the manufacture of large-diameter, long-length underwater cables, the army sent a deputation to call on Wylie Brown. One executive, William Dunbar, recalled being summoned to Brown's office and finding there a stony-faced group that included a brigadier general, a colonel, and a captain, and Phelps Dodge's own plant manager and chief engineer from Yonkers.[6]

Brown turned to the army men and said, "You state for my people here exactly what you want us to do." The chairman of the military delegation emphasized by way of introduction that the secret project they had in mind would have "every priority that the government can give you; we've got to have it as fast as possible." Here he paused before stating, "We want a hollow submarine cable."

"What d'you want it for?" interrupted the always irascible Brown.

"Top secret," came the reply. "You'll never know until after the war."

"Well," responded Brown, "we can make a hollow cable. We can make [it] with nothing in it, no guts in it, no wires—nothing. . . . It'll be a tube with nothing inside. It'll be lead." Here he paused and turned to his Habirshaw works manager, Earle A. Mitchell, and barked in his customary tone, "Mitchell! You can do this?"

Having commanded a navy submarine chaser during World War I, Mitchell understood the urgency of Brown's request. "Yes, sir," he responded quickly.

"How're you going to do it?" Brown fired back.

Mitchell responded as if thinking aloud, "I'm going to [hesitates] . . . going to draw [extrude] a tube, and I am going to put some antimony in the lead to harden it, but you'll still be able to bend it. I'm going to put jute [coarse fiber] around the outside of it. . . . I am going to put steel armor wires on that jute—the jute will give the armor wires a bedding—and I'm going to put jute over [everything] as additional protection."

"How're you going to ship it? How're you going to move it?" Brown knew that the usual communication cable contained several hundred telephone wires and that its wire core supported

the weight of the cable when coiled for shipment. Hollow cables handled that way would collapse under their own weight. Mitchell's proposed solution to that problem was to fill each cable with water during shipment to support the weight of the coil above it.

The military people grew increasingly interested, Dunbar recalled. "The general says, 'Your plant at Yonkers . . . is on the river. It's navigable. Deep water. I'm going to put Liberty Ships along your dock . . . and as this cable is manufactured and goes on these reels, it goes right into the hulls. . . . As one ship fills up with these cables, the next Liberty Ship will pull in.'"

Military engineers brought their specifications to Phelps Dodge, which on April 8, 1944, launched construction of an entirely new plant at Yonkers (since demolished) to handle the mammoth order. Its own engineers rushed to design, build, and install special machinery required to perform all manufacturing operations simultaneously. Those giant machines applied the many separate layers of protective coverings in a single continuous operation to create a sheathed conduit that was then carried outside where it was coiled before being loaded into specially converted cargo ships alongside the plant's deepwater dock.

Three other companies—General Electric, Okonite, and General Cable—shared in closely guarded Operation Pluto, but Phelps Dodge Copper Products Corporation bore the main responsibility. Each section of pipeline was four inches in diameter but designed to withstand high-pressure pumping. None of its competitors could build the required conduits in sections of more than two thousand feet, yet Phelps Dodge not only brazed those lengths together at Yonkers but itself fabricated over two-thirds of the total required distance of 140 miles.

First delivery took place on July 5, 1944, and exactly 162 days after Phelps Dodge Copper Products Corporation first tackled its unique assignment, the new Yonkers plant shipped the last foot of its quota on September 16, 1944. The conduit was soon in place to provide fuel to Allied armored divisions under General George Patton, then driving toward the German border. The Habirshaw Division received a coveted Army-Navy "E" Pennant for excellence in production of war materials. It was one of many commendations that Phelps Dodge people earned, each a token of appreciation from the armed forces to the nation's copper workers who contributed so much to achieve victory.

From War to Peace

THE GENERAL ALEXANDER ANDERSON was one of the first big troopships to return from Europe after fighting ended there in the spring of 1945. As it sailed up the Hudson River past Phelps Dodge's Habirshaw plant, it seemed suddenly to list slightly to starboard as many of the 5,519 soldiers aboard rushed to return greetings from a group of women production workers who had gathered on the dock. As the plant whistle sounded long and loud to welcome the troops home, still more women spontaneously left their jobs to cheer as the ship passed by. Phelps Dodge workers had much to celebrate, including their own vital role in winning the war.

Due in large measure to the massive expansion at Morenci, Phelps Dodge at the end of World War II in August 1945 employed about fifteen thousand people, or about double the number it had just before the war. During the conflict, production of copper continued around the clock despite the loss of employees who volunteered or were drafted for military service, though War Department officials later furloughed a number of service personnel with mining experience to return to their former jobs. A total of 5,950 Phelps Dodge employees entered the United States armed forces during the war, and more than 100 of them died in service.

The pace of production did not slow appreciably even after the return of peace in 1945 because, although military demand for copper dropped abruptly, pent-up demand for civilian goods of all kinds created a vigorous postwar market for copper. Apart from a short pause after 1945 because of dislocations associated with government decontrols, inflation, and labor militancy, there followed a protracted level of affluence that caused America's copper industry to prosper as it had not done in nearly thirty years. Phelps Dodge copper production rose to nearly five hundred million pounds in 1948 to help meet unsatisfied demands that had accumulated during the war years.

That same year Phelps Dodge purchased outright the expanded plant facilities at Morenci and El Paso and the tube mill in Los Angeles, all of which Uncle Sam had financed during the war. Then came the cold war, and only five short years after the end of World War II, the Korean War again increased the military demand for copper. Phelps Dodge was ready, because even before the guns of World War II fell silent, its executives were already looking ahead to anticipate the needs of a postwar world.

As a result, they planned and implemented many improvements and expansions in existing plants, and they established a completely new division of Phelps Dodge Copper Products, the Indiana Rod and Wire Division, with a $4.5 million plant in Fort Wayne. Its construction began in August 1945 at a site directly across New Haven Avenue from the company's Inca Manufacturing plant. When the Indiana rod mill commenced production late in 1946, it was the most modern facility of its kind in the world—as well as one of the first major industrial projects completed in the United States after World War II. The output of the fabricating divisions of Phelps Dodge Copper Products kept pace with the increased activity of the nation's wire and cable industry as a

whole. By 1949, in fact, the corporation ranked as the nation's largest rod and wire manufacturer.

On the mining side, the economic health of Phelps Dodge remained synonymous with production of copper from Bisbee, Ajo, Jerome, and Morenci. Day after day the mine at Morenci remained busy, and the landscape there constantly evolved as the great pit grew wider and deeper and as disposal of rock and concentrator tailings filled up old canyons and created impressive new mountains. Somewhat the opposite was true for the United Verde Branch where World War II output represented a last valiant effort by a once great producer. The end was clearly at hand by the late 1940s. Every annual report speculated that because it had made no major discoveries, the mine could operate only two or three additional years. Zinc ore mined deep underground helped to postpone the United Verde Branch's inevitable day of reckoning until 1953.

A MODERN CHANGE HOUSE *for women was opened at the Morenci Branch in 1944. To deal with wartime labor shortages, Phelps Dodge recruited Navajo and Hopi Indians from Arizona reservations, Jamaicans, and women, many of them wives of its Morenci employees. That was the first time women worked in the company's mining and milling operations, not just its offices. By the end of 1943 some 25 percent of its concentrator workforce was female.*

Dealing with Uncle Sam

AFTER 1937, Phelps Dodge found itself working ever more closely with federal officials, and that was truly a departure for the venerable corporation. In contrast to the titanic legal battles waged by Montana copper barons at the turn-of-the-century, which made national headlines, Phelps Dodge had long sought to maintain a quiet, inconspicuous political stance. Exceptions were rare, such as in 1864 when William E. Dodge, who thought New York City's mercantile community needed a competent spokesman in the House of Representatives, ran successfully on the Republican ticket. He declined renomination, however, after a single term in Congress dampened his interest in politics.

During the early 1870s, the partnership's traditional distaste for politics increased because of unfounded accusations to the effect that the company had defrauded the federal government of large sums of money through undervaluation of metal imports. A Dodge family biographer concluded that the company perhaps "should have taken the case to court and fought the charges" but that it was "intimidated by the arbitrary methods of the customs officials," whose authority was backed by the Port of New York collector Chester A. Arthur as well as by the Empire State's powerful United States Senator Roscoe Conkling. To members of the city's notorious Tweed Ring it was all "honest graft," but to some members of the press it was blackmail "by a conspiracy of petty official rogues." When old Daniel James, a model of business rectitude, learned from an English newspaper that the federal government was planning to sue Phelps Dodge for $1,750,000, "the shock well-nigh killed him."[7]

Phelps, Dodge & Company, concerned for the firm's good name, settled the case out of court for approximately $250,000. The episode left William E. Dodge furious. He sought to avoid political maneuverings of every type—except to congratulate his former Republican congressional colleagues Rutherford B. Hayes and James A. Garfield upon their respective elections as president in 1876 and 1880 and to applaud Lucy Hayes's refusal to serve any wine or liquor in the White House.

The corporation's links to the federal government remained minimal throughout the 1920s. Only after the Great Depression tightened its grip on American industry did Cates and other Phelps Dodge leaders take an activist stance by pressuring Congress for a tariff on imported copper, and later they prominently associated themselves with New Deal agencies like the National Recovery Administration by helping shape policies intended to assure fair competition within the nation's industrial and manufacturing sectors. Phelps Dodge's need to work closely with Uncle Sam increased dramatically during World War II, and a growing array of interconnections with the federal government ended the laissez-faire political and economic climate that copper producers took for granted during much of their business lives.

During the early 1940s, an expanding bureaucracy in wartime Washington formulated a maze of regulations unprecedented in scope and number. Two federal agencies most directly affected Phelps Dodge. One was the Defense

Plant Corporation, which financed the factories, mills, and equipment that enabled the corporation to expand production at its Morenci, El Paso, and Los Angeles facilities. Another was the Metals Reserve Company, which acquired so many million tons of domestic and imported copper and other metals for the war effort that it became sole buyer of all such commodities from domestic mines—and thus gained the power to set their prices. Uncle Sam became of supreme importance to Phelps Dodge's "bottom line."

Invariably, wartime copper prices reflected opposition by the Roosevelt administration to any repeat of the industry's high World War I profit margins. That opposition explains why Uncle Sam arbitrarily pegged the price of copper at twelve cents per pound, compared to the market price of as much as twenty-six cents per pound during World War I, and permitted only minor adjustments at locations where the cost of production was demonstrably higher than the national norm.

President Cates wanted shareholders to understand why around-the-clock wartime production and record sales would not automatically equal soaring profits. Between 1941 and 1945 the corporation produced nearly twice as much copper as it had during 1936 to 1940, he explained, but severely limiting Phelps Dodge profits were the taxes that averaged 356 percent higher than during prewar years and payrolls that climbed 227 percent. True enough, Phelps Dodge earned sufficient money to maintain its quarterly dividend of forty cents per share, but the general purpose of government controls such as high taxes and ceiling prices for copper and other products, emphasized Cates, "was to restrict earnings as nearly as might be possible to a pre-war basis."[8]

LOUIS CATES RESIGNED the presidency of Phelps Dodge in 1947. The board of directors subsequently elected him chairman and chief executive officer of the corporation. By that time, he had had the opportunity to oversee nearly eighteen years of extraordinary growth and dizzying change, but anyone who thought that he would relax and enjoy the fruits of past labors was wrong. He remained chief executive officer until 1954, when he finally did relinquish that responsibility, but he continued as chairman until his death in 1959. Until just days before he passed away at age seventy-seven, Cates remained actively involved in day-to-day decisions at the corporation's offices in New York. Virtually everything that made Phelps Dodge the large and complex organization it had become by the end of the 1950s could be traced back in some way to his efforts. In all, it had been quite an extraordinary era, and Cates must certainly rank with Dr. James Douglas as one of the titans of Phelps Dodge history.

When the formal change of command came in the 1950s, Phelps Dodge was in superb financial condition. In 1954, for instance, it earned slightly more than $41 million (after a depletion allowance on the sale of more than 444 million pounds of copper); and it invested $27 million in new capital equipment and operations. Phelps Dodge Corporation had money in the bank, a total of $131.5 million in cash and securities, plenty of copper in the ground, and absolutely no long-term debt! Seldom had the future looked brighter.

~ Chapter Ten ~

IN SEARCH *of* EASY STREET

O N AUGUST 23, 1966, a massive explosion ripped apart the Fort Wayne offices of Phelps
Dodge Copper Products Corporation. Surrounded by a lush lawn and big evergreens, the
modern, two-story administrative building formed the centerpiece of a neatly kept oasis of
beauty and quiet on the city's industrial east side. Directly across busy New Haven Avenue was the
huge Inca works, from which wafted the faintly medicinal smell of magnet wire insulation, while on
the opposite side was the sprawling plant of Indiana Rod and Wire, another division of Phelps Dodge
Copper Products. What most people recall of that Tuesday morning was the sudden rumbling boom
like a distant cannon as the office building vanished in a billowing cloud of white smoke and dust.

They recall next that shredded paper floated down like confetti in a ticker-tape parade, that
desks and chairs lay tossed about like toys, and that tons of twisted steel and broken concrete dropped
into the basement. The floors of the remaining offices tilted at crazy angles. The force of the blast
stunned Walter Ainsworth, an Inca vice-president, and ripped the front off his second-story office, but
otherwise left him unharmed. He climbed down a ladder put up by employees and immediately joined

*~ THE SPRAWLING INCA plant in Fort Wayne stood on the opposite side of New Haven Avenue from where the Phelps Dodge Copper
Products Corporation office building, in this view not yet constructed, blew up in 1966. Tragic as the explosion was, it had no long-term
impact on wire and cable manufacturing.*

223

the rescue effort. Ainsworth recalled that dazed and injured colleagues lay all over the front lawn. One person who had been blown through a front window was disheveled but apparently not seriously hurt.

Rastus Powell, an employee since George Jacobs founded Inca in 1929 and who knew everybody by name, did a quick inventory of personnel. In all, five people died in the explosion, a majority in the center section of the building, and two more died of heart attacks attributed to the blast. The death toll would likely have soared had the blast not occurred just before noon, during the lunch hour. Twenty people were injured, and the property loss was about $5 million.

Thirty minutes prior to the blast, employees smelled gas. A quick check of the building revealed nothing amiss, although in the basement private contractors worked to install a new gas main. They confidently assured everyone the odor was perfectly normal, but a coroner's inquest blamed the outside firm for the needless tragedy. For the next two or three years, every Fort Wayne office worker at Phelps Dodge Magnet Wire spent a few hours a week on "salvage duty" attempting to piece together barrels of tattered records recovered from the blast site.[1]

THE FORT WAYNE explosion notwithstanding, the 1950s and 1960s were generally good years for Phelps Dodge Corporation, an era of cautious and quiet growth. In fact, during the decade 1947–57 the company's stockholders enjoyed a string of fat dividends not seen since copper's boom times before World War I; the annual

payout was at least four times greater than during the preceding decade. In addition, Phelps Dodge accumulated an ample financial reserve that during the best postwar years totaled more than $100 million, or nearly half the firm's net worth. *Forbes* magazine reported in 1958 that "in an industry famed for erratic winds of fortune, conservative, cash-rich Phelps Dodge makes its own luck by hewing strictly to the straight and narrow path of solid fiscal virtue."[2]

Robert G. Page, who succeeded Louis S. Cates as president of Phelps Dodge in 1947 and in later years as chief executive officer and chairman of the board, believed that because copper mining was inherently a cyclical business the only way to survive was to follow a conservative debt and cash policy. An additional reason to keep so much money in the bank was to build up the financial muscle needed to develop or buy new deposits of copper. "When a new venture comes up," Page told *Forbes* in 1958, "you've got to have ready cash if you want to get in on it." He recalled that Phelps Dodge had "borrowed $20 million in 1937" to finance its Morenci mine, "which was a lot of money in those days. But we have not had any need to borrow since."[3]

During the Page years the company's copper business generated profits year in and year out seemingly without heroic efforts. Analysts in the mid-1950s noted that Phelps Dodge could profitably produce copper for eighteen to twenty cents a pound at a time when it sold for close to thirty cents a pound. Furthermore, as Page noted in 1958, while mining, smelting, and refining provided a steady but relatively low margin of profit, fabricating copper into various industrial products generally paid much better, particularly when a firm controlled its own source of supply, as Phelps Dodge did. In sum, the corporation's various branches and divisions meshed well in the 1950s and 1960s to ensure that one prosperous year followed another.

During the thirties and forties, Phelps Dodge Corporation had weathered the Great Depression and the breakneck pace of World War II production. Now it appeared basically content to enjoy the hard-won fruits of earlier efforts, notably when it integrated copper production from mine to market and brought the modern open-pit mine at Morenci into full operation. With copper seeming to take care of itself—in 1950 its mines yielded some 490 million pounds of red metal and substantial additional amounts of gold and silver—the time seemed right for Phelps Dodge to venture cautiously into oil and aluminum, two highly seductive industries compared to tried-and-true copper. In fact, between 1952 and 1962 Phelps Dodge actually spent more money exploring for black gold than searching for new deposits of copper.

Although the 1950s and 1960s were unusually prosperous years, and although Phelps Dodge often gave the impression that it had at last found easy street, chance misfortunes like the explosion at Fort Wayne reminded executives that they could take nothing for granted. Who in 1950, for example, could have predicted that sixteen years later the United States government would arbitrarily cap the rising price of copper at thirty-six cents a pound during a war in Vietnam (a distant place that on 1950 maps was still labeled French Indochina), or that thousands of miners from several different companies would remain on strike for eight months in 1967–68 in the longest industrywide work stoppage in the history of American copper?

Easy street had some bumpy stretches. That was to be expected in any business, but the rough places were neither extensive nor serious enough to harm Phelps Dodge's long-term prospects. Only later, in the late 1970s and early 1980s, would the combination of runaway inflation, copper prices sagging even lower in real dollars than during the worst years of the Great Depression, and strident environmental regulation stir up a whirlwind of troubles that at times completely obscured the road ahead.

New Leadership: From Cates to Page

IT WAS IN October 1947 that Phelps Dodge directors resurrected the office of chairman of the board for Louis S. Cates, who for a few more years also served as chief executive officer. Board members selected Robert Guthrie Page as administrative and operating head. Unlike Cates in background and temperament, Page in his own way came to be respected as much as his older, more dynamic colleague. One major difference between Cates and Page lay in the fact that Page was a lawyer, not an engineer. Born in Columbus, Ohio, in 1901 and a graduate both of Yale University and Harvard Law School, Page was the first member of the legal profession to preside over Phelps Dodge. The heads of the old partnership had been merchants; following them was a succession of three presidents renowned in mining and metallurgy. Just as hiring Cates marked a major turning point in the company's history—bringing in the first outsider to run Phelps Dodge—so too did hiring Page. It began a forty-year era of lawyers at the helm of the corporation.

Old-timers at Phelps Dodge often used the word *brilliant* when they recalled Page. He was chief editor of the *Harvard Law Review*, and upon graduation in 1925 (with the highest grade point average in the history of Harvard Law to that time) he taught law for a year at his alma mater. One executive remembered that Felix Frankfurter, then a member of Harvard's law faculty, regarded Page as one of his most outstanding students. The future Supreme Court justice probably had a hand in securing his appointment in 1926 as secretary to Justice Louis D. Brandeis, a position that was a plum in any legal career. After a year, Page went into private practice, specializing in corporate organization and finance. In the mid-1930s he served a year as regional administrator in New York City for Uncle Sam's recently created Securities and Exchange Commission before becoming a senior partner at the firm of Debevoise, Stevenson, Plimpton and Page, general counsel for Phelps Dodge. In that capacity Page acted as special adviser to the corporation in financing its massive Morenci enterprise and subsequently represented Phelps Dodge almost full time in increasingly complex dealings with federal agencies during World War II. In all his work in Washington, Page demonstrated a formidable intellect.

After World War II, when Phelps Dodge directors recognized that their corporation's recent close association with federal administrators was no passing phase, they decided that top management should include a person who possessed the legal knowledge and experience the new business climate demanded. Page was their natural choice. Accordingly, in June 1947, they invited him to join the board of directors, and the following October they elected him president of Phelps Dodge. Page became chief executive officer in 1955 and chairman in 1966, and he continued in both positions until retiring in 1969.

The corporation's new administrative head— diminutive, urbane, witty, and a true gentleman of the old school who would always bow to ladies—was as wise as he was brilliant. He realized early that mining was scarcely his area of expertise, yet he gladly joined Cates on lengthy tours of company properties to acquaint himself with Phelps Dodge operations and personnel in the Southwest. He studied correspondence courses on mining, milling, and smelting, and afterwards whenever he went into company facilities, he posed numerous questions to individual operators. Always self-effacing, Page never tried to show off his newly acquired knowledge. In the end, he left policy on mining matters mostly to Cates and to his managers out West. In areas of his personal expertise—finance, diversification possibilities, and copper products—he always tended toward conservatism. Like a chess player carefully plotting strategy, he invariably made his move only after long and careful deliberation.

We speak somewhat glibly of the Page era, but in truth it was difficult to define with certainty when the era of Page began and that of Cates ended. Even after Cates's death in 1959,

several talented men supervised ongoing mining operations and sustained his vision. Two key administrators, Charles R. Kuzell from United Verde and Walter C. Lawson from Calumet and Arizona, owed their appointments at Phelps Dodge to mergers that Cates pursued. Those men essentially "ran the west" during the 1950s and 1960s, when the corporation still maintained a modest administrative staff at its New York headquarters and delegated considerable day-to-day authority to general managers based in Douglas at Western Operations offices. Both Kuzell and Lawson, his successor, officially reported to Page or Cates, but New York exercised control mainly by making the big decisions that affected production rates, marketing, and pricing and by handling the purse strings in major financial matters. Kuzell was a metallurgical engineer by training and a genuine innovator in smelting technology, a field to which he devoted a lifetime. Lawson's specialty was open-pit mining, and it was he who was largely responsible for overseeing Morenci's breathtaking wartime growth. A square-jawed stereotype of a mining engineer, he was known as Laconic Lawson for his stinginess with words.

Others, notably the Phoenix lawyer Bill Evans, unquestionably deserved an ample measure of credit for the ongoing success of the corporation's vast mining complex in Arizona and New Mexico. In the quest for water to run the company's mills in Morenci and Tyrone, the names Evans and Phelps Dodge were closely linked. Behind all the developments of those years, however, stood Louis Cates, "King Louis," as some called him beyond his hearing. For certain, Phelps Dodge paid Cates generously for his many contributions to the corporation. In 1947, for instance, his salary totaled $163,750, a substantial sum for the era (Wylie Brown ranked next at $118,500), and growing wealth allowed Cates to live regally. He never emulated the showiness of Daniel Jackling, his mentor, who once paid to redecorate a full floor of San Francisco's gracious Saint Francis Hotel as his living quarters, but Cates enjoyed touring Europe in imperial style. He ranked high among a vanishing breed of copper industry grandees.

The late 1940s and early 1950s also witnessed the emergence of new leadership at Phelps Dodge Copper Products Corporation. Two years after Page arrived at the parent corporation in 1947, the indomitable Wylie Brown relinquished day-to-day management responsibilities to become chairman of its manufacturing sector. The man who for two decades was synonymous with the industries side of Phelps Dodge Corporation retired in 1950. Filling his shoes was not easy. The first person to succeed Brown as president was Whipple Jacobs, a thirty-year employee of Belden Manufacturing Company in Chicago, but in 1952 at age fifty-five he suffered a fatal heart attack. Replacing Jacobs was Howard T. Brinton, who when he retired as chairman in 1965 had given more than half a century of service to Phelps Dodge Copper Products Corporation or its predecessor companies.

Regardless of its new leadership, Phelps Dodge Corporation in comparison to the other big producers, Anaconda and Kennecott, probably changed the least during the quarter century after World War II. If anything, the period of relative calm provided a welcome opportunity to catch its breath and quietly position itself to seize future opportunities. Of course, changes did occur at Phelps Dodge during the years between 1945 and 1970, but few innovations in technology or changes in its basic mix of businesses were as dramatic as those the company experienced during the preceding fifteen years—or would experience during the 1980s.

During the postwar era it made no significant alterations to the way it extracted ore. Except for 1949, a year of industrywide decline, Phelps Dodge expanded production each year to average almost five hundred million pounds of copper annually between 1947 and 1951, totals that compared favorably to the approximately three hundred million pounds it produced during each of four years preceding United States entry into World War II. Its consolidated net income rose accordingly, to approximately $40 million a year, or roughly six times higher than 1937–41 figures. The basic trend continued during the 1950s, when copper prices increased slowly from slightly more than twenty-one cents per pound in 1950 to nearly thirty-three cents in 1960.

Because yearly profits remained so robust for so long, Phelps Dodge felt no need to redefine its business strategy in any major way. It remained a model of prudence. When oil exploration excited other major copper producers, Phelps Dodge joined the search only cautiously by forming partnerships first with Carter Oil, then Continental Oil, and finally with the petroleum division of Cyprus Mines. Together they looked for black gold along the Gulf Coast of Louisiana and Texas during the 1950s, but without real success. Phelps Dodge did better through passive diversification, such as when it turned a profit on a 1952 purchase of two hundred thousand shares of Amerada Petroleum common stock.

Even if the 1950s and 1960s were not years of sensational growth and change like the 1930s and 1940s, several mileposts in the corporate landscape during the Page era were impressive nonetheless. It was then that Phelps Dodge noticeably increased its mineral exploration activity, added a large underground ore body at Safford, Arizona, and new open-pit mines at Bisbee and Tyrone, gained interests outside the United States in several wire- and cable-making businesses and a one-sixth stake in the Southern Peru Copper Corporation, expanded Phelps Dodge Copper Products Corporation, and entered the alluring new field of fabricating aluminum products. In the late 1960s, toward the close of the Page era, Phelps Dodge even joined the race to acquire uranium mining and milling property in Wyoming, supposedly the newest road to wealth.

By the time Cates died in 1959 he had

bequeathed to Page a financially robust Phelps Dodge. Its entire capitalization consisted of a relatively small ten million shares of common stock. It had issued neither preferred stock nor bonds, and it had no long-term debt. At that time its common stock sold for around $55, and behind every share were valuable and highly negotiable assets.

"Unless Further Discoveries Are Made"

WITHIN THE UNITED STATES during the 1950s and 1960s, Phelps Dodge copper production came mainly from open-pit mines at Ajo and Morenci and both underground and open-pit mines at Bisbee. On the other hand, at Jerome both the quantity and quality of ore declined significantly following World War II. "Unless further discoveries are made," warned the 1947 *Annual Report*, Phelps Dodge expected to halt production there in the near future. In 1951 Jerome contributed slightly less than 20 million pounds of copper to the corporation's record total that year of 501 million pounds. Not surprisingly the United Verde Branch ceased all production two years later.

All in all, the Jerome gamble paid off surprisingly well. During eighteen productive years after Phelps Dodge acquired the mine in 1935, the United Verde Branch yielded an impressive quantity of metals, including nearly 920 million pounds of copper, 15 million pounds of zinc, 53 million ounces of silver, and 400,000 ounces of gold. Even in the late 1990s, Phelps Dodge continued to own the property and

employ a small staff to maintain it. No one could say exactly what the future held, but in mining, as in movies, a sequel of some sort was always possible. The value of any remaining ore would depend on future mining technology and the all-important price of copper and zinc.

South of the international border, Nacozari and the Moctezuma Copper Company seemed destined to share the fate of Jerome and the United Verde Branch. In truth, the Mexico property would probably have remained shut after its 1931 closure had it not been for a subsequent rise in demand for copper. Operations resumed there in September 1937, and during World War II production at the Pilares mine and Nacozari concentrator was increased substantially at the request of the United States War Department. But when the conflict ended, production again dropped off sharply. Phelps Dodge ceased underground copper mining in Mexico in 1950, but it continued leaching operations for another decade. In 1979 a Mexican court dissolved the Moctezuma Copper Company and formed instead the Minera Pilares, S.A. de C.V. Phelps Dodge retained a 49 percent stake in the property.

Along with United Verde and Moctezuma, Phelps Dodge's coal mines at Dawson, New Mexico, steadily declined after the World War II boom. For employees with long memories, it was sad to watch the decades-long downward trend at Dawson. Its black diamonds had energized Phelps Dodge smelters, powerhouses, and railway locomotives, but blast furnaces gave way to more efficient reverberatory smelters, and by the mid-1920s, oil from Texas and later natural gas rapidly elbowed aside coal and coke for many industrial uses. One major coal customer was the Southern Pacific, but the railroad increasingly burned oil in its steam locomotives. During the decade after World War II, railroads of the West rapidly converted to diesel-electric locomotives, which were far more efficient than the picturesque but maintenance-intensive steamers.

In its final years, Dawson had only one mine in production, and earnings of the Stag Cañon Branch were at best anemic. Dawson was so obviously dying that no one could have been surprised when Phelps Dodge permanently ceased production there on April 28, 1950, after workers ran a trip of cars out of the mine and unloaded the last of some thirty-three million tons of coal dug there during forty-six years of operation. Still, for some of the old-timers the shutdown hurt. On May 11 the opera house witnessed one final public gathering: Dawson High School's Class of 1950 graduated. By then the town's population had dwindled to about twelve hundred residents, and it was hard for anyone not to feel mixed emotions when living in a community composed mainly of empty houses.

For tax reasons, Phelps Dodge quickly disposed of most aboveground structures to the National Iron and Metal Company of Phoenix for half a million dollars, and the new owner razed or sold or moved to a new location about two thousand different structures. Before long, Dawson's only remaining dwellings consisted of sixteen houses used by cowboys who turned the fifty-two thousand acres of coal and timber property into Phelps Dodge's own Diamond D Ranch. Nearly two decades later, in 1968, the corporation leased the property to the historic CS Cattle Company of Cimarron, New Mexico,

for grazing and for elk hunting. In the late 1990s, one of the remaining houses had been enlarged and remodeled to serve as a base for sportsmen.

Many former residents still gather every other year at what once was Dawson for a Labor Day–weekend reunion. They voice no bitterness because they understood the harsh realities of coal production. "Phelps Dodge did right by its employees, and there's a loyalty like you never find anywhere else." So noted Carol Myers, a 1946 graduate of Dawson High School who helped run the reunions. In the late 1990s they still attracted as many as eighteen hundred people.[4]

By that time, just about the only way for old-timers to reconstruct Dawson mentally, apart from sharing their collective memories, was to examine faded photographs or personally trace the pattern of eroded and weed-obscured streets and sidewalks. And in March 1997, most evidence of the urban landscape disappeared along with long-abandoned coal ovens and smokestacks in a Phelps Dodge remediation program. Old Dawson simply ceased to exist. It vanished even as a ghost town. Ironically, in the end the town's most permanent feature was its cemetery. Old William E. Dodge could probably have drawn a theology lesson from that.

As for the once mighty Copper Queen Branch in Bisbee, during the half decade following World War II it too declined as a big contributor to the corporation's bottom line, not only because long-reliable shafts, notably the Campbell, Cole, and Junction, lacked the rich ore of former years but also because underground operations required skilled miners. That category of worker became increasingly rare during and

SPECTATORS WATCH
as the grand finale blast officially
opens Bisbee's Lavender Pit on
August 7, 1954.

after the war. Yet in the early 1950s, the Bisbee operation avoided the fate of the United Verde and Stag Cañon branches and, remarkably, enjoyed one more copper boom.

In 1947, Phelps Dodge purchased the last of its major rivals in Bisbee, the Shattuck-Denn Mining Company. At about the same time it began evaluation of the Bisbee East ore body, a large low-grade copper deposit located just south of its long-closed Sacramento Pit. Workers in the

early 1950s removed the first of some forty-six million tons of rock to launch development of what soon became the Lavender Pit. The name honored Harrison M. Lavender, the Phelps Dodge vice-president in charge of Western Operations, although he did not live to see completion of the project. The hard-driving man died early in 1952 of leukemia at La Jolla, California. Compared to the Sacramento Pit, which actually formed the northwest section of

the Lavender Pit, Bisbee's newest mine excavation was approximately ten times larger than its predecessor, though its ore was of significantly lower grade.

Ore from the Lavender Pit first reached the new concentrator on June 24, 1954, and production continued for the next twenty years. Phelps Dodge itself financed the $25 million mine and processing complex. Guided by Charles Kuzell and then Walter Lawson, the Copper Queen Branch quickly resumed its former standing among Phelps Dodge's major mine properties by outproducing Ajo's New Cornelia Pit in the mid-1950s to rank second only to Morenci.

The concentric rings of the expanding Lavender Pit spiraled gradually outward to consume some of Bisbee's earliest suburbs, notably the neighborhood of Lowell. The main highway through town ran next to the rim of the chasm, and that offered motorists a spectacular view of daily mining operations. They could watch as workers drilled holes sixty feet deep in the concentric benches and carefully loaded each one with about 450 pounds of explosives. Soon the Arizona earth trembled as simultaneous blasts fractured tons of rock. A cloud of dust drifted away with the wind as if to raise the curtain on the next act of an ongoing drama. Electric shovels scooped six cubic yards of muck at a time into twenty-five-ton diesel trucks, which hauled it a short distance to fill side-dump railroad cars that diesel-electric locomotives slowly pulled up to the rim. The impressive show was quite a tourist draw. Children raised in nearby Douglas remember when Sunday afternoon entertainment consisted of family

outings to the Dairy Queen in Bisbee followed by a stop to watch the afternoon blast at the Lavender Pit.

The New Cornelia Branch continued to be a dependable producer, although news from Ajo rarely warranted more than a paragraph or two in the company's annual report, and what coverage it did receive often amounted to perfunctory news of a couple of new trucks, or modern drills, or an innovative explosive. Probably the two most significant developments occurred in 1946–47 with electrification of the railway line extending down into the pit and in 1950 with completion of Ajo's first smelter, as distinct from the vacuumizer and other technological frauds of earlier years. Phelps Dodge invested $8 million in the new facility "blown in" at Ajo in June 1950, the same

A PHOTOGRAPH FROM *December 1947 shows the New Cornelia Pit after recent electrification of its haulage railroad. The substantial nature of the support structure shows that this was truly a first-class upgrade. When the Ajo mine launched production thirty years earlier, the principal motive power consisted of oil-fired steam locomotives.*

month it closed its venerable smelter at Clarkdale. Thereafter, the New Cornelia Branch rapidly scaled back and soon eliminated regular rail shipments of concentrate to Douglas. The Ajo smelter was strictly an in-house job, its design influenced by early-day operations at Clarkdale where some of its equipment originated and where its chief architect, the assistant general manager Kuzell, worked earlier.

The bellwether of Phelps Dodge copper mines during the Page era was still Morenci, the branch that produced as much as three hundred million pounds of copper a year starting in the late 1940s. Its annual profits rapidly repaid the corporation's earlier investment of nearly $100 million. Not content with its record production, the Morenci Branch further stepped up production by enlarging its concentrator to

A RETROSPECTIVE
LOOK *at ongoing operations at the*
El Paso refinery in 1953, where
workers insert the starter sheets
used to make copper cathodes.
Here anodes from Phelps Dodge's
three Arizona smelters were
electrolytically refined to produce
virtually pure copper cathodes.

handle sixty thousand tons of ore per day and by constructing a large precipitation plant to recover copper leached from dumps of copper-bearing material of grades too low for processing through the big concentrator. As had happened during World War II, expanded operations at Morenci increased the need for water and required that Phelps Dodge enlarge its already extensive hydraulic network.

During the 1950s and 1960s, Phelps Dodge personnel made almost continuous refinements to equipment and processes that at individual properties increased capacity and reduced costs. Among the projects were larger grinding mills and development of the leach-precipitate-flotation process at the Morenci concentrator. Near its Morenci complex in 1953, Phelps Dodge initiated exploratory drilling in the old King-Metcalf area where earlier it had abandoned several underground mines and where in the future it would launch the new Metcalf open-pit mine.

Phelps Dodge did not limit new construction during the post–World War II decades to industrial facilities. In September 1948 it completed a new recreational center for employees at Ajo that included a modern 660-seat movie theater, a library, and a community recreation room. The latter facility was used for dancing or banquets or as an auditorium. In 1953 Phelps Dodge provided 1,287 additional homes for employees and other eligible residents in the Morenci-Plantsite-Stargo-Clifton area. In the early 1960s, as Phelps Dodge expanded its open-pit mines at Ajo, Morenci, and Bisbee, it removed and replaced several mine buildings and

processing facilities, but nothing it did was as dramatic as relocating the entire community of Morenci starting in 1965. The old town stood in the way of the big pit as it expanded south, and thus during the course of several years the company systematically relocated Morenci to a new site about three miles south of the mine. Opening a new high school in 1982 concluded the move.

Not Sticking Close to Home

ONE SIGNIFICANT DEPARTURE for Phelps Dodge during the quarter century after 1945 was the corporation's growing involvement in mining and manufacturing activities outside its familiar sphere of operation in the United States and Mexico. Overseas commerce of the old New York–based mercantile partnership had extended to Great Britain and continental Europe. That changed after 1881 when Phelps Dodge's growing involvement in copper in the Southwest caused it finally to focus its full attention on that one remote corner of North America. In the early 1930s its angle of vision expanded once again to encompass fabricating plants located in various parts of the United States, primarily the Northeast.

By midcentury, Phelps Dodge remained the only one of America's Big Three copper producers to avoid any direct involvement in mining properties outside North America. In fact, apart from the Moctezuma Copper Company and a few small properties in northern Mexico and a short-lived part-ownership of a

Canadian refinery in Montreal gained with acquisition of Nichols Copper, the corporation continued to restrict its mining and manufacturing operations to the United States.

To some extent, the narrow geographic focus resulted from deliberate policy. In the words of *Forbes* magazine, Phelps Dodge did not intend for "its fortunes to hinge, as Anaconda's may, on the outcome of an election in Chile, or as [Cerro de Pasco's] may, on the consequences of a military coup in Peru." That course had long seemed wise. For United States copper producers that had expanded into international operations, *Forbes* observed, "the rewards have been anything but commensurate with the risks in recent years," whereas by "sticking close to home, New York's Phelps Dodge has consistently led the copper industry in profitability."[5]

Ironically, by the time *Forbes* published those comments in 1964, Phelps Dodge had taken its initial steps toward greater international involvement. To outsiders they must have appeared rather timid, such as when Cates's postwar trip to Africa led Phelps Dodge to buy four million dollars' worth of stock in the American Metal Company (later AMAX)—or about a 10 percent interest. Because of American Metal's sizable investments in a northern Rhodesia copper venture, the stock purchase meant that Phelps Dodge became an interested party as well.

During the 1950s, Phelps Dodge executives cautiously scanned distant horizons in search of new mining opportunities, as distinct from passive investments abroad. They spotted an especially attractive prospect high in the Peruvian Andes near the borders with Chile and Bolivia, where several American mining companies

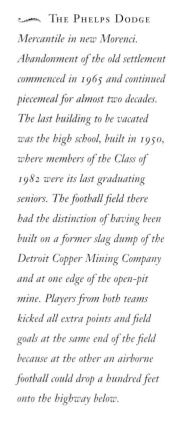

THE PHELPS DODGE *Mercantile in new Morenci. Abandonment of the old settlement commenced in 1965 and continued piecemeal for almost two decades. The last building to be vacated was the high school, built in 1950, where members of the Class of 1982 were its last graduating seniors. The football field there had the distinction of having been built on a former slag dump of the Detroit Copper Mining Company and at one edge of the open-pit mine. Players from both teams kicked all extra points and field goals at the same end of the field because at the other an airborne football could drop a hundred feet onto the highway below.*

discovered large deposits of copper ore. Phelps Dodge became involved in the isolated locale mainly because American Smelting and Refining (ASARCO), a company hitherto interested almost exclusively in metal processing and sales, obtained a promising ore body known as Toquepala. Nearby lay a second large copper deposit, Quellaveco, while about twenty miles northeast lay yet another, Cuajone, the latter owned jointly by two mining companies, Cerro de Pasco and Newmont. All together, the three deposits contained more than a billion tons of ore assaying around 1 percent copper.

Exploratory drilling in the late 1940s and early 1950s revealed that the deposits were amenable to open-pit mining. However, two difficulties stood in the way of easy development. One was the magnitude of the project and hence the high cost that development would incur. A second was the fact that the property-owning companies lacked adequate experience in open-pit technology. Phelps Dodge participation would contribute much to solving both problems. The owners, needing to secure an Export-Import Bank loan to develop their Peru properties, carefully sounded out Cates. He suggested that Phelps Dodge might be willing to join a consortium, but only if American Smelting and Refining, Cerro, and Newmont agreed to develop the properties jointly with a common infrastructure that encompassed Toquepala, Cuajone, and Quellaveco. The four companies turned the novel proposition over to their staffs, and after what one participant later referred to as "the battle of the burning slide rules," they joined forces to create the Southern Peru Copper Corporation early in 1955. That same year, the newly formed enterprise began the formidable task of stripping away 130 million tons of overburden and laying out a landscape of production at Toquepala.

The highly successful Peru consortium was largely Cates's brainchild. In some ways it was his last major contribution as Phelps Dodge's supreme authority on mining matters. The corporation also brought to the table the basic knowledge and mining skill it had previously gained from successful open-pit operations at Morenci and Bisbee. All this was important because the Southern Peru deposits were located in exceptionally difficult places, and they required even more massive commitments than Phelps Dodge had made in Morenci. The man who directed operations at Toquepala was Warren Smith, former superintendent at Bisbee's Lavender Pit.

Eventually, after minor realignments, the owners of Southern Peru divided the enterprise as follows: Phelps Dodge 16 percent, Cerro 22.25 percent, Newmont 10.25 percent, and American Smelting and Refining 51.5 percent. As an added inducement to Phelps Dodge, all parties agreed that when Southern Peru paid its first dividends, Phelps Dodge would recover part of its investment in advance of the other companies—which it did in 1966. Together with annual payments of $7 million during each of the subsequent five years, the money was sufficient to liquidate the corporation's loans of approximately $20 million to Southern Peru Copper. Further, they equaled a return on investment of nearly 100 percent a year on Phelps Dodge's remaining equity! So promising were the operations of Southern Peru Copper that its partners began in the early 1970s to reinvest their dividends in a companion project, Cuajone, which was even larger than Toquepala.

The new landscape of production at Cuajone began operations in September 1976. Although it cost a whopping $750 million in United States dollars, as compared to $238 million spent to open the Toquepala complex sixteen years earlier, Cuajone proved a good investment because even a quarter century later it still processed many tons of copper ore each day. As for Quellaveco, its third major deposit of red metal, Southern Peru Copper ceded ownership to the Peruvian government in 1971.

In all, Southern Peru Copper retained two excellent properties where modern equipment and an efficient management team combined to make production costs among the lowest in the world. As a result, it was one of very few copper-mining companies able to produce a positive cash flow during the disastrous downturn of copper prices during the early 1980s. That was when Southern Peru Copper emerged as Peru's largest single producer, accounting for about two-thirds of that nation's output of red metal. Phelps Dodge's minority share of Southern Peru Copper added millions of dollars to its balance sheet over the years.

During the 1950s and 1960s, as construction work in Peru pounded ahead, Phelps Dodge entered into several wire and cable manufactories outside the United States. That was when a dapper young Dane named Edward H. ("Ted") Michaelsen planted the initial seeds of what later flowered as Phelps Dodge International Corporation. (A detailed story of the global enterprise that resulted from one man's unusual angle of vision appears in chapter 16.)

A "Full Market Basket" Includes Aluminum

DURING THE DECADE from 1952 to 1962, Phelps Dodge followed the lead of its major competitors and invested in oil exploration. But after ten disappointing years that yielded at best meager returns on investments, Phelps Dodge exited oil, at least temporarily, and turned its attention to aluminum, the "wonder metal" that America's big copper companies found all but irresistible in the early 1960s.

Aluminum represented the latest in a series of attempts by Phelps Dodge to extend its technological expertise from copper to other primary metals. For years it had recovered gold, silver, lead, zinc, molybdenum, and several other nonferrous metals as by-products of copper. On at least two occasions between 1927 and 1950, Phelps Dodge proclaimed itself a serious entrant into the field of lead-zinc ores, and it constructed lead-zinc concentrators in Bisbee and a sizable lead smelter in Douglas. By 1950 its on-again, off-again relationship with those two white metals was definitely off. Copper once more became the primary focus of the company's attention and the metal that remained the foundation of its business.

Robert Page's own prudent disposition, together with declining earnings in America's nonferrous metals industry during the late 1950s, inclined him to pursue a conservative course for

Phelps Dodge. Nonetheless, as one publication observed, Page "has never hesitated to spend when he could see a profit in the offing." He fully supported overseas investments in mining and manufacturing; and despite gloomy prognostications by many observers in the industry, he remained upbeat about copper's long-range future. "With the exception of the three postwar recessions," he liked to remind any naysayers, "the use of copper has been determined only by its availability."[6]

As Phelps Dodge entered the 1960s, the price of copper remained unprecedentedly stable at approximately thirty cents a pound through 1963, a fact that vexed speculators but was welcome news to the cautious Page. He realized that this unusual steadiness could not last indefinitely. As he had foreseen, tight supplies of red metal began driving up its price early in 1964. In the context of the 1960s that was not entirely good news.

During that decade, persistent interference by the United States government with copper prices once ostensibly determined by a free market agitated the economic waters, but Phelps Dodge was fortunate in having a man of Page's political dexterity at its helm. The problem grew most troublesome during the presidency of Lyndon B. Johnson. In a doomed effort to fund both the war in Vietnam and Great Society social programs, and all the while keep inflation in check, Uncle Sam strong-armed American copper producers into "voluntarily" selling their metal within the United States at substantially lower than world prices. That was when political instability in Africa, strikes in Chile and elsewhere, and fighting in Vietnam combined to boost the price of copper. When London Metal Exchange spot copper (that is, for immediate delivery) soared to more than seventy cents per pound, federal officials sought to "persuade" major American producers to hold the line against inflation by selling their copper mined in the United States at thirty-six cents per pound. Later, when voluntary restraints proved inadequate, on January 20, 1966, they outlawed export of American copper in any form.

The ongoing manipulation disgusted the usually composed Page. His great concern was that government attempts to hold the line against inflation in the near term would cause the price of copper to soar later. Because of the rapidly rising competition between copper and aluminum that was not a good thing for Phelps Dodge. Rising copper prices could easily cause the company's wire and cable customers to switch to lower-cost aluminum or stainless steel products. Once substitution became standard in what formerly had been mainly copper-based products, erstwhile customers might well be lost permanently to Phelps Dodge. The threat of substitution was a primary reason the company entered the aluminum business in 1964.

Aluminum to meet civilian needs rose sharply after World War II. Would-be manufacturers easily secured ample supplies, often with strong support from Uncle Sam, and because of its intensive efforts to meet war needs, the aluminum industry came to know more about the metal's special properties than ever before—and that included how to substitute it for copper in familiar products. Added to that was aluminum's seeming price edge over copper.

～ Precious Aluminum ～

UNTIL THE LATE *nineteenth century, aluminum was so difficult to produce that it posed absolutely no competitive threat to copper. For years, laboratory methods yielded aluminum in tiny amounts that cost about $545 per pound, a price equivalent to perhaps $8,000 per pound in late 1990s dollars! A French scientist in 1854 succeeded in cutting the cost of aluminum to a mere $90 a pound, but even with further reductions, the price of aluminum remained almost as high as that of silver at a time when copper sold for mere pennies a pound. The white metal's great value and durability were why builders of the Washington Monument capped it in 1884 with six pounds of precious aluminum.*

Nothing much changed until February 23, 1886, when young Charles Martin Hall, a recent graduate of Oberlin College, discovered an electrolytic process (known as the Hall-Heroult process today) that offered a cheap way to produce pure aluminum, which at that time sold for about $8 per pound. His laboratory breakthrough gave rise to the newest primary metals industry in the United States. Hall established a plant in Pittsburgh, where its first mass-produced aluminum product in 1889 was the common teakettle. In 1907 the pioneering firm changed its name to the Aluminum Company of America, better known as ALCOA. As late as 1939 ALCOA remained the sole American producer of aluminum, but during World War II aircraft manufacturers required huge quantities of the lightweight metal, and several new makers successfully challenged the ALCOA monopoly.

Phelps Dodge almost joined the aluminum business then when Wylie Brown announced plans in mid-1943 to manufacture parts for planes and other implements of war. He had a plant partially built in Hammond, Indiana, before the War Production Board pulled the plug on the venture. It was a false start for a company that later became one of the biggest manufacturers of aluminum wires and cables in South America. However, within the United States the relationship between Phelps Dodge and the white metal was one of hope mixed with frustration.

When, for example, during a fifteen-month span in the mid-1950s the price charged by United States copper producers leaped 53 percent (from thirty cents to forty-six cents a pound) while aluminum held rock steady at about twenty-four cents a pound, copper users howled in protest.

Many of them rushed to substitute aluminum as well as steel and plastic for copper, which rapidly lost ground in many common uses.

"Engineers," Page worried aloud, "were told to work designing copper out of products, particularly in the auto and electric utility markets." Copper subsequently bounced back in many applications, but in several markets formerly dominated by copper and bronze, such as screen for windows and doors, aluminum held its newly gained ground. In other industries, engineers kept near at hand their recently drawn blueprints for cutting down copper consumption in the event that red metal prices spiraled upward anew.[7]

For all such reasons, during the years from 1956 to 1962, per capita consumption of copper in the United States declined 12 percent while that of aluminum increased by an identical amount—and plastics jumped by 40 percent. That was why the allure of aluminum overwhelmed major copper producers. At Phelps Dodge the man perhaps most eager to enter the white-metals business was Edgar Dunlaevy, president and chief executive officer of Phelps Dodge Copper Products Corporation. An engineer trained at the famed Massachusetts Institute of Technology, he understood far earlier than most people at Phelps Dodge that the future of overhead lines for long-distance transmission of high-voltage current belonged to steel-core aluminum cables. By being efficient conductors of electricity (having about two-thirds the conductivity of copper) but weighing less than half as much as copper, aluminum cables were already well on the way to becoming the industry standard by the late 1930s—and that worried Dunlaevy. He grew even more alarmed when the communications giant American Telephone and Telegraph switched to aluminum for some of its long-distance lines during the early 1950s.

Repositioning Phelps Dodge to gain it a competitive edge in the manufacture of

aluminum cables and other metal products did not appear to be difficult at first. Aluminum fabrication, after all, seemed to closely parallel that of familiar copper in rod rolling, wire drawing, tube drawing, and insulating and cabling wire. On the other hand, there were plenty of good reasons not to join the rush to aluminum, especially when long-term success seemed to require a new-fashioned and integrated landscape of production wholly distinct from copper's; and Phelps Dodge would need both expertise and confidence to compete head-to-head with three established giants, ALCOA, Reynolds, and Kaiser, the trio that after World War II largely controlled production in the United States. There were several alternative strategies that beckoned to America's three biggest copper producers, which in 1956 ranked among the nation's top twenty-five money earners of all industrial companies and may have been overconfident as a result.

Kennecott launched a program of portfolio investments in aluminum and other metals as early as 1949, and by the end of the 1950s it had gained a position in Kaiser Aluminum valued at almost $100 million. Far more aggressive was Anaconda, which launched its own aluminum subsidiary in 1953 and boldly sought to position itself as the fourth member of the ruling aluminum oligopoly. Its $65 million quest, which meant building a new aluminum smelter at Columbia Falls, Montana, and fabricating plants elsewhere, ended in financial disaster that weakened the once great copper producer.

Phelps Dodge too responded to the allure of aluminum, but in its own carefully reasoned way. Unlike Anaconda, it wisely did not sink millions of its dollars into the costly smelting and reduction stages of aluminum production. Phelps Dodge based its entry solely on the need to complement its primary line of copper products. That is, if it made its own aluminum wire and cable, it could serve its customers better and thus strengthen its overall marketing position. To that end, Phelps Dodge in 1963 built a fabrication plant in Du Quoin, a town in southern Illinois, and acquired a small extrusion, or shaping, facility at nearby Murphysboro. Neither plant required a huge outlay of capital, but together they formed a solid foundation upon which to erect Phelps Dodge Aluminum Products Corporation.

As Page explained, "We wanted to get into aluminum fabricating first, to have a so-called full market basket. And, second, it was a chance to diversify and grow in a field with which we were somewhat familiar." By making only this modest commitment, Phelps Dodge hoped to remain fully competitive in its usual lines of business "with the smallest possible capital expenditure." Whatever fields the company chose to expand into, emphasized Page, "we have hoped and still hope to finance our capital program internally without borrowing."[8]

Threatening to upset his conservative calculations were large outlays of money required for Phelps Dodge to guarantee itself a steady supply of aluminum to fabricate. That meant a troubling new level of commitment. "You can't just start with a smelter," Page explained. "You've got to start with alumina, which means if you're going to make any sense, you've got to start with bauxite," the raw material mined in Australia, the Caribbean, South America, and elsewhere. In simplest terms, mining four pounds of bauxite, a reddish substance composed of aluminum oxide and water, yielded two pounds of alumina, which, in turn, was refined to make one pound of finished aluminum. Only at that point did Phelps Dodge hope to enter the stream of production as a fabricator and thus avoid the mind-boggling financial commitment that successful aluminum mining and processing required.[9]

To that end Phelps Dodge developed an ingenious strategy. Having learned from Anaconda's horrific miscalculations, it charted a far safer course for itself by lending money to two well-established Dutch and Swiss firms, mining and industrial giants, to enlarge their productive capacity, and then it negotiated long-term contracts to assure a steady supply of aluminum. N. V. Billiton Maatschappij agreed to supply alumina from its bauxite mines in Dutch Guiana, and the Consolidated Aluminum Corporation (CONALCO), an American subsidiary of Swiss Aluminium Ltd. (Alusuisse), refined the alumina at its plant in New Johnsonville, Tennessee. Phelps Dodge guaranteed to take a minimum of thirty thousand tons of aluminum a year starting in 1966 to supply its growing number of fabricating facilities. So promising did the future seem for the lightweight metal that Phelps Dodge increased its annual commitment to forty thousand tons. The Billiton and CONALCO deals together cost about $30 million, a bargain price for the steady supply of aluminum that Phelps Dodge needed. "We don't make as much money this way," explained Page. "But we don't have to put in nearly the amount of money we would otherwise have to do."[10]

By the late 1960s Phelps Dodge Aluminum

Dramatizing Phelps Dodge's role as a major manufacturer of aluminum products is this 1979 image. To place its commitment to the white metal in perspective, during the years 1964 to 1970 Phelps Dodge invested fully $112 million in aluminum production, $195 million in copper mining, $15 million in refining, and $79 million in domestic copper fabrication.

Products Corporation comprised seventeen plants, located chiefly in the mid-South, that in addition to supplying aluminum needed for power and communications cables also fabricated such items as wire screens, fasteners, furniture, and ladders. By the end of Page's tenure in 1969, however, serious doubt had arisen within the corporation concerning its aluminum venture. No matter what anyone did or said, Phelps Dodge never realized the profits it originally anticipated, and substitution never became quite as serious a threat as some people feared earlier in the 1960s. Clearly, the corporation had to do something about what had become less-than-alluring aluminum. Phelps Dodge, though never suffering the great losses of Anaconda, still spent more than $100 million in its bid to fabricate aluminum between 1964 and 1970, money that might have yielded appreciably greater returns if invested elsewhere. With that perspective in mind, company executives took corrective steps in December 1970 to merge Phelps Dodge Aluminum Products Corporation with Consolidated Aluminum Corporation and the Gulf Coast Aluminum Corporation, two subsidiary firms of Swiss Aluminium Ltd. In return, Phelps Dodge gained a 40 percent stake in CONALCO.

With the advantage of hindsight, some people may look back at Phelps Dodge's aluminum venture as a mistake. Certainly, early on the prospect of substitution had seemed far more menacing than it ever became in fact. Meanwhile, Phelps Dodge's investments in Amerada Petroleum and American Metal Climax Company appreciated handsomely, but the 195,000 shares of New Jersey Zinc it acquired in

OPPOSITE: INSIDE THE *Habirshaw Division's new Nepperhan plant in Yonkers in 1956, which had commenced production a year earlier. This factory specialized in production of building wire.*

INSIDE THE HOPKINSVILLE, *Kentucky, plant of Phelps Dodge Magnet Wire Company, a facility completed in 1967 to double the firm's manufacturing capacity, at that time concentrated in Fort Wayne. An innovation called catalytic fume burners kept air inside the plant clean.*

1956 for $10.3 million performed poorly. That was a small setback. Phelps Dodge still earned attractive profits by selling its copper.

Reshuffling the Deck at Phelps Dodge Industries

DURING THE 1950s and 1960s, Phelps Dodge Copper Products Corporation experienced its own unique combination of advances and setbacks. It was unhappy to report in 1960, for example, that Fidel Castro's new government had nationalized the four-year-old wire and cable plant in Cuba and thus forced Copper Products to write off one of its most promising joint ventures abroad. At times it appeared that Page grew discouraged with the entire fabricating side of Phelps Dodge: "Today we could do just as well selling mine products

without fabricating facilities," he explained to one interviewer in 1962. "Primarily we thought we could make money in fabricating, and now we don't like to give it up. Maybe if we hadn't spent that money we might be out of fabricating." At the time, hardly any participant in the greatly overbuilt fabricating sector made a real profit either. Page's occasional pessimism notwithstanding, Phelps Dodge Copper Products stayed the course, though on several occasions it reshuffled the component parts of its fabricating operations in hopes of finding a winning combination.[11]

By 1959, the four plants of the Habirshaw subsidiary were leaders in making defense products such as coaxial transmission cables essential to the Distant Early Warning (DEW) radar line, a cold war electronic shield extending across Arctic Canada and Alaska, and special high-frequency copper-sheathed cables installed at the Minuteman Missile Wing VI site at the Grand Forks air force base in North Dakota. To better serve its commercial customers, Phelps Dodge Copper Products added a mammoth new tube mill at South Brunswick, New Jersey, in 1958, to the one it already operated in Los Angeles, and Habirshaw constructed a plant in Fordyce, Arkansas, in 1966 to enlarge its production of telephone wires and cables.

The Inca Manufacturing Division in Fort Wayne positioned itself in 1967 to become the world's largest producer of magnet wire for electric motors and various types of electronic and automation equipment when it formally adopted the more descriptive and modern-sounding name Phelps Dodge Magnet Wire Corporation and opened a large state-of-the-art plant in Hopkinsville, Kentucky, primarily intended to make yoke coils that wrapped around the necks of television picture tubes, then a booming business. From executive offices that remained anchored in Indiana, President S. Allan Jacobs oversaw the expanding operations.

In addition, by the end of the 1960s Phelps Dodge had established more than twenty joint manufacturing ventures abroad, and it greatly multiplied its fabricating facilities within the United States after a buying and building spree earlier in the decade. Most of the acquisitions and expansions were not of major consequence, but the aggregate still looked impressive. Among the new properties was Edlen, Inc., a maker of electrical cable connectors, which the parent firm moved from New York City to suburban Yonkers soon after its acquisition in 1960. Another early 1960s addition formed the basis for Phelps Dodge Electronic Products Corporation, established in 1962 to make specialty antennas and high-end broadcast equipment in North Haven, Connecticut, and Marlboro, New Jersey. The following year Phelps Dodge gained Lee Brothers Foundry Company of Anniston, Alabama, an old-line manufacturer of brass and bronze cast fittings and valves for plumbing and refrigeration. All those 1960s acquisitions had in common that they were significant end users of copper and its alloys.

Following still more additions, the management of Phelps Dodge acknowledged that the expanding family of fabricating subsidiaries had outgrown the old organizational structure. That was the main reason it formed a new wholly owned subsidiary, Phelps Dodge Industries, Inc., on December 16, 1966. Edgar Dunlaevy, who had

been with Phelps Dodge Copper Products Corporation since its creation back in the early 1930s, served as first president of what was essentially a management company. Phelps Dodge Industries provided legal, technical, marketing, and other services to several distinct operating subsidiaries having a total of twenty plants in the United States and another fifteen abroad.

At the time, the structure of the manufacturing sector seemed reasonably clear and simple. Phelps Dodge Corporation occupied the topmost rung of the organizational ladder, and one rung down stood Phelps Dodge Industries, which from offices in New York City provided various forms of management assistance to eight separate subsidiaries: Phelps Dodge Copper Products Corporation, Phelps Dodge Aluminum Products Corporation, Phelps Dodge Electronic Products Corporation, Phelps Dodge Magnet Wire Corporation, Lee Brothers Corporation, Mackenzie Walton Company, Inc., Phelps Dodge Puerto Rico Corporation, and Phelps Dodge International Corporation. There seemed little that Phelps Dodge Industries and its far-flung fabricating plants couldn't do with copper, brass, or bronze.

They alloyed, cast, extruded, drew, rolled, formed, welded, and tin coated and insulated copper to make an impressive array of insulated cable and wire, magnet wire, telephone and communications cable, tubes, and plumbing fixtures. In short, apart from not making copper sheets or strips, nothing was missing at Phelps Dodge Industries—except robust profits!

Because the financial performance of Phelps Dodge Industries during its initial years was so

TIME FOR SOME *harmless* *fun at Phelps Dodge Magnet Wire: "Miss" Poly-Thermaleze. The product name referred to the first in a line of high-temperature, dual-polyester-coated magnet wires designed to perform well in difficult applications. Over the years those products included Armored Poly-Thermaleze® 2000, a series of Bondeze® products, and Thermaleze® Q⁵ (Quantum Shield™ for inverter-driven motors).*

unimpressive, the parent corporation in 1968 promoted Dunlaevy to chairman of its manufacturing sector and chose Ted Michaelsen, a young and energetic executive vice-president, as his successor. It gave the new president of Phelps Dodge Industries full responsibility to do something about its sickly earnings, which had declined each year after its formation. Michaelsen, who had most recently served as president of Phelps Dodge International Corporation, his first love, and where he remained as chairman another decade, reshuffled the deck several times before he at last drew a winning hand for Phelps Dodge Industries.

The structural details are unimportant. The real question was whether restructuring would make any real difference—whether the 1971 incarnation of Phelps Dodge Industries would perform any better than the 1966 and 1969 models. It certainly seemed to. In fact, even before Michaelsen administered a second strong dose of organizational medicine, observers noted signs of a recovery. The parent corporation reported record sales and earnings for 1970, helped in part by improvements in its once-lagging manufacturing sector. Among the stellar performers Michaelsen singled out in 1972 was Phelps Dodge Magnet Wire. He warned, however, that he would quickly prune away any plant that failed to show satisfactory profit or marketing potential. Only recently he had closed plants in England and Germany for those reasons, and "it was I who started the plants," he added wryly in one interview.[12]

End of the Page Era

GROWTH OF PHELPS DODGE'S manufacturing arm had been in large measure the result of Page's personal interest in expanding what Cates began. That did not mean, however, that the eastern-educated lawyer neglected the vital extractive sector. Page understood well that copper mining remained the company's core business, and in time he emerged as an acknowledged leader of America's copper industry. He played a major role in the new Copper Development Association, which he served as its first chairman, and in the International Copper Research Association, of

FOR A SINGLE DAY in February 1963, New York's Times Square celebrated the source that lit its famous lights by acquiring a descriptive new name: *Electric Square. The man on the left is William K. Dunbar, vice-president of Phelps Dodge Copper Products Corporation and longtime associate of Wylie Brown; and on the right is Armand D'Angelo, the city commissioner of water, gas, and electricity.*

which he was chairman at the time of his death. For several years he was a director also of the American Mining Congress. But perhaps Page attempted too much. He was anything but mercenary, having refused several proposed pay raises by professing that he "was being paid enough now." By habit he was inclined to overwork. He even answered his own office phone. "It was hard, in a way, to help him," remembered his successor George Munroe. "He was such a bright guy, and grasped things so easily, that a lot of people around him tended to come to Bob with every problem."[13]

Munroe summed up Page's preoccupation with Phelps Dodge by recalling a Christmas Eve when Page uncharacteristically had forgotten to purchase a gift for his wife. After rushing to Cartier's just before it closed, he presented her with a lovely piece of jewelry. Noticing that she looked bewildered, he asked if she liked it. "Why, it's beautiful, Bob," she replied. "But it's the same thing you got me last year."

A heavy smoker much of his life, in his later years Page suffered from such serious emphysema that he kept a special oxygen breathing apparatus in his office. A slender man with a steady cough, he died of acute bronchitis and cardiac arrest on Christmas Day in 1970, exhausted no doubt from twenty-three years of heavy responsibilities at Phelps Dodge, twenty-two of them as president or chairman. He was sixty-nine.

During the 1950s and 1960s, he had guided with resourcefulness and vigor the company's continued growth in mining and refining copper and its expansion into aluminum manufacturing. Page was a man of numerous talents and broad interests, dedicated to the well-being of Phelps Dodge, its employees, and its shareholders. For all his caution, Phelps Dodge Corporation during his tenure generally ranked as the most profitable major producer in the industry. At one point in the mid-1960s, it recorded a 14.7 percent return on equity as compared to 12.2 percent for Kennecott and 7.1 percent for Anaconda. In turn, Page gained the respect and affection of all his associates. His passing was a sad event for Phelps Dodge.[14]

Chapter Eleven

A DECADE *of* UNCERTAINTY

THE MODERN OPEN-PIT mine complex that took shape at Morenci between 1937 and 1944 always ranked as one of the most complicated and expensive projects Phelps Dodge ever undertook, but a project requiring perhaps even greater commitments of money in real dollars and even more breathtaking technological leaps of faith was the Hidalgo smelter it opened in New Mexico in 1976. To that end Phelps Dodge acquired a land area larger than the state of Rhode Island, built a completely new town called Playas, and trusted that the technical experts could design and operate a plant with unprecedentedly low levels of atmospheric emissions and an ingenious way of causing the sulfur in sulfide ores to ignite and burn hot enough to conserve precious fuel during a time of sharply rising energy costs.

Building a "flash smelter" had never before been attempted in the United States. Certainly it was Phelps Dodge's boldest response to the numerous bewildering uncertainties, both economic and environmental, it faced during the 1970s. The Hidalgo smelter complex was, all things considered, a daring move by Page's successor, George Munroe, and his generation of Phelps Dodge leaders and the corporation's board of directors.

~ A FORM OF INDUSTRIAL ART: *inside a reverberatory furnace, possibly in Douglas, before the newly reworked smelter was fired up in 1915.*

Facing the
Environmental Whirlwinds

To understand what was at stake at Hidalgo, let us recall briefly the formative years of smelter technology, when glassmakers in the early 1600s became the first artisans to use reverberatory furnaces—so named because the arched ceiling of the furnace deflected rising flames and caused intense heat to "reverberate" downward toward the hearth. This fuel-saving innovation helped to conserve England's rapidly dwindling forests. Woodcutters chopped down one stand of trees after another to provide fuel enough for an earlier invention, the household fireplace. Because the fireplace made warming individual rooms of a home easy, it greatly increased popular demand for window glass, which, in turn, required a steady supply of wood for charcoal to heat the glassmaking furnaces.

Parliament in 1593 outlawed all such glassworks within eight miles of England's rivers to safeguard the supply of wood required by the British navy. About two decades later, someone invented an efficient glassmaking furnace that metalworkers soon adapted to their trade as well. By burning England's plentiful coal instead of its increasingly scarce wood, the reverberatory furnace was popularly hailed as environmentally friendly technology.[1]

That bit of medieval history was unfamiliar to most people living in the 1960s and 1970s, when protection of the environment became a mainstay of American politics and the nation's vital mining industry a convenient whipping boy. Much earlier in the century, small groups of influential citizens had worked to focus popular attention on specific conservation or preservation concerns, such as protecting remnants of the great bison herds that once roamed the Great Plains, but never before had so many Americans been interested in so broad a range of environmental issues—or, more specifically, so inclined to punish copper producers because the tall stacks of their reverberatory furnaces vented plumes of smoke into the atmosphere during the smelting process.

It was a rare individual who understood that reverberatory furnaces originated as environmentally friendly technology or sought to learn why copper smelters continued to use them. People angry about environmental pollution, both real and perceived, upheld federal and state regulations that deluged America's mining industry and sent it reeling.

To put events of the 1960s and 1970s into historical perspective, the first federal environmental regulatory law in the United States was probably the Migratory Bird Treaty Act of 1918. Not until nearly half a century later, in 1963, did Congress pass a Clean Air Act, which with subsequent amendments implemented the first federal air pollution laws. At the same time, federal and state legislators added measures intended to protect water quality, control lead emissions, or outlaw the popular pesticide DDT. The list of lawmakers' environmental concerns grew longer each year. The most encompassing federal enactment was probably the National Environmental Policy Act of 1969, which President Richard M. Nixon signed into law on January 1, 1970, to create the Environmental Protection Agency.

Four months later environmentalists celebrated their first Earth Day, and more vocal adherents popularized environmentalism on college campuses and in street demonstrations. Television sets multiplied images of Americans protesting environmentally callous practices; and buzzwords like *conservation*, *ecology*, and *environment* became part of the public vocabulary.[2]

For at least a century prior to the 1970s, Phelps Dodge and its various subsidiary companies had responded to legitimate concerns voiced by individuals or communities about the impact of mining and smelting on local air and water quality. It acted according to accepted practices of the time. In one case, back in the fall of 1900, a rancher living south of Clifton lodged a complaint against both the Arizona Copper and the Detroit Copper companies by alleging that concentrator tailings they discharged into the San Francisco River had damaged his orchard and garden. The two producers not only settled the rancher's claims but also agreed to supply him with ample fresh water by means of a specially constructed pipeline. At about the same time, they dammed Chase Creek (which flowed through the heart of their industrial landscape) to prevent tailings from reaching the San Francisco River and thence the Gila River, where downstream farmers and ranchers had complained about damaging water pollution. Phelps Dodge also worked to prevent mine water from discharging into nearby rivers.

After Edward H. Robie, associate editor of the *Engineering and Mining Journal*, visited Douglas, Arizona, in 1928 he wrote that "its group of smelter stacks, belching into the atmosphere all the sulphur contained in the Bisbee ores, make a distinguishing landmark." All the smoke, however, created a landmark headache for copper companies even in that environmentally innocent age. Phelps Dodge regularly paid farmers in the Douglas area for any sulfur dioxide damage to their crops. Personnel from the corporation's Smoke Department toured farms on both sides of the international boundary to assess damages, if any, and operated a greenhouse in which they exposed common farm plants to controlled concentrations of sulfur dioxide gases to provide them benchmarks for settling claims. Corporation scientists learned that certain crops were very susceptible to excess sulfur dioxide in the air, while others were far less so. In fact, they discovered that sulfur dioxide in some instances acted as fertilizer. That intriguing discovery prompted the general manager Charles R. Kuzell to quip that the farmers should pay Phelps Dodge for smelter smoke.[3]

In truth, there was once considerable uncertainty among copper industry veterans about any deleterious effects caused by sulfur dioxide smoke. Feeling a cold coming on, some old-time employees actually asked to go into the smelter to inhale deeply the sulfurous fumes, and smelter workers did tend to have fewer colds, or so insisted the onetime president of Phelps Dodge Corporation Richard T. Moolick. "That was well known," he emphasized during a 1997 interview. Warming to his controversial subject, he hastened to add that New England housewives once burned pure or elemental sulfur to disinfect homes during spring housecleaning.[4]

Such claims caused the corporation's retired engineering vice-president Richard Rice to observe wryly, "I don't know. I worked at the Hidalgo smelter for five years and don't remember that stuff ever helping my colds." He recalled mainly the sulfurous smell *and taste* of smelter smoke that wafted through Morenci when he grew up there, but "people tolerated it because it represented their jobs." Indeed, until the late 1960s and early 1970s, smoking stacks continued to symbolize progress, industry, and jobs to a good many Americans. That was especially true in Pittsburgh and many other gritty centers of heavy industry, even if the smoke-filled air required executives to change their white shirts at least twice a day and sometimes brought darkness at noon.[5]

In the late 1940s an employee in the smelter office at Morenci wrote a poem of praise that summed up sentiments shared by many of her coworkers. Published in the *Prospector,* a branch newsletter of the time, its verses perhaps sound odd in today's environmentally conscious age, but undoubtedly they came from the heart. "The Smelter" begins with these words:

> *Out into open spaces, drifts this mighty cloud,*
> *Belching forth its magnitude of which we're*
> *   very proud.*
> *It symbolized copper produced by many hands*
> *And solves the metal needs our nation now*
> *   demands.*[6]

No one doubted that smoke equaled jobs in turn-of-the-century Bisbee, but during the late 1890s, as sulfide ores increasingly replaced oxide ones, people became aware of emission problems in the confines of the mountain canyons. This image shows Bisbee smelters and power plants before the Copper Queen moved its reduction works to Douglas in 1904 to form a massive new landscape of production.

J.O. 2563
COPPER QUEEN
LOOKING WEST

The image of stacks belching forth smoke causes many Americans today to cringe, but in its historical context it reminds us why any attempt to judge and condemn practices commonly accepted until the 1960s and 1970s is a bit like condemning our own grandparents because they had no emission controls on their automobiles. No one did. The earliest such device, incidentally, was the positive crankcase ventilation valve first mandated in the United States on 1964 models. Quite simply, smoke remained of little concern to most people until recent times, although a few residents of Swansea, Wales, complained about dense concentrations of fumes from smelter furnaces as early as the 1600s. Those were voices crying in a wilderness of world indifference.

In twentieth-century America, the reverberatory furnaces favored by copper producers were low, hulking structures into which operators blew oil, gas, or pulverized coal to create heat intense enough to melt the concentrate. It was in the early 1970s that environmental activists claimed copper smelters were a major cause of atmospheric contamination and successfully promoted state and federal laws forcing them to install high-priced pollution-control equipment—or else shut down. Everyone

In an early *quest for clean air,* Phelps Dodge constructed *a large and expensive dust collector as part of a major renovation project at its reduction works in Douglas during the mid-1920s. By the 1970s the Douglas works had become the corporation's oldest smelter and, because of changing environmental standards, its most troublesome environmental headache until plant closure in January 1987.*

wanted clean air, but retrofitting an aging reverberatory smelter to limit its sulfur dioxide emissions was a costly procedure, and some mandated modifications were at best of marginal or even unproved effectiveness.

All the while, the ranks of an earnest but often scientifically and technologically inexpert or naïve cadre of regulators, both state and federal, kept growing. To mining company executives, themselves worried and uncertain about what the future held, it sometimes appeared that the goal of government was to raise the bar of industry compliance to impossible heights—and then set it higher still. "Every time we'd do something," recalled the onetime Phelps Dodge vice-chairman L. William Seidman, "there would be three more things" needing our attention.[7]

For example, the Arizona legislature, which decided that federal standards were not strict enough, enacted a tough new air quality law in May 1970 that set state pollution standards and regulations for copper smelters. It required them, as of 1974, to recover at least 90 percent of all sulfur released from concentrate during smelting, though no known technology could be applied to existing plants in order to meet the standard— only new plants could achieve a 90 percent recovery rate. What was effectively a death sentence for many Arizona smelters became a subject of enormous concern for producers because in 1972, of the fifteen copper smelters in the United States, seven were located in Arizona. There was only one each in eight other states. Most troubling for Phelps Dodge in the near future, the 90 percent requirement, if not modified by state or federal regulators, meant permanently closing the big Douglas Reduction

Works, which would cost hundreds of workers their jobs and create major logistical headaches as the corporation scrambled to find (or build) other facilities to smelt its concentrates.

Various pollution-control devices then in existence were at best able to recover 70 percent of the sulfur dioxide emissions from Phelps Dodge's reduction works at Douglas, Ajo, and Morenci. By the dawn of the 1970s most smelters had long had some kind of dust collection system in place to recover particles that might otherwise vanish up their stacks. At Douglas in 1940, dust-filtered sulfur dioxide gas drifted into desert air from a 572-foot smokestack, recently raised to become the tallest in the state. White and blue smoke when diluted with clean air simply trailed off into the distance instead of collecting above the city and its surrounding ranches. Engineers had claimed that the tall stack's lighter, filmier smoke was nearly free of noxious gases and dust. Back in 1942 the company added an electrostatic precipitator to its recently completed Morenci smelter and a baghouse where discharged gases passed through a series of large porous bags (not unlike those used in home vacuum cleaners) to collect dust particles from which were recovered copper and other minerals.

By the early 1970s copper producers could no longer address the smelter emissions problem merely by adding another baghouse or further raising the height of stacks to dilute smoke to harmless levels before it reached the ground. Quite simply, there was no easy way to deal with sulfur dioxide smoke because it easily passed through the big bags used to collect dust.

Back in the early 1920s, when both Phelps Dodge and the Calumet and Arizona Mining

Company wrestled with the problem of smoke damage to crops around their Douglas smelters, someone suggested that fumes might be transformed into sulfuric acid. Colonel John Greenway of the Calumet and Arizona apparently quipped that "his smelter alone might furnish all the acid needed in western America." Despite the limited market, only by capturing sulfur dioxide in a sulfuric acid plant or as elemental sulfur could smelters keep any portion of it from reaching the atmosphere. In fact, by the time Phelps Dodge acquired the Calumet and Arizona smelter in 1931 the facility actually contained two acid plants.[8]

During the early 1970s Phelps Dodge spent approximately $25 million to construct acid plants at both Ajo and Morenci, sites far more remote than Douglas, in an effort to bring those two big smelters into compliance with very stringent standards established by Arizona's new air quality law. The "solution" created other problems. Although sulfuric acid was an important raw material for a number of industries, all markets for it near Ajo and Morenci remained almost as limited as in Greenway's day. Further, it was difficult to store at the site of production as well as expensive to ship to customers. Sulfuric acid soon glutted national markets, and Phelps Dodge found itself producing far more than it could sell at a profit. Further, concluded the senior vice-president and general manager Arthur Kinneberg, "the enormous pollution control expenditures . . . have not increased our production capacity by even one pound of copper!"[9]

The estimated cost of bringing Phelps Dodge's three Arizona smelters into full compliance with state demands was astronomical—at least $240 million, a sum that represented fully one-third of the corporation's total capitalization. Thus did the mounting whirlwinds of environmentalism test Phelps Dodge's new generation of leaders. They worked diligently to save the three Arizona smelters, preserve the jobs involved, and remedy emissions problems.

The emergence of environmentalism, as embodied in various recently fashioned laws and mandates, and what it portended for Phelps Dodge was foremost among the several uncertainties George B. Munroe faced after he succeeded Robert Page—and *uncertainty* was truly the operative word for the 1970s from the corporation's perspective. Those years were, in addition, a time of skyrocketing inflation, when wages, energy costs, and interest rates climbed sharply. The corporation recorded numerous accomplishments during the decade, but always seeming to overhang everything was a sense of uncertainty, even malaise, about the world in which it operated, uncertainty that at times seemed to spin almost out of control. "There is going to be the devil to pay with the environment," Page confided to Munroe. "And, George," he emphasized shortly before his death in 1970, "I'm not sure we can get through this one."[10]

New Leadership for Difficult Times

UNTIL THE LATE 1960s, Phelps Dodge had continued to enjoy a surplus of cash. Then troubles started to loom large on the horizon.

The corporation's innate conservatism helped it weather the early 1970s when the industry leader, Anaconda, suffered a one-two punch delivered by Chile's expropriation of its lucrative Chuquicamata mine complex and by its own ill-starred diversification into aluminum, which *Forbes* called "a bomb." At that time too, Kennecott incurred mountainous legal expenses trying to defend its recent acquisition of Peabody Coal, and still it lost the antitrust fight. In 1972 the Federal Trade Commission ordered Kennecott to divest itself of Peabody (although in the end it came out slightly ahead financially from the forced sale).

Of far more concern to America's Big Three copper companies, as the developing nations won progressively greater control of metal mines within their borders, they flooded world markets by large increases in production. Curiously, in the near term the price of red metal skyrocketed. That was because the major industrial countries stockpiled copper in 1973 in the misguided belief that world supplies would be tight. Realizing their blunder, they then unloaded massive tonnages of the metal, glutted markets, and briefly caused the price to plummet. Even when buffeted by these strange gyrations, Phelps Dodge continued to take a long-range view. During the decade of the 1970s it invested approximately $1.5 billion to replace exhausted mines, add modern smelting, refining, and fabricating facilities, and install expensive emission controls.

Guiding the corporation during those sometimes troubled but always challenging times was a new and soon-to-be-battle-tested generation of leaders. As president of Phelps Dodge, Robert Page had recruited two especially

able assistants: George Barber Munroe was an Ivy League attorney from Page's former law firm in New York; Warren E. Fenzi was a civil engineer educated at the California Institute of Technology who, since joining Phelps Dodge at Morenci, had shown a flair for management. Munroe from the East and Fenzi from the West would play out their respective leadership roles at Phelps Dodge and in the process become close friends.

The Illinois-born Munroe was an intellectually and physically ambidextrous person: as a Dartmouth student he earned both a Phi Beta Kappa key and popular acclaim as one of the Ivy League's star basketball players. Though standing only slightly more than six feet tall, he helped power the Dartmouth team to the 1944 National Collegiate Athletic Association finals. During World War II Munroe served in the Pacific as an officer aboard the battleship

Maryland. Afterwards he played professional basketball for the old Saint Louis Bombers and then for the Boston Celtics while attending Harvard Law School.

Following graduation, Munroe worked for three years in the Office of the United States High Commissioner for Germany, first on the legal staff in Bonn and later, at age thirty-one, as the youngest judge on the United States Court of Restitution Appeals in Nuremberg. In that capacity he worked to recover property stolen from the victims of Nazi persecution—"everything from jewelry in the pawnshops to the Hermann Goering Steelworks," he recollected.[11]

Then Munroe returned to New York City, where he joined Debevoise, Plimpton and McLean, the firm that served as outside counsel for Phelps Dodge Corporation. There in 1958 he came to the attention of Page, who offered him a job as assistant to the corporate secretary John E. Masten. That was an unexpected turn of events, and because he knew relatively little about Phelps Dodge—apart from having been briefly associated with the personable Ted Michaelsen while working out legal details for a wire and cable venture in Venezuela—Munroe took three weeks to educate himself about the corporation before he accepted Page's offer. Although he had no formal title at first, he soon became assistant to the president.

Fenzi, a California native, had joined Phelps Dodge some twenty years before Munroe. That was in 1937, a portentous year both for young Fenzi and Phelps Dodge. Just two days after he graduated from the California Institute of Technology, his father, Cammillo, died in an automobile accident. A quarter of a century earlier, the elder Fenzi had prepared landscape plans for Walter Douglas and the El Paso & Southwestern system, and in more recent years he had beautified the grounds of the Pasadena home of Louis D. Ricketts, a big supporter of Caltech and its engineering graduates. It was the legendary Ricketts, a member of the Phelps Dodge board, who interceded with management during the job-tight depression era to help Fenzi find work at the Morenci Branch just as it embarked on its huge open-pit venture. There he had worked for Walter C. Lawson, chief engineer, and for a time supervised drilling and blasting. Apart from service as a navy Seabee officer in the Pacific during World War II, Fenzi spent two decades at Morenci, where he became mine superintendent and then general superintendent.

Whatever his title, Fenzi proved an immensely popular leader. Appointed assistant to the general manager of Western Operations at Douglas in 1957, he relocated to New York only two years later as assistant to President Page. Even in the button-down bastion of corporate power, Fenzi did not discard his big Stetson hat, long a symbol of the wild American West. Employees recalled fondly that Fenzi had an impressive ability to remember their names and those of their family members—even of their dogs.

It was after the 1961 retirement of Cleveland E. Dodge, a longtime vice-president and the last direct living link between management and the Phelps Dodge founders, that Fenzi and Munroe moved up to vice-presidencies. Munroe succeeded Page as corporation president five

years later and as chief executive officer in August 1969. It was significant that Phelps Dodge replaced one lawyer as chief executive with another. Training in the law, Munroe once observed, "is helpful in overseeing a complex organization in three ways. It sharpens mental discipline. It helps one communicate effectively. And it teaches one to steep himself in any field."[12]

Those attributes, plus the fact that the soft-spoken and thoughtful Munroe possessed a generous measure of a largely indefinable quality called presence, proved especially helpful during the environmental hearings held in the 1970s. Charles A. Burns, Phelps Dodge's retired vice-president for government relations, once quipped that, when he and Munroe made "Hill Calls" to visit members of the Arizona and New Mexico delegations in Congress, it was "sort of like having Cary Grant with you." Munroe, a skilled debater and public speaker, often attended meetings where big and contentious crowds of people wanted to argue about whether smelters should be shut down, but he was not at all flustered. He did much to determine the corporation's positive response to mounting environmental political pressures.[13]

Between them, Munroe and Fenzi amassed a wealth of training and experience by the time they attained top leadership positions, and that was fortunate. They needed it all during the tumultuous 1970s, when whirlwinds of change obscured the road ahead and presented major American copper producers with uncertainties totally different from anything Cates or Page had faced just a generation or two earlier. Above all, the new leaders of Phelps Dodge Corporation would be good stewards of the environment—if only someone would provide a reasonable definition of what that meant.

Uranium Frenzy

EVEN AS IT responded to a variety of vexing challenges during the 1970s, Phelps Dodge continued to search for a winning formula to make its ventures outside copper profitable. The company maintained a 40 percent stake in the Consolidated Aluminum Corporation, the nation's number four producer and fabricator (behind ALCOA, Kaiser, and Reynolds) into which it merged its several aluminum products plants in 1971. But no longer was its heart in aluminum, and the return on its investment in CONALCO was not satisfactory.

Early in 1977, when substitution once again tested Phelps Dodge's commitment to copper, it waged a million-dollar advertising campaign to stress copper's advantages over substitute materials such as plastic for household plumbing and aluminum for wiring. It placed advertisements in the *Wall Street Journal, Business Week, Forbes,* and some trade journals and magazines to emphasize the nation's self-sufficiency in copper and copper's recyclability and scrap value. The hard-hitting series usually ended with the tag line, "Today, more than ever, there's no real substitute for copper."

Significantly, among several advantages that Phelps Dodge claimed for copper in the advertisements was that it required roughly one-third less energy to produce than an equivalent amount of aluminum. No doubt that was an important selling point during a decade when sharply rising energy prices dismayed and frightened producers and consumers alike. In response to energy shortages that resulted from an Arab embargo of oil shipments to the United States in the spring of 1973, when fuel prices escalated almost daily, Americans endured long lines and limited supplies at service stations and slowed their automobiles to the newly imposed nationwide speed limit of fifty-five miles per hour to conserve precious gasoline. For its part, Phelps Dodge established a small venture that could provide gas and oil for its operations in the West, and it stepped up pursuit of uranium, which at the time was an alluring mineral that promised an alternative to fossil fuels.

The road to success in uranium, like that in aluminum, proved anything but straight. In each metal, Phelps Dodge expected to make a sizable profit. And why not? Several nuclear-powered

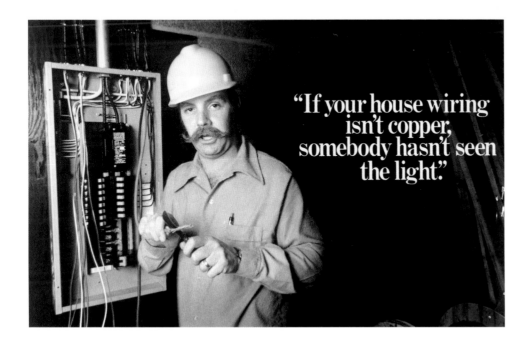

"If your house wiring isn't copper, somebody hasn't seen the light."

electric generating plants were already in operation, and many more were on the drawing boards. All those installations required uranium oxide for fuel, and utilities paid well for the radioactive substance. Alas, Phelps Dodge's uranium venture, though an interesting tale, does not end happily. It shows how all business calculations involve some element of risk, and it underscores the wisdom of Phelps Dodge's innately conservative approach to all such alluring prospects. Caution kept the corporation from being swept along in what one commentator labeled an episode of "uranium frenzy" in mining circles or from committing more than a minor portion of its financial resources to pursuit of the exotic metal.

The story of Phelps Dodge's interest in uranium dates from the mid-1950s when a largely forgotten fad spurred Americans, often people of limited means and aided by little more than a clicking Geiger counter, to crisscross the Mountain West in search of telltale signs of uranium deposits that they hoped would make them rich. As was also true of its first investments in Arizona copper in 1881, Phelps Dodge entered the picture mainly after prospectors had located and started to develop some of the more promising finds.

Exploration, whether for uranium or copper, was not a corporate priority until the early 1950s. Only after America's other big copper producers implemented multimillion-dollar programs did the general manager of Western Operations Harrison Lavender and his assistant Charles R. Kuzell set up a separate office in Douglas in 1949 to assist with search activity that until then was scattered and localized. The Geophysical

Research Department was in due time headed by Elmer E. ("Red") Maillot, a "swashbuckling" geologist whom Phelps Dodge sent to the Colorado School of Mines to take a six-week-long crash course in geophysics. Even with professionalization of exploration, members of the company's Douglas personnel were initially not welcome at the Morenci Branch, which preferred to use its local staff to search for metal deposits. In 1957, Herbert Z. Stuart became the first manager of exploration based at headquarters in New York. Six years later, because of increased emphasis on mineral exploration, the Geophysical Research Department was renamed the Western Exploration Office of Phelps Dodge Corporation.

Apparently spurring Phelps Dodge to increase the pace and sophistication of its metal exploration activity was the corporation's growing interest in uranium during the 1950s. At Robert Page's urging, Walter Lawson, now general manager of Western Operations, launched a program to find uranium mines. That led in due course to Robert W. Adams, a short and barrellike man who was a pioneer in the frenzied pursuit of the radioactive metal on the high plains of Wyoming. The storied romance of the frontier had earlier lured him west to seek his personal fortune. Adams worked as a short-order cook in Rawlins, Wyoming, a job that did not seem likely to lead to success, but during the excitement of the mid-1950s he acquired several leases at Crooks Gap and the Browns Park area in central Wyoming and won financial backing from C. W. Jeffrey (after whom Adams named Jeffrey City, the ramshackle company town at the heart of his would-be uranium empire).

At one time Adams prospected the Wyoming backcountry from the air using a small airplane equipped with a Geiger counter. With a fourteen-year-old boy as his pilot, he identified promising sites by dropping sacks of white flour. Keen-eyed rivals on the ground who rushed to the marked locations to stake claims ahead of him forced Adams to deploy numerous decoy bags. Such was the unbridled nature of competition in the early days of the uranium frontier.

The initial mine of Adams's Lost Creek Oil and Uranium Company began production in June 1957 and the next month its mill, with a capacity of four hundred tons per day, rumbled to life, processing uranium into yellow cake, or uranium oxide (U_3O_8). After expanding operations north into Wyoming's Gas Hills area to develop uranium reserves large enough to warrant a contract with the United States Atomic Energy Commission, Adams reorganized his firm in 1959 as Western Nuclear, Inc.

A decade later, because the future of uranium seemed so promising, Phelps Dodge paid almost $13 million in cash and a like amount in stock to obtain Western Nuclear, which in 1971 became a wholly owned subsidiary based in Denver. From his Colorado offices, Adams presided over mine and mill operations centering on the high plains of Wyoming and over exploration activities in several far-flung locations. In that way Phelps Dodge expected to profit from an anticipated boom in uranium driven by Uncle Sam's cold war defense program and the world's growing nuclear power industry.

That was the plan. "I thought nuclear was going to be the answer to the environmental problem," recalled L. William Seidman, a onetime vice-chairman of the corporation. "I thought we had a good thing in nuclear." In reality, Phelps Dodge lost money on uranium almost from the start. Yet so complete was its confidence in uranium's long-term prospects that the corporation seemed resigned to near-term losses while it slowly nurtured its new subsidiary. In time some executives within Phelps Dodge reluctantly concluded that Adams was far better at finding uranium deposits than in making any profit from them. Some recalled him, not always fondly, as a great entrepreneur "who could sell ice to the Eskimos."[14]

Buoyed by its faith in the future of uranium and nuclear power, Phelps Dodge continued to commit development money to uranium projects located in central Wyoming, in western New Mexico near Grants, and in eastern Washington on an Indian reservation about forty miles northwest of Spokane. The latter development, known as the Sherwood Project, it based largely on fully priced contracts entered into with the Washington Public Power Supply System. Western Nuclear also gained a 50 percent joint-venture interest in the Beverley deposit unearthed in South Australia.

Western Nuclear at last turned a small profit in 1978, and again the following year when production rose to almost 2.1 million pounds and the company made deliveries on the last of its low-price contracts that retarded earnings. Now with additional contracts in hand that provided for substantial purchases of uranium oxide by Washington Public Power Supply System and Union Electric Company of Saint Louis at good prices, Phelps Dodge seemed on the verge of realizing its long-deferred dreams for Western

Nuclear. Its future never looked brighter than in 1979. Then the unexpected happened. No one could have predicted the Three Mile Island disaster, but within the context of the 1970s it seemed timed almost perfectly to provide a sensational finale to the decade of uncertainty. Early on the morning of March 28, 1979, at a Metropolitan Edison power plant south of Harrisburg, Pennsylvania, a valve failed, and water required to cool the atomic reactor stopped flowing because of confusion in the facility's control room. That resulted in a partial meltdown of the uranium core and release of radioactive gas within the big containment building.

Three Mile Island was an accident that was never supposed to happen, and the unfolding drama focused the nation's attention as never before on the safety of nuclear power. The reactor core did not completely melt and burn through the earth toward China, a scenario feared by some overwrought imaginations, and only a tiny amount of radiation ever escaped into the atmosphere, but financial fallout from Three Mile Island was immense. It spread like a black cloud to blight the landscapes of uranium production across the West.

As the market for uranium oxide contracted rapidly, prices plummeted from around $40 to less than $25 a pound in 1980, a figure well below Western Nuclear's production costs. Soon prices dropped still lower. The Phelps Dodge subsidiary suspended Wyoming operations in 1981. The parent corporation hoped that the uranium market would improve during the mid-1980s, but it never did. Ironically, the only substantial money Phelps Dodge made from

uranium came from settlements in multimillion-dollar lawsuits against two utilities that defaulted on their commitments to buy uranium from Western Nuclear. It received almost $25 million from Washington Public Power Supply System in 1984 and approximately $47 million from Union Electric the following year. Even so, by 1985 Western Nuclear owed its parent more than $85 million, and that was at a time when rock-bottom copper prices imperiled the entire corporation. Hard-pressed, Phelps Dodge decided to close its uranium operations and sell off whatever assets it could. In the end, dollars spent to reclaim mine and mill sites in Wyoming and Washington far exceeded any money Phelps Dodge earned from its great uranium misadventure. Fortunately, after the allure of uranium and aluminum faded, it still had copper.

Growing the Corporation

From the mid-1940s, after Morenci's open-pit mine came on line, until the late 1960s, when it still had money in the bank and no long-term debt, Phelps Dodge worked hard to maintain the steady production of copper, often by rearranging parts of its landscape of production. For instance, even before the first pound of red metal came from Bisbee's Lavender Pit in the early 1950s, the corporation had constructed a new smelter at Ajo to handle New Cornelia concentrates and free its venerable Douglas Reduction Works for anticipated production from Bisbee.

During the 1960s, Phelps Dodge planned for the day when the historic mines of Bisbee no longer produced copper. That meant stepped-up exploration both in Bisbee and elsewhere. Searchers found an especially promising site in New Mexico beneath old Tyrone, the long-deserted company town designed by Bertram Goodhue. Back in the 1950s, Phelps Dodge geologists had revisited the remote Burro Mountains and bored more than seven hundred holes to delineate a large low-grade copper deposit amenable to mining by open-pit methods. On September 2, 1966, Page, who was in his final years as head of Phelps Dodge, formally announced that the corporation would launch construction of a modern mine and mill complex in New Mexico. The news rated front-page coverage from New York to Phoenix. The *Wall Street Journal* noted old Tyrone's impending conversion into an enormous pit; a headline in the *New York Times* reported, somewhat erroneously, "$100 Million Bringing Western Ghost Town to Life"—though, in fact, the old ghost town would soon be gone.

It was newsworthy too, in contrast to the debt incurred during the 1970s, that when Phelps Dodge announced its $100 million Tyrone project, it did not need to borrow any money to finance the new mine and mill. Those were the days! The corporation explained that it needed three or four years to complete the project, after which it would ship copper concentrates from Tyrone to its smelter at Douglas. In that way it proposed to phase out ore from the Copper Queen Branch, where the end of mining was in sight.

Phelps Dodge launched development of the Tyrone pit in 1967. During the next two years, crews worked with rotary drills, electric shovels, and a fleet of eighteen eighty-five-ton trucks just to remove fifty million tons of overburden. They sculpted an impressive canyon with fifty-foot steps winding to its ever-deepening floor. Eventually the pit grew to nearly two miles in diameter and thirteen hundred feet deep. Tyrone's new concentrator loomed large among the mine structures, even if it seemed diminutive next to the pit and rock dumps. Nearby stood settling tanks that conserved and reused precious water from the Gila River.

Water was crucial to successful production at Tyrone. That was why as early as 1955, Phelps Dodge arranged with the Phoenix law firm Evans, Kitchel and Jenckes and with its own Pacific Western Land Company subsidiary to purchase thousands of acres of agricultural and grazing land along the Gila River—and in that way gained rights to fourteen thousand acre-feet of water a year. That was a monumental achievement not unlike building its extensive water-gathering system in Arizona. Phelps Dodge completed a diversion project that redirected its share of Gila River water into Bill Evans Lake, a sixty-five-acre reservoir named for the lead attorney of the Phoenix law firm and the water wizard whose vision helped make the mining and milling operations at Morenci and Tyrone possible. When completed in 1968, the big reservoir stored water that was then pumped more than twelve miles uphill to the Tyrone mine and concentrator site.

At the mine site itself, giant earthmovers smashed through dilapidated pastel-tinted, tile-roofed dwellings. That was all that remained of Goodhue's model company town. Phelps Dodge built a newer, but less picturesque Tyrone in

Pipeline Draw about five miles northeast of its bygone gem in the Burro Mountains. Before building some two hundred modern homes, construction crews contoured a hillside landscape covered with live oaks and bear grass into lots and streets that gradually ascended from the base to the top of the slope. Unlike residences in old Tyrone, all new ones included running water, sewers, electricity, and a garage. Dirt and noise problems vanished because all the streets were paved and curbed and because there were no industrial operations close by. Because virtually all employees now owned cars, some of them chose to live outside Tyrone in other parts of Grant County.

Under Munroe's leadership, Phelps Dodge formally reopened its onetime underground operation at Tyrone as an open-pit mine. In all, the corporation had successfully integrated various parts of a complex new industrial landscape. During Tyrone's first full year of production in 1970, the mine and mill complex designed to produce 110 million pounds of copper a year turned out almost 120 million pounds, and that helped boost the corporation's total production to a record 627 million pounds. The results were so encouraging that Phelps Dodge announced it was proceeding to expand Tyrone's capacity to 180 millions pounds, at an additional cost of $38 million.

By July 1972, when its expansion was substantially complete, Tyrone easily ranked as a truly world-class copper producer. During 1973, it produced more than 208 million pounds of copper, compared to a little less than 240 million

pounds for Morenci, the corporation's longtime pacesetter. Phelps Dodge again set production records, turning out a total of 639 million pounds of copper that year.

Crews seeking to replace Bisbee's dwindling supply of copper also found a promising site across the canyon from the Morenci open-pit mine and above the abandoned town of Metcalf, where drilling revealed substantial tonnages of low-grade copper ore. Development of the Metcalf open-pit mine and concentrator complex began in 1969, two years after Tyrone. In importance it was fully equal to Tyrone, though it did not receive the same kind of national media attention. The Metcalf project required removing no less than seventy-five million tons of overburden, building both primary and secondary crushing plants and a concentrator capable of processing forty thousand tons of ore per day, adding a power plant and housing for employees, and acquiring considerable mining and related equipment. Phelps Dodge's capital investment

THE JUSTICE COURT *is one of a handful of buildings that survive of Tyrone, the town that Bertram Goodhue designed. Although saving all its architecturally valuable properties has not been practical, Phelps Dodge has done so where possible—for example, the former Copper Queen Mine headquarters now houses the Bisbee Mining and Historical Museum, and the attractive plaza complex in Ajo stands intact.*

in Metcalf totaled $194 million, including the premerger mine development expense of $28.5 million.

The Metcalf mine and concentrator launched production early in 1975, just about the time the last mines of Bisbee shut down. During its first year of operation, Metcalf produced slightly more than 62 million pounds of copper; during the next year, 1976, it more than doubled that amount to almost 140 million pounds.

Addition of the Metcalf concentrator to other installations at Morenci gave Phelps Dodge greater flexibility in processing ores from both the Morenci and Metcalf mines. By the end of the 1990s it was Metcalf, in fact, that formed the heart of Morenci's still-expanding mine complex that originated more than half a century earlier with the Clay ore body, now effectively exhausted.

As for Bisbee, where the Copper Queen

～～ EARLY METCALF, *a town that gradually disappeared after miners abandoned its underground workings in the early 1930s. Four decades later this was the location for Phelps Dodge's open-pit Metcalf mine, shown on the following page.*

Branch was once the corporation's proud flagship property, a century of production ended with the closing of all its mines. In late 1974, Phelps Dodge ceased production at the Lavender Pit and laid off 488 workers. It continued layoffs at a slower pace through the winter and spring of 1975 until in June it finally shut down underground production as well and discharged a final 360 workers. As of June 13, 1975, only a skeleton crew of approximately fifty employees remained where a thousand people had worked a short time earlier. "You always knew there was going to be an end, but nobody would believe it," recalled Jack Ladd, Phelps Dodge's former director of labor relations, who helped relocate Bisbee employees to the new Tyrone and Metcalf mines, the Hidalgo smelter, and other company operations.[15]

For the first time in nearly a century, mining did not set the pace of life in Bisbee. Just as low-cost copper from mines in the western United States had earlier closed high-cost operations in England and northern Michigan, so cheap copper from low-cost mines in Africa and South America humbled the once mighty Copper Queen as its reserves dwindled. For a time, gloom and depression gripped Bisbee. Several downtown businesses closed their doors for good, including the corporation's venerable Mercantile in 1976.

Phelps Dodge launched development of its Safford project in 1971 in anticipation of the day when copper demand and prices would rise again. Also within Arizona, the Copper Basin near Prescott looked more promising than it had in years. Overseas, Phelps Dodge devoted an increasing amount of attention to both its mining and its manufacturing operations. During the

1970s, the company intensified the search for ore deposits, and not just for copper, but also for metals such as molybdenum, gold, silver, lead, zinc, cobalt, tin, and tungsten.

From the mid-1950s until the late 1960s, Phelps Dodge had spent a paltry $2 million a year on exploration activity. That financial commitment (along with the pace of exploratory work) stepped up noticeably after George Munroe grew unhappy with Phelps Dodge's relative lack of success in finding major new mineral deposits and hired William K. Brown to head its corporatewide exploration program. Brown, an English mining engineer educated at Cornwall's prestigious Camborne School of Mines, would leave after a decade to become president of Gold Fields Mining Corporation, yet he brought Phelps Dodge a wealth of experience gained in the mining industry around the world. He and Munroe decided that the time was right for the corporation to expand its exploration activities beyond the United States and Canada to Australia and South Africa, which were two of a handful of nations where the threat of government expropriation was low and thus offered a safe field for mining investments.

During the three years from 1969 to 1971, Phelps Dodge's exploration budget grew to almost $9 million per year—and unlike Anaconda and Kennecott, the company maintained its commitment to exploration even during times of low copper prices and operational cutbacks. During those years too, its exploration teams located promising ore bodies at Woodlawn, near Canberra in Australia, and at a remote location north of Cape Town in South Africa, which became the site of the Broken Hill Mine of the

IN JUNE 1969, *Phelps Dodge initiated development of the Metcalf complex (located a short distance north of the Morenci mine). The removal of tons of overburden began early in 1970, not long before this picture was taken, but later that same year Phelps Dodge decided to defer the Metcalf project and launch a $438 million expansion project at Tyrone. The first ore was delivered to the new Metcalf concentrator in 1975. Phelps Dodge later combined the Metcalf and Morenci pits.*

Black Mountain Mineral Development Company, Ltd. When it went into full production late in 1979 it became a major producer of lead and zinc along with copper and silver.

The Quest for a Nonpolluting Smelter

UNDER THE LEADERSHIP of Munroe and Fenzi and their lieutenants in Western Operations, Phelps Dodge by the mid-1970s developed the capacity to produce nearly 20 percent of all copper mined in the United States, and that was at a time when it phased out mines at Bisbee and brought new ones into production at Tyrone and Metcalf. Moreover, partially in response to the growing environmental challenge, the corporation built a state-of-the-art smelter that was the cleanest in the United States—and a whole new company town to go with it. It was perhaps the company's boldest response to the uncertainties of the 1970s.

Phelps Dodge initially responded to passage of the federal Clean Air Act and subsequent state regulations by slowing production at its Morenci smelter in order to reduce sulfur dioxide emissions. Even so, the corporation still needed a way to process the vast output of copper concentrates coming from Morenci as well as those anticipated from the new mines at Tyrone and Metcalf. Retrofitting its existing "reverbs" to capture sulfur dioxide emissions seemed both a costly and only a temporary solution at best. That was why Bill Evans, the indisputable mastermind behind Phelps Dodge's modern land and water

acquisitions in the desert Southwest, suggested building a large, ultramodern smelter in the remote and sparsely populated boot heel of southwestern New Mexico.

To do that Phelps Dodge needed to purchase most of the Playas Valley, a vast and isolated landscape covered with mesquite, wild grasses, yucca, and greasewood, or creosote bush, that extended north from the Mexican border. In total, the expanse of rangeland was seventy-five miles long and twenty-five miles wide. Because the sprawling valley generally received less than ten inches of rain a year and had virtually no surface water, Phelps Dodge needed to acquire many thousands of acres of land solely for their water rights, because without water it could not transform troublesome sulfur dioxide into beneficial sulfuric acid. One big question was, why would a long-established and tightly knit ranching community sell land and water rights to an industrial outsider who wanted to construct something so alien to agriculturists as a copper smelter and acid plant?

Once an avid outdoorsman, Evans was confined to a wheelchair by multiple sclerosis, but his mind remained sharp. He formed the Great Lakes Land and Cattle Company and two other firms and arranged for a Phoenix real estate man, Milton D. ("Bud") Webb, to run them all. Webb spent several lonely months on the road quietly purchasing more than six hundred thousand acres of land for Phelps Dodge, although no seller ever knew who the real buyer was. Because all checks were drawn on the Harris Bank of Chicago, rumors circulated up and down the Playas Valley. One claimed that Cuban refugees sought land to build a giant casino complex to replace the ones in Havana lost to Fidel Castro; another asserted that the purchaser was the Mormon church. Webb's only contact with Phelps Dodge was through brief and infrequent conversations with the Western Operations general manager John A. Lentz. To exchange information the two men usually huddled in Tucson in the bleachers amidst the noisy crowd at University of Arizona football games.

So well guarded was their secret that nearly everyone was caught by surprise when Phelps Dodge announced in March 1972 that it planned to build a $100 million smelter complex at a remote site in Hidalgo County, New Mexico. It was a surprise for two reasons. First, few people, not even employees of Phelps Dodge, understood fully how the copper giant had quietly amassed its new landed empire; and second, leaders of the copper industry assumed that no United States producer of sound mind would build a new smelter while emission standards remained unresolved. "If we can't build a modern, low-emission smelter in one of the least populated areas, then there will never be another copper smelter built anywhere in this country," Munroe boldly proclaimed in the pages of *Forbes* magazine. According to the law as Phelps Dodge understood it, at that time federal officials could not monitor stack emissions until they wafted beyond the edge of private property, but the corporation went one step farther and determined that little sulfur dioxide should be emitted in the first place.[16]

Construction of the Hidalgo smelter began in March 1973, and production commenced

during the second quarter of 1976. Rising out of a seemingly desolate landscape inhabited mainly by jackrabbits, coyotes, and rattlesnakes, Phelps Dodge's first new reduction works since the Ajo smelter twenty-five years earlier was specifically designed and constructed to comply with a growing body of pollution-control regulations. In fact, Hidalgo was the first copper smelter in the United States to meet established standards. It was also the first in the nation to use flash-furnace technology, a process that partially offset the large amount of energy required to control sulfur dioxide emissions. Developed by Outokumpu Oy of Finland to overcome that country's lack of fuel, flash smelting appealed to Phelps Dodge because, of all technologies it evaluated, it was the one that permitted the greatest degree of air pollution control. The process had already proved itself since its introduction in Finland in 1949. To that nation's mining outback newly hired Leonard R. Judd headed with his family in early 1973 to learn everything he could about the fiendishly complex process he soon would be overseeing in New Mexico.

The flash furnace combined the roasting and smelting of a conventional roaster-reverberatory series into one step. The fine concentrate—a mixture consisting principally of copper, sulfur, and iron—was combined with preheated air and flux and dropped through a reaction shaft where it released nearly enough heat to sustain the smelting process. The secret was that it literally burned the concentrate's own sulfur along with a small amount of fuel oil or coal to produce the intense heat needed to melt copper-bearing material. That was important in a country like

THE NEWLY BUILT *Hidalgo smelter. Here and at all its mine facilities, Phelps Dodge fashioned a new landscape of production around the smelter. For many years, beef cattle owned by its subsidiary Pacific Western Land Company grazed nearby.*

❧ PLAYAS, THE NEWEST COMPANY TOWN ❧

PHELPS DODGE OFFICIALS *decided not to locate the new town of Playas next to the Hidalgo smelter because past experience taught them that people living within earshot of a smelter could hear it breathe, as if it were a living presence. When it shut down unexpectedly during the night the silence seemed ominous, and workers often lay awake wondering what had happened. The company instead decided to locate Playas nine miles north of the smelter to give its employees a sense of separation between their jobs and their leisure-time activities. John Lentz's assistant general manager, William W. Little, ultimately determined exactly where the new town was to go.*

❧ ILLUSTRATING HOW *Phelps Dodge planned Playas to be attractive as well as functional, this aerial photograph shows a pattern of gently curving residential streets closely following the natural terrain. Company housing in the pueblo and territorial styles, each home separated from its neighbors by lot-sized buffer zones of undisturbed desert vegetation, maintained privacy and a feeling of spaciousness. The Hidalgo smelter site is in the distance beyond the top edge of the photograph.*

Pat Scanlon, a retired vice-president, recalled that Little came into his office one day and asked if he wanted to help plan the new townsite. As the two men rode across the empty range in a jeep, they discussed whether to design a town that had an obvious hierarchy—a geography of power—that placed the manager's house at the most desirable site and arranged everything else down slope (and down scale) from it. In the end, they decided to maintain that kind of traditional relationship. Playas was, after all, a company town.

When Leonard Judd became smelter superintendent (the top management position at Hidalgo) in 1973, Playas consisted of six prefabricated houses, most of them used by security people. At the smelter site itself, where his job was to oversee Phelps Dodge's costly new technological wonder, he found workmen putting in the initial foundations. All roads were dirt. Judd recalled that when he, his wife, Gloria, and their two young children first arrived at the main gate, the sole entrance to both the town and the smelter and located just off the primary state highway, the guard excused himself to shoot a pesky rattlesnake before he checked them through. When they reached their new living quarters, Gloria turned on the faucet at the kitchen sink and out came a two-inch-long scorpion.

She did not complain, but later, after a rough-and-ready community dance in an abandoned schoolhouse in nearby Hachita, she commented, "Len, I promised to follow you to the ends of the earth, and I think you've found them." To buy groceries, she regularly drove 120 miles each way to stores in Douglas. For weekend recreation the Judds occasionally motored around the vast Phelps Dodge outback, dug cactus plants, and replanted them in their yard—where they invariably died. Judd recalled that in early Playas you were grateful for any recreation opportunities, even digging cactus.

A quarter century later, in the late 1990s, Playas consisted of 256 three-bedroom homes, 25 apartment units, a business plaza, a medical clinic, recreational facilities, and a population of about nine hundred people. At its heart lay a green common of grass and shade trees that made the company town look very much like the campus of a modern community college. Social life centered on the two churches and on the Copper Pins Recreation Center, which boasted six bowling lanes, billiard room, game room with twelve different electronic games, snack bar, multipurpose room with kitchen facilities, and hair-styling salon. Opened in 1981, the recreation center was operated by Phelps Dodge Mercantile Company. For outdoor recreation there was a junior Olympic–sized swimming pool complex, softball fields, tennis courts, a children's playground, and basketball courts. Equestrian-minded residents of Playas built corrals for their horses.

The town did not have its own school, but Playas children were educated in Animas, New Mexico, twenty miles to the west. The logistics of medical care were even more complex. When a Playas resident was sick or injured, the local clinic was linked by two-way television to the Med Square Clinic in Silver City, New Mexico. Physicians there could converse with the patient and a physician's assistant in Playas to diagnose an illness or injury and prescribe needed medication and treatment, as if everyone were together in the same examining room. Implemented late in 1975, the Telehealth system at Playas was the first privately funded system of its kind in the nation.

(Based on interviews with Leonard R. Judd, Paradise Valley, Arizona, March 20, 1996; Matthew P. Scanlon, Phoenix, Arizona, March 22, 1996; and a telephone interview with Leonard Judd, September 7, 1998.)

Finland which lacked fuel resources, and no less so in the energy-conscious United States during the 1970s. Some industry observers prematurely sounded the death knell for smelting, but smelting experts outsmarted them with their clever new technology.

The flash smelting process captured 97 percent of all sulfur dioxide emissions, which Phelps Dodge then treated in an acid plant to produce sulfuric acid. In fact, Hidalgo produced nearly four times more sulfuric acid by weight than anode copper. One-third of the cost of the huge facility went for air pollution–control equipment, of which the acid plant was a major part. The remaining steps of the fiery process needed to transform copper concentrate into heavy anodes of 99.7 percent pure copper paralleled those in a conventional smelter.

The new smelter site was advantageous not just because it offered Phelps Dodge abundant land and water but also because it was located only a moderate distance from other key features of the landscape of production. It was easy to build a thirty-eight-mile industrial railroad to link the Hidalgo complex with the Southern Pacific main line east of Lordsburg. That enabled trains to arrive at the smelter with hopper cars filled with copper concentrates from Morenci and Tyrone and outgoing trains to depart with leased tank cars full of sulfuric acid and flatcars of copper anodes destined for the electrolytic refinery in El Paso. Hidalgo typically shipped out twenty to thirty tank cars of sulfuric acid each day, seven days a week. Some of it traveled to other Phelps Dodge operations in the Southwest and some went as far as Florida to produce fertilizer from phosphate rock.

To handle additional supplies of copper, Phelps Dodge soon enlarged the smelter's capacity from fourteen hundred to nineteen hundred tons of concentrate per day, an amount that ranked it among the largest reduction works in the world. In all, it was capable of producing 500 tons of anode copper, 2,500 tons of sulfuric acid, and 1,350 tons of slag every twenty-four hours.

Initially, Phelps Dodge's own Pacific Western Land Company held title to all rangeland it had acquired in the Playas Valley and raised approximately seventy-five hundred head of cattle and crops of grain, sorghum, corn, alfalfa, potatoes, onions, beans, and chilies. By the late 1990s, though local water was still used predominately for industrial purposes, cowboys from other ranch outfits herded cattle across large expanses of grazing lands leased from Phelps Dodge. The big smelter complex, which once employed about five hundred people, provided work for a number of former cowboys. Hidalgo hired about 40 percent of its original workforce from its sparsely populated corner of New Mexico, and no more than 5 percent had ever worked in a smelter.

A Gathering Storm

By the time the Hidalgo smelter came on line in 1976, its final cost, including industrial railroad and townsite, topped $360 million (or probably $1 billion in today's money). By contrast, the last smelter Phelps Dodge had built—at Ajo between 1948 and 1950 with about half the capacity of Hidalgo—cost slightly more than $8 million. The difference was not merely a result of greater capacity or runaway inflation, though both added to Hidalgo's eye-popping price tag. It was that about a third of Hidalgo's capital cost could be attributed to pollution-control equipment required to meet state and federal air quality standards. To that end it was, in the words of Len Judd, who knew it intimately, one of the "most complicated processing plants on the planet." To the future president and chief operating officer of Phelps Dodge, it placed the corporation on record as having made a major commitment to clean air.[17]

Apart from being by far the cleanest smelter in the United States in terms of atmospheric emissions, Hidalgo also relieved some of the load carried by Phelps Dodge's three aging smelters in Arizona, all of which waged a long struggle to meet federal and state air pollution rules and ultimately lost. Hidalgo's success involved a major trade-off, however. More than 40 percent of all energy it consumed went for pollution abatement.

Originally, some observers thought Phelps Dodge was crazy to think about building a new smelter. They mistakenly regarded environmentalism as only a passing fad or, conversely, believed that copper companies could do nothing to satisfy highly vocal critics of smelter emissions. For a time the surge of environmentalism seemed to leave the American mining industry dazed and uncertain which way to turn.

George Munroe recalled his own rude introduction to the strident side of modern environmentalism during a trip to Tyrone in 1969 to dedicate Phelps Dodge's new mine. The corporation had invited five thousand guests to a barbecue to celebrate the opening. Afterwards at a smaller dinner for the press at the rustic Buckhorn Saloon in Pinos Altos, near Silver City, Munroe made a few remarks to the assembled reporters and offered to answer their questions. One journalist shot back, "When are you going to quit poisoning the citizens of Arizona?" Munroe responded that Phelps Dodge was working hard to control emissions but did not have the technology to meet a 90 percent sulfur removal requirement then being considered by the Arizona legislature, and if it enacted and enforced such a regulation, the company's existing smelters would have to close. Because his words created a furor, Munroe, who was new to his job as chief executive officer, worried that he might have sunk the corporation by his hasty but honest remarks.[18]

In some ways, Munroe and Phelps Dodge, like the American mining industry in general, were unprepared for the environmental onslaught. During the decade of the 1970s, despite its expensive commitment at Hidalgo to virtually pollution-free technology, the corporation often appeared to fight a rearguard action on environmental matters so as to continue producing copper as usual. What to do about its smelters at Ajo, Morenci, and Douglas was a continuing problem. The Douglas smelter was especially troublesome because, unlike the other two, it employed roasters to drive off sulfur from concentrates, especially those from Bisbee, where local ore contained far more sulfur than ores mined in Ajo or Morenci.

The 1970s posed uncertainties for Phelps Dodge quite apart from environmental regulations, which the corporation, after

⁓ PHELPS DODGE IN WASHINGTON ⁓

ONE REASON PHELPS DODGE *directors hired Robert Page was that he knew his way around Washington during World War II and the cold war that followed. The need for the corporation to work closely and effectively with Uncle Sam grew even more compelling after the rise of the environmental movement in the late 1960s. In early 1973, George Munroe formally recognized that federal policy makers had become a permanent addition to the management scene at Phelps Dodge by hiring Charles A. ("Charley") Burns to be the corporation's first director (later vice-president) of government relations. Having earlier been administrative assistant to a North Carolina congressman and in charge of legislative matters for the American Mining Congress, Burns, an amiable Southerner who loved politics, knew his way around Washington.*

That was helpful to Phelps Dodge, the last of the big American copper companies to open a permanent office in the national capital, where it became Burns's duty to present the corporation's views on pending legislative and regulatory matters. After 1974, when Congress changed the law affecting corporate political contributions, he also headed Phelps Dodge's federal political action committee.

More than once it crossed Burns's mind that his lobbying Congress to protect the corporation's interests during the difficult years of the late 1970s and early 1980s might make the difference between success and failure for Phelps Dodge. "That's a thought that's terrifying," he later confided. Burns served Phelps Dodge for twenty years. Succeeding him in 1993 was Linda D. Findlay, among the first women to serve as vice-presidents in the history of the corporation.

(Based on a telephone interview with Charles A. Burns, October 1, 1998.)

recovering from the initial shock, learned to address in a positive way. There were escalating fuel costs—a primary reason for the decade's runaway inflation; and with inflation came the sharply rising interest rates that only added to the growing burden of long-term debt that Phelps Dodge incurred mainly to meet air quality standards at its four smelters.

In total, Phelps Dodge spent a third of a billion dollars on pollution-control devices intended to address the constantly changing rules set forth by federal and state bureaucracies. Operating that antipollution technology contributed to the sharply rising cost of producing copper. Because current earnings covered only part of those capital expenditures, Phelps Dodge spent its surplus funds in the bank and then in 1967 borrowed a modest $2.7

million. After more than eighty years of pay-as-you-go mining, the company bowed to changed rules.

By the end of the 1970s its total long-term debt had climbed to $606 million, of which about half could be attributed to compliance with environmental regulations. Even that sum created no undue alarm as long as the all-important price of copper remained in harmony with the cost of production. Alas, as the corporation headed into the 1980s that was not the case. During the years 1981–84, instead of wrestling with uncertainty in terms of environmental remediation and inflation, the nation's Big Three copper producers ran headlong into a genuine economic disaster.

As if to portend the deluge ahead, a tailings dam at the Tyrone Branch ruptured on October 13, 1980, and covered nearly three hundred acres of Mangas Draw, mostly owned by Phelps Dodge Corporation, with varying thicknesses of finely ground rock. The flood, which ripped out power lines and broke water mains, shut down the mill for about three weeks. Fortunately, a series of check dams along the dry streambed contained the mess, and there were no personal injuries. Phelps Dodge spent months moving spilled tailings, covering the affected area with soil, and fertilizing and seeding to reclaim the land. Grass soon blanketed the ground, and the rapid regrowth of native vegetation provided good forage for the cattle and deer. Not all challenges of the 1980s were so easily met.

~~~ *Chapter Twelve* ~~~

RED METAL BLUES

C OPPER IS USUALLY referred to as the red metal, but in solution it is blue, about as blue as the mood around Phelps Dodge during much of the early 1980s. "A very heavy dependence on one commodity gives you some real ups and downs," Chairman George B. Munroe once confided to *Forbes*. "I'm not saying that's bad. But the extent and suddenness of the swings take some getting used to." Because the time-honored corporation had a history of adjusting successfully to swings in the price of copper, what was so different about the four-year period from 1981 through 1984?[1]

The early 1980s were by far the worst of times for Phelps Dodge, when its troubles multiplied like breakers across the face of a stormy sea. In the second quarter of 1982 it paid its last dividend until December 1987. During part of that period it suffered a string of money-losing quarters unprecedented in its history—thirteen quarters in all with the lone exception being the second quarter

---

~~ BECAUSE THE PRICING *of copper is difficult to illustrate, the photographs on pages 280 through 291 offer a retrospective look at copper production during the four prosperous decades between 1940 and 1980. In the 1940s, no less than in the 1970s and 1980s, rows of anodes awaiting shipment from the Douglas Reduction Works to the El Paso refinery symbolized the heavy dependence of Phelps Dodge on the price of copper.*

281

of 1983 when it enjoyed modest net income totaling $300,000. Not even the decade of the Great Depression was so bad for so long.

In the pages of *Phelps Dodge Today*, an in-house magazine, the senior vice-president and general manager Arthur H. Kinneberg worried aloud in early 1983 about the previous year's losses of $74.3 million. He could not know that during 1983 the corporation would again lose huge sums of money—and still more the year after that. In all, losses for the three-year period amounted to a whopping $400 million (including a nonrecurring pretax charge of $195 million in 1984 that reflected an asset-restructuring program). Any way you calculated it, those were dreadful years for Phelps Dodge and other American copper producers—and for other major industries too, like the nation's automobile giants, brought low by swift-footed Japanese competitors.[2]

The copper crisis of the 1980s, though it appeared to leap suddenly into prominence, was actually several years in the making. It emerged only after certain long-noticed trends converged unexpectedly to form a combination both frightening and dangerous, much as happens when charcoal, sulfur, and saltpeter unite to form gunpowder, an explosive mix far more volatile than any of its constituent parts. America's copper crisis took shape when the distinct legacies of events occurring during the 1970s—national-ization of mines overseas, double-digit inflation at home, and long-term debts that producers shouldered to meet changing emission standards—coalesced in a menacing new way.

Starting in 1982, one money-losing quarter followed another with depressing regularity until

employees of Phelps Dodge wondered whether their corporation had a future. It still had plenty of copper in the ground, but as the vice-president Richard Pendleton noted in the *New York Times* in April 1982, "the more we sold the more we lost." No wonder some executives quietly pondered whether to dust off their résumés and seek new jobs outside the copper industry. Pay cuts for salaried employees and sales of major assets helped to fuel dark rumors of impending bankruptcy. Some members of middle management were certain that the corporation's chairman already had the necessary legal documents drawn up and waiting in his desk.[3]

Not so, emphasized Munroe during a 1998 interview at his New York office. "There was never any danger of bankruptcy," he recalled, and any claims to the contrary he branded a "total falsehood." Phelps Dodge was "in bad shape. No question about it," he agreed. "There were plenty of dark days." But Munroe was quick to cite the many positive measures that he and his lieutenants took to navigate Phelps Dodge around all mortal dangers. By the end of the 1980s, in fact, it had become a stronger and far more nimble enterprise than just a few years earlier. Improbably, company observers could judge the last two or three years of the decade the best of times. Perhaps that was only the natural giddiness and sense of relief that come from surviving a life-threatening situation, but certainly there was a noticeable, even profound, mood swing at Phelps Dodge after early 1985 when the corporation started making real money once again. To place the decade's crucial events in proper perspective, Munroe adds, one need only recall that among the firm's longtime rivals

Anaconda did not survive the 1980s copper crisis and Kennecott emerged only a shadow of its former self. For Phelps Dodge, the events of that tumultuous era shaped the corporation's future course every bit as much as those of the 1880s, when it first acquired a stake in Arizona copper.[4]

More Dark Days Ahead

PRESIDENT WARREN E. FENZI in New York typically talked by phone to the Arizona-based general manager of Western Operations every day. And when Art Kinneberg held that post he flew back to New York at least once a month to attend meetings of the senior management committee. Further, it was a long-standing Phelps Dodge custom for top executives based in the East to visit corporation properties in the Southwest each January. That was why in early 1980, Munroe and Fenzi headed west to tour the refinery and new rod mill in El Paso, the mines and smelters in Arizona and New Mexico, and the new La Caridad mine and mill near Nacozari, Mexico. Along the way they renewed old acquaintances and made a special effort to talk with as many members of branch personnel as possible. The annual January tour by New York executives provided western mine and mill employees an occasion to roll out the red carpet. However, the mood this time around appeared slightly less upbeat than usual.

During that swing through the West, Munroe addressed luncheon meetings of business and civic leaders in Phoenix, Albuquerque, and Tucson to report on the current status of the copper industry and discuss its outlook. In his

remarks the Phelps Dodge chairman hammered home an important point: although American copper producers had enjoyed a reasonably successful year in 1979, their profits were not excessive if evaluated in terms of inflation.

From 1974 through the mid-1980s, except for one year, Phelps Dodge saw the price of copper remain at uninspiring levels even as the corporation spent an ever-increasing number of dollars to operate its facilities in compliance with environmental regulations. The lone exception, 1979, was a good year because strong demand for copper and its products by the electric utility, construction, and automobile industries resulted in the highest production figures yet recorded by the corporation: 686 million pounds. If evaluated in constant dollars, however, Phelps Dodge's 1979 earnings of nearly $111 million actually fell below those recorded in half of the previous twenty-five years. Still, 1979 offered a happy contrast to the dark days that soon followed.

The average price of a pound of copper increased from thirty-eight cents in 1967 to ninety-three cents in 1979, but after correcting for inflation it equaled only forty-two cents a pound in 1967 dollars, or an anemic gain of just four cents. In other words, when runaway inflation was taken into account, copper prices barely crept upward while production costs virtually skyrocketed. That dramatic variance equaled trouble ahead. So too did the fact that during the decade of the 1970s Phelps Dodge found it harder to reduce its long-term debt than company experts had anticipated. "Normally we would have paid that debt down rather rapidly," explained Munroe. But those were not normal times. The corporation and its several

competitors, at least those that remained independent, entered the 1980s carrying a heavy and unaccustomed load of debt. That burden nearly crushed American copper producers when the price of red metal sagged even lower in terms of constant dollars than it had during the hard times of the 1930s.[5]

Consider that the nation's nine leading copper companies recorded aggregate debt of $250 million back at the end of 1964. By the mid-1970s that figure had ballooned to $3.2 billion, a twelvefold increase attributable for the most part to capital expenditures for environmental controls. And the total continued to climb.

In all, federal and state environmental programs added about fifteen cents per pound to the cost of producing copper. That was the unavoidable legacy of the previous decade's heavy borrowing and spending for new facilities, especially for environmental technology that promised to clean up smelter emissions even as it greatly increased operating costs. Further, because of the way terminal markets priced copper (explained later in this chapter), Phelps Dodge was unable to pass increased costs along to customers in the form of higher prices, unlike electric utilities, for example, which simply added the cost of emission controls at their power plants to rates they charged customers. A copper company could pass those costs only to its stockholders through reduced profits—if it had any profits.

In January 1980 the price of copper in constant dollars remained below the level needed to justify any new mine development. However, Munroe predicted to listeners during his annual tour that, because demand for copper exceeded

supply, copper prices should move up to the level needed to attract investment in new productive capacity by the mid-1980s. He expected that the opening years of the new decade would be good ones for the copper industry, assuming reasonable federal and state policies regarding taxation, the environment, and various regulatory matters. Those were big assumptions, however, and as it soon turned out, no one in the American copper industry accurately foresaw the time of trouble that lay just ahead. Since its founding in 1834, Phelps Dodge had weathered many periods of hard times, but nothing in its lengthy history compared to the disastrous combination that swamped and nearly sank it during the first half of the 1980s. In fact, from hindsight, the New York management's January 1980 tour of the Southwest actually looked like a rare moment of sunny optimism amidst gathering gloom.

In September 1980, President Fenzi retired (although he continued to serve as a member of the board of directors until 1984). In a move that was a first for Phelps Dodge, George Munroe then added the title of president to that of chairman and chief executive officer. The previous June he formalized a senior management committee composed of himself, three vice-chairmen (a newly created post), and the general manager of Western Operations. During the two previous years, members of the group had gathered on an unofficial basis in an effort to break down barriers that had long separated the corporation's mining and manufacturing sectors.

The trio of executives that Munroe, age fifty-eight, had named vice-chairmen were Edward H. Michaelsen, sixty-two, who was responsible for Phelps Dodge Industries; Richard

T. Moolick, fifty-nine, who had climbed through the ranks of western mining to gain responsibility for the mining and smelting side of the company; and L. William Seidman, fifty-nine, a senior vice-president and the corporation's chief financial officer. Seidman, who had been Munroe's roommate at Dartmouth and more recently President Gerald Ford's economics adviser (a member of the so-called Michigan Mafia at the White House), joined Phelps Dodge in 1977. He added to his stature as a money wizard when he later headed the Federal Deposit Insurance Corporation and the Resolution Trust Corporation that brought order out of the savings and loan chaos of the early 1990s. The final member of Munroe's newly formed team was Arthur H. Kinneberg, fifty-eight, a senior vice-president.

Listing the men's ages in news accounts of the management restructuring was important because it seemed to signal that no one from within the senior group was likely to succeed Munroe when he retired at sixty-five. The chairman after him would probably come from the corporation's next younger tier of executives. Talk of succession might have been dismissed as corporate politics, a subject mainly for speculation around the watercooler, but within the context of the monumental problems Phelps Dodge soon faced, questions about the next generation of leadership came to loom far larger than they would in ordinary times. What no one foresaw in 1980 were events culminating four years later in what Seidman recalled as an "internal takeover" unprecedented in the annals of American business.[6]

Meanwhile, it was Munroe and his senior managers who must have sometimes turned and tossed at night wondering what new troubles the dawn of a new day would visit upon Phelps Dodge. The nagging recession that started during the second quarter of 1980 moderated somewhat during the centennial year of 1981, when the corporation commemorated how it first looked west for its future. That year the combined output of all Arizona producers was a record 2.3 billion pounds of copper, or nearly 68 percent of the total production in the United States. Then the price of copper crashed hard during the first half of 1982.

Because unlike most industries copper producers cannot set prices, they had to fight back with sharp pencils to pare away any unnecessary costs. When the cost of producing red metal far exceeded its selling price, fourteen of Arizona's twenty-four mines suspended operations for an indefinite period, while others operated on a reduced work schedule. In total, some twelve thousand miners lost jobs between December 1981 and September 1982, and that predictably had a negative impact on the state's economy—not to mention the household budgets of the miners themselves.

Phelps Dodge spotted trouble ahead as early as the fall of 1981. At that time it became common for employees to speak of hard times, particularly after corporation officials announced production cuts at all branches. They did that by adding five days to the usual Christmas shutdown at Tyrone, six days at Ajo, and twelve days at the Morenci concentrator in addition to an eight-day shutdown previously scheduled for the entire Morenci complex. As hard times persisted, especially after the fourth quarter of 1981, the drain on corporation cash grew so severe that in April 1982 Phelps Dodge closed all its mines in the United States and its three Arizona smelters for an indefinite period. Only the Hidalgo facility in New Mexico continued to operate to fulfill smelting commitments to other companies. The $1.4 billion Phelps Dodge Corporation entered a state of suspended animation when it laid off about thirty-eight hundred of its six-thousand-member workforce and simply sold copper from its inventory.

"We deeply regret the hardship this action will cause the employees and communities affected," observed Munroe of the shutdown, "but the current recession has driven copper prices to their lowest levels in real terms in more than thirty years." He emphasized that during the early months of 1982 the company lost money virtually each day until its April shutdown, the first such suspension in fifty years. Phelps Dodge's "current predicament," observed *Business Week* in a mid-1982 article that many people considered unduly critical, "is a far cry from its glory days of the 1960s, when it had virtually no debt and was flush with cash."[7]

Munroe preferred to emphasize that "sooner

⌐ CASTING COPPER *wirebars at the Laurel Hill Refinery in the early 1950s. From here they typically traveled aboard barges to the Bayway, New Jersey, manufactory, where heavy machinery hot-rolled each of the 265-pound bars into rod used for drawing wire.*

THE NEXT STEP: receiving wirebars at Phelps Dodge's Bayway facility in New Jersey in the early 1950s, when it was a leader in rolling wirebars. Bayway at that time formed the heart of the corporation's manufacturing landscape.

or later" the country would work its way out of recession. Meanwhile, a pall of gloom and uncertainty hung over the copper towns. No one was sure what the immediate future held. Phelps Dodge sought within its newly limited means to aid laid-off employees. It continued their life insurance benefits for sixty days, medical and dental benefits for ninety days, extended limited weekly credit at Phelps Dodge Mercantile stores, and allowed those who lived in company housing to defer the payment of one-half their rent. As a result of a past labor agreement, it also supplemented unemployment compensation by fifty dollars per week from a trust fund into which it had contributed for the past seventeen years.

Unfortunately, mining and smelting operations did not resume on June 1, 1982, as many at Phelps Dodge had hoped, and because of the copper country's extended depression, medical, dental, and life insurance benefits expired. Perhaps worst of all, the trust fund for supplemental benefits ran dry for New Mexico. Approximately two thousand layoffs at Morenci plus their impact on the local merchants pushed unemployment in all of Greenlee County to 64 percent in May 1982. A series of community workshops called "How to Survive on $165" a week sought to help out-of-work families in Clifton.

In a May tour of Phelps Dodge properties, Munroe attempted to explain the continuing copper crisis to employees and their families. By that time the message should have been familiar: "Phelps Dodge is experiencing one of the worst economic depressions in the company's history." He emphasized that the price the company received for a pound of copper, in real terms, was lower than it has been at any time in the twentieth century, except for very brief periods following the world wars and a somewhat longer period during the Great Depression of the 1930s.[8]

Despite austerity measures at every level, Phelps Dodge suffered a net loss of $38.5 million during the first six months of 1982; a year earlier, it had earned a net income of $54.3 million for the same period. In view of the continuing economic slump, directors omitted the dividend on common stock for the third quarter of 1982. Arthur Kinneberg assured employees in Western Operations that there was "no chance" that Phelps Dodge would "go out of business. We have premier copper mining properties, and we have good people—we expect to be in business a long time, and we expect to be mining copper for a long, long time." He was unwilling to predict, however, when operations would resume. "The timing depends upon Phelps Dodge's cash flow position—whether we would lose less money if we remained shut down than we would lose if we resumed operations."[9]

Not surprisingly, when the corporation offered its employees an opportunity to retire early if they met certain age and service qualifications, the number of retirements jumped sixfold during the spring months of 1982. Morenci did finally reopen the following October, Ajo in February 1983, and Tyrone in early May 1983, but with only 70 percent as many salaried and hourly employees as had been on corporation payrolls in April 1982. Times remained difficult and threatened to grow even worse.

From a national perspective, the United States in the early 1980s was in a recession, but for commodity producers in Arizona, New Mexico, and other western states, it was appropriate to speak of full-blown economic depression. In addition to the world glut of copper, the exceptionally high value of the American dollar vis-à-vis other world currencies put United States producers at a global disadvantage. At home, interest rates rose so high that consumers delayed purchases of big-ticket items like automobiles and appliances, all of which were sizable users of copper. The ongoing copper crisis prompted a great deal of soul-searching at Phelps Dodge as people sought to understand its causes and find a cure.

The Gospel of Efficiency

THE ONLY THING for certain in the copper industry is change, and in the late 1970s and early 1980s there was plenty of that. For old-timers, it was sometimes hard to think of Morenci and not Bisbee as the company's copper flagship. With each passing year Morenci grew bigger and more important, until by the late 1990s it alone produced about one-third of all red metal mined in the United States. When

THE OLD WAY.
After workers hot-rolled
wirebars into rod, they welded the
short segments together before
shipping coils to wire and cable
manufacturers. Unfortunately,
each weld increased the chance
of breakage during the wire-
drawing process. After the modern
continuous-casting process
overcame that difficulty, the
loaflike wirebars became a novelty.

Phelps Dodge purchased the Western Copper property for $10 million in 1981, it consolidated virtually all viable copper claims in the Morenci area under its control and completed a process of acquisition that began exactly a century earlier when it first invested in the Detroit Copper Mining Company. Acquisition of the eighty-one Western Copper claims enabled Phelps Dodge to increase efficiency at its Morenci and Metcalf open-pit mines by removing any physical obstacles that separated them. And efficiency in all forms became the corporate gospel in the early 1980s.

Efficiency, sometimes called the challenge of productivity, included everything from the use of larger haul trucks and computerized dispatching

systems in open-pit mines, beginning at Tyrone in 1980, to automated welding machines and modern leaching and precipitation facilities at Morenci. For a time it appeared that greater efficiency alone—improved technology and good management—would enable Phelps Dodge to weather the runaway inflation of the early 1980s and compete successfully with low-cost mines outside the United States.

The emphasis on achieving greater efficiency through improved technology was not actually new. For decades, Phelps Dodge had pursued research and development programs that sought to modernize mine production and better its manufactured copper products. That enabled the corporation to remain at the forefront as one of the lowest-cost producers in the copper industry. At times, however, it seemed all efforts to achieve greater efficiency meant only that Phelps Dodge ran ever harder just to stay in place.

During the decade of the 1970s, the price of virtually everything Phelps Dodge required to produce copper and copper products escalated. The dramatic rise was spurred in part by the Organization of Petroleum Exporting Countries (OPEC) oil embargo of 1973 and a second oil price "shock" in 1979 that lifted the price above $30 a barrel. While most Americans could only fume about fuel shortages and sharply rising prices, Phelps Dodge early worked to gain a secure energy supply of its own. To that end, in 1974 it formed Phelps Dodge Fuel Development Corporation, a wholly owned subsidiary intended to explore for oil and gas in the continental United States. That first year it participated in twenty-five exploration wells, all located in Texas. During the remainder of the 1970s, Phelps

Dodge Fuel Development explored for oil and gas in various parts of the country, though with only moderate success. By the early 1980s, nonetheless, several of its wells supplied fuel for Western Operations. Beginning in 1982, Phelps Dodge also substituted pulverized coal for much of the fuel oil burned in the Hidalgo smelter.

Unfortunately, not all inflation-related problems were so easily solved. Consider, for example, that the cost of each brake shoe used on railroad dump cars at its Morenci mine jumped 95 percent, a rotary drill bit 131 percent, and a grinding ball, of which it used thousands in its mills, 141 percent. The cost of each refractory brick used in smelters ballooned by an astronomical 979 percent over a ten-year period. Often, it seemed that the more basic the item the more its price jumped. All kinds of similarly inflated numbers could be cited for everything from building lumber and pickup trucks to chemical reagents and explosives.

But that was not all. As Kinneberg explained to readers of *Phelps Dodge Today* in 1980, labor costs—that is, the total Phelps Dodge paid for employees' wages and benefits—had ballooned 171 percent since 1970. Further, back then there were few environmental costs; "yet now, only ten years later, the Company pays tens of millions of dollars every year to comply with a maze of environmental laws and regulations."[10]

The Dismal Science

To REPEAT ECONOMICS lesson one in the copper classroom of the early 1980s: when the price of copper was expressed in 1981 dollars

it had decreased 34 percent during the previous decade, while Phelps Dodge's total production costs had increased by about 50 percent. The three principal reasons for the corporation's increased production costs, explained George Munroe, were the skyrocketing price of energy, the additional burden imposed by pollution controls, and the impact of automatic cost-of-living adjustments (COLA) on wages in the industry. The question of COLA was something about which Phelps Dodge employees would soon hear a great deal.

In the short run, the corporation emphasized still greater utilization of technology to achieve still greater efficiency to reduce production costs. But greater efficiency alone had its limits. It simply could not offset the financial damage done when copper prices lagged behind inflation—and copper remained the backbone of Phelps Dodge Corporation. In 1981, for instance, when the price of copper dropped by as little as a single cent, it reduced the corporation's annual earnings by as much as $5 million. But as important as red metal was to Phelps Dodge, the corporation remained at the mercy of copper prices because the world, not just the United States, had become its marketplace.

The continuing copper crisis focused attention as never before on how copper was priced, and if people wondered why economics was often called "the dismal science," they certainly didn't wonder after hearing American copper's continuing complaint. "A person unfamiliar with our industry," explained Munroe in early 1982, "might suggest that we merely increase the price we charge for our copper to cover any increases in our labor costs." Actually, supply and demand determined the price that Phelps Dodge and other metal producers received for a pound of copper over the long term, and the obvious imbalance of the early 1980s placed the mechanism of copper pricing in the spotlight.[11]

A fundamental axiom of economics is that the more there is of anything the less it is worth. To apply that principle to copper, imagine a pricing mechanism somewhat like the inflow and outflow of water for a common bathtub. If the tap of production was opened full but the drain of demand was closed or partially plugged, the tub would fill up. The water level in the tub was analogous to the world supply of copper, which would likely decline in price as the level, or supply, rose. Around the world, everything from industrial accidents, earthquakes, and floods to wars and rumors of wars influenced the price of copper.

Phelps Dodge typically sold its copper to buyers by setting its own "producer price," but by the early 1980s it had tied that price closely to terminal markets in London and New York, both of which were in turn influenced by supply and demand. Like corn, wheat, and dozens of other commodities, copper was traded on the futures market. The older of the world's two main metal markets is the London Metal Exchange, or LME, which dates back to 1877 when metal traders set up their modest exchange over a hat shop in Lombard Court. Its influence diminished temporarily after the outbreak of war in 1939, at which time it closed its doors and kept them locked to copper traders for the next fourteen years. The other was the Commodity Exchange, or COMEX, in New York. Established in 1933, it is today located in the World Financial Center near the landmark twin towers of the World Trade Center. In contrast to an atmosphere of British formality that prevails on the London Metal Exchange, life at the COMEX mirrors the more raucous demeanor typical in the United States, as a circle of leather-lunged buyers and sellers bellow orders at one another and litter the trading floor with scraps of paper. These clamorous exchanges are called, appropriately enough, open outcry sessions.

Activity at the Commodity Exchange or the London Metal Exchange also illustrates how in the real world the relationship between copper's supply and demand is complicated by speculators and floor traders who have no direct interest in production or consumption of copper. They have no intention of taking physical possession of any of the copper they buy or sell, but nonetheless they can create rapid change or movement in prices and cause difficulty for Phelps Dodge as it seeks to establish a fair and consistent price for its copper customers. Finally, the availability of scrap copper further affects the price of newly mined metal.

As has already been noted, copper is a commodity subject to substitution. If its price per pound climbs too high, users might replace it with other metals or plastics in the manufacture of certain products. For example, plastic pipes might replace copper ones if they became too expensive. All that was old news by the early 1980s, but a reminder of the lurking threat of substitution was near at hand in the form of the common "copper" penny, which in recent years evolved into a slug of zinc electroplated with a thin outer layer of copper. It was clear that Phelps Dodge fared best when copper was priced low enough to discourage substitution but high

⟶ The Pulse of Copper ⟵

Anyone who ever *imagined the nation's mining frontier to be a boisterous place should visit the floor of the COMEX on a busy day. For better or worse, and though nary a pound of copper is in sight, it offers a way to visualize what is really meant by the often-repeated phrase, "it all depends on the price of copper." The frenzied pace of activity a person usually witnesses on New York's metal exchange may seem a world away from the carefully orchestrated landscape of production at Morenci or the hush of Phelps Dodge offices in Phoenix, but they're all intimately joined together. In fact, at its headquarters building the company maintains a trading center where large wall clocks give the time in Phoenix, New York, London, Tokyo, and Santiago, five cities vitally important to Phelps Dodge's up-to-the-minute assessment of the pulse of copper.*

enough to provide the corporation a reasonable annual profit.

Forbes magazine observed as early as 1958 that the copper mines of the world could not "legally regulate or practically rationalize production among themselves. Yet if the world's formidable array of producers mine so much as 4% more copper than the market needs, the price can plummet by nearly 50%." That, incidentally, was at a time when private enterprise dominated world copper production, but things became vastly more complex during the early 1980s. The problem that Phelps Dodge faced at that time was that recently expropriated copper companies in developing nations seemed determined to leave the tap of production fully open in defiance of world demand. Thereby they overflowed the tub and left the world awash in low-cost copper. All of this, of course, translated into depressed prices for everyone. Perhaps nothing more effectively illustrated how different the world of copper had

become since World War II or how major political or social changes affected the price of red metal.[12]

With each passing month, Phelps Dodge executives saw ever more clearly how radically their industry had evolved during just the past ten or fifteen years. In former times, companies in the United States produced about one-quarter of the western world's newly mined copper. Likewise, private enterprise owned and operated all major foreign mines. In fact, until the late 1960s just six companies basically controlled world copper production. Those giants were Anaconda, Kennecott, and Phelps Dodge in the Americas, and Anglo American, Rhodesian Selection Trust, and Union Minière du Haut Katanga in Africa. Their managements typically responded to market surpluses by cutting production and thereby kept supply and demand in reasonable balance.

The pricing mechanism went haywire during the late 1960s and early 1970s because expropriation of big mines in Chile, Zambia,

Zaire, and other countries effectively jammed open the tap of production. By the end of the 1970s more than 42 percent of all copper coming from the non-Communist world originated in mines owned or controlled by governments of developing nations. Chile, often aptly described as the Saudi Arabia of copper, annually produced a whopping 17 percent of the world's supply; Zambia 9 percent; Zaire 8 percent; Peru 5 percent; and Mexico 4 percent.

Overproduction during the recession years of 1982 and 1983 forced extra copper into a market where it found few buyers. As inventories built up, the price fell still farther, to seventy-four cents in 1982, a figure that when expressed in constant dollars just about equaled the fourteen cents that copper sold for in 1946. Normally, a price decline such as Phelps Dodge experienced in the early 1980s resulted in reduced copper output. But those were not normal times—at least not in the old sense of normal. Where privately owned copper companies had once responded to market forces, state-owned companies continued to produce all the red metal they could, apparently in the mistaken belief that flat-out production would result in maximum foreign exchange for their countries and full employment for their people. Because of social policy they did not cut production.

Government-controlled Third World copper producers competed in markets with copper coming from Phelps Dodge and its counterparts in the private sector, who were forced to bear the entire burden of trimming production to meet demand. Because copper was an international commodity—that is, a pound sold in Brazil was the same as a pound sold in Bangladesh or Boston—Phelps Dodge was wholly dependent on world economic events, not just those taking place in the United States where only about 25 percent of the world's copper was consumed. No wonder *Forbes* magazine as early as 1978 labeled the American copper industry "sick."

During the recession of 1982, United States producers reduced output by 25 percent and laid off nearly 40 percent of the nation's copper miners. That same year, Chile's state-owned CODELCO *increased* production by 15 percent to an all-time high and maintained full employment in its mines. CODELCO, in Munroe's words, was the "King Kong of Copper." When newly expropriated mines did not reduce production, they thus transferred the burden for doing so onto the private sector, specifically the United States copper industry, and that meant big production cuts, layoffs, and personal hardship.[13]

For nearly a century, from 1883 until 1980, the United States remained the world's leading copper-producing nation—except for 1934, when depressed economic conditions curtailed production—but beginning in 1980, and especially after 1982, the United States lagged steadily behind Chile, and over time the gap widened. In Chile, where copper was expropriated in the late 1960s and early 1970s, output increasingly reflected the wishes of politicians who wanted their country's mines to produce copper in great quantities and thus provide steady jobs in defiance of all known laws of economics. Chile's red metal production continued to increase each year between 1980 and 1984, as did the amount of copper it exported to the United States. As a result, in the early 1980s both the United States and world markets found themselves drowning in a sea of low-priced metal. When the copper glut combined with global recession and runaway inflation, Phelps Dodge found itself groping through the darkest days in its long history.

Because Phelps Dodge could not effectively fix its own price for copper, if it hoped to earn a profit during times of low prices it needed to reduce its costs of production, and by 1982 it had already embarked on a quest for greater efficiency. The corporation might also turn to Uncle Sam for short-term protection from low-cost imported copper. To that end, copper producers participated in a successful lobbying campaign mounted by the nation's smokestack industries, all of which suffered from the prevailing high interest rates, a strong dollar, and economic recession, to persuade Congress and members of the Ronald Reagan administration to grant them special tax credits needed to weather the economic storm. At the same time, however, Phelps Dodge and its American competitors failed to persuade Uncle Sam to impose tariffs or import quotas, at least during periods of recession, to protect them from extreme market distortions caused by the subsidized actions of state-owned mines. By 1983 imports had taken 26 percent of the domestic market for refined copper, up from only 10 percent as recently as 1979.

Failure to obtain tariff relief seemed particularly unjust because in the early 1980s, overproduction by expropriated or government-controlled copper mines in Chile and other developing countries continued to be under-written by financing agencies like the World Bank and the International Monetary Fund to which

United States taxpayers were the largest contributors. In short, and as bizarre and unjust as it seemed to American producers, Uncle Sam indirectly provided financial assistance to compensate Chile for losses it sustained while maximizing production at the expense of its American competitors. Not only that, financing agencies also granted loans at very low interest rates to permit further expansion of foreign copper production facilities.

On one occasion, Chile applied for and received $32 million from the International Monetary Fund to offset a "loss" in export revenues primarily attributed to low copper prices, prices driven down, ironically, by Chile's own overproduction! "It is discouraging to see U.S. dollars going overseas to underwrite the expansion of competitors whose predatory market behavior is the largest single cause of our copper industry's difficulties," observed Kinneberg in March 1984. Even worse, money for those baleful loans came from the pockets of American taxpayers, including Phelps Dodge's own copper miners in Arizona and New Mexico who lost jobs because of United States–subsidized copper imports. "The cure for this intolerable situation" fumed the normally mild-mannered Munroe, "is to get the international financing institutions to better understand the dynamics of supply and demand in a free commodity market and to apply that knowledge to their financing decisions."[14]

That was not all. In addition to using dollars from American taxpayers to bolster the over-production that was flooding world copper markets, those same developing nations generally had richer ore bodies to mine and less stringent and hence less costly environmental regulations to meet than their United States competitors, and they paid their workers less. That was one major reason the ABC television program "20/20" spoke in July 1984 of a "once proud industry brought to its knees" by foreign copper.

No End of Challenges

FROM THE 1920s until the early 1980s, Phelps Dodge did not change significantly the way it produced its copper or managed its far-flung operations. It was a remarkable record of stability disrupted only when the price of copper nose-dived and the corporation lost money on every pound it sold. The ensuing distress lasted four years and unleashed a tidal wave of structural and personnel changes.

A shuffle in top Phelps Dodge management in 1982 only heightened public concerns for the future of the corporation. *Business Week* observed in midyear that L. William Seidman, the man Munroe brought in to modernize management and accounting procedures, was leaving that September, and the impending retirement too of Edward H. Michaelsen, who at age sixty-five became managing director of New York's Carnegie Hall, "will mean that Richard T. Moolick, the remaining vice chairman and a tough and abrasive mining engineer, is likely to be named president." The question was, would he be too tough and abrasive given the patrician image cultivated in recent years by the corporation's top management in New York?[15]

Moolick, who served as vice-chairman for mining operations from 1980 until 1982, the year he became president of Phelps Dodge, did

implement a variety of programs intended, in his words, to make the company "lean and tough." Earlier, he himself had acquired a reputation for toughness, an attribute he greatly relished because he saw himself as embodying the American frontier spirit. His grandfather was a miner and his father an oil driller. To friend and foe alike, he seemed to have the tenacity of a prickly pear cactus and sometimes needed to be approached with the same degree of wariness the cactus required. Moolick and Munroe were studies in contrast: whereas the chairman was cerebral and inclined to work through difficult matters with a willingness to compromise, the president was visceral and inclined to give no quarter. Their differences in style became most apparent during the great copper strike of 1983 when Moolick locked horns with nearly three dozen unions long entrenched at Morenci and other Western Operations properties.

Back in the mid-1970s, after he went to Phelps Dodge's New York headquarters as a vice-president, Moolick made a lengthy firsthand study of Western Nuclear that caused him to worry that the pace of the uranium company's mining activity and its metal reserves bore no relationship to its sales; and it was he who set in motion a process that led Robert Adams to resign as president of Western Nuclear. That gained the parent company direct control over its ailing subsidiary, especially after Fenzi appointed Moolick president of Western Nuclear.

As the Western Nuclear episode illustrated, Moolick saw himself as a man of action. Take the Metcalf mine as another example. Originally estimated at $100 million, the cost ballooned because of inflation to twice that amount by the time the mine came into production in the mid-1970s. After that it was plagued by large amounts of overburden, especially when copper prices were low, and the solution extraction–electrowinning technology had not yet been embraced by Phelps Dodge. In 1980, the mine produced only 11,400 tons of copper, yet it had to move 7.6 million tons of overburden to get to 2.4 million tons of ore, an unsatisfactory ratio of over three to one. "My first action as vice chairman," recalled Moolick, who assumed that position in 1980, "was to shut down the Metcalf Mine, and to feed the Metcalf Mill from the Morenci Pit. That resulted in a saving of approximately $25 million a year."[16]

By 1982, belt tightening all down the line was the order of the day at Phelps Dodge, but the history of the corporation during the early 1980s is not entirely a story of retrenchment. Phelps Dodge Copper Products Company opened a continuous-cast rod mill in Connecticut in July 1981 and another in Texas a year later. Each of those plants covered an area bigger than a football field, and together they secured Phelps Dodge's position as the world's leading producer of copper rod. Those facilities, furthermore, represented vital connecting links between Phelps Dodge Mining Company and its customers, which included several manufacturing plants of its sister Phelps Dodge Industries.

Also on an upbeat note, late in 1982, Phelps Dodge opened a precious metals recovery plant at the El Paso refinery. At about the same time the newest unit of Phelps Dodge Industries—the recently formed (1978) Phelps Dodge Solar Enterprises—moved into spacious new quarters in California. Well attuned to the energy-conscious decade, it produced copper solar panels for domestic hot-water installations, such as for swimming pools and spas, a market that was big in the Southwest. Branching out into that area marked a "dramatic departure from Phelps Dodge's past attitude toward manufacturing," which was just to move copper. Previously, the company didn't want to worry about small-volume, high-margin products.[17]

Moolick as president did not simply close mines and trim away perks. He sought to lessen the company's dependency on massive open-pit copper mines by fostering the Small Mines Division to exploit the kind of precious metals deposit that Phelps Dodge had previously snubbed. Dating from 1981 and based at Safford, Arizona, the Small Mines Division developed an operating silver property in southeastern Arizona and recovered precious metal ores from its underground copper mines in Bisbee, which by 1982 had been officially idled for seven years. It even hoped to use its small stakes in oil and

Located inside the El Paso complex at the aptly named intersection of Slimes Alley and Gold Drive is the precious metals recovery plant opened in 1982. It is from slimes created as a by-product of the electrolytic refining of copper anodes that Phelps Dodge recovers gold, silver, and other valuable metals at this high-security facility.

Where Mining Meets the Market

For years, plants *in Bayway, Fort Wayne, and El Paso had made Phelps Dodge a leading supplier of hot-rolled copper rod, the basic feed material for drawing copper wire. The technology's main drawback was that, in order to produce the longer lengths desired by wire and cable manufactures, workers had to weld several shorter rods together, and each weld carried a risk of breaking when machines drew the rod into wire. The surface of hot-rolled copper rod also tended to oxidize, and this impurity had to be shaved off mechanically. Continuous-cast rods, a technology born in Europe during the 1970s, overcame both problems, but Phelps Dodge already had a huge investment in traditional rod-making equipment, and so it held back until 1979 when the new technology had at last matured. When Phelps Dodge engineers gave their approval, the corporation announced plans to spend $67 million to install two state-of-the-art continuous-cast rod mills, one in Norwich, Connecticut, to serve the eastern market, and another next to its refinery in El Paso. This spelled the end of rolling mills at Bayway and Fort Wayne.*

Because industrial Connecticut had for years suffered from an exodus of cotton mills and other manufactories to the South, the word mill *was anathema, so Phelps Dodge purposely named its new facility the Norwich Rod and Wire Plant. When the Norwich plant opened, the corporation announced that it had come home. Indeed it had. Not only were two of its founders, Anson Phelps and William E. Dodge, Connecticut Yankees, but Dodge had as a teenager worked in a cotton mill his father ran in Bozrahville, a village located just down the road from Norwich Rod and Wire.*

uranium to expand further in energy production. However, the Small Mines Division was never able to surmount low metal prices that caused it to close in 1984, though not before it acquired a small mine in Chile called Ojos del Salado, an unassuming property that unexpectedly led exploration personnel to Candelaria, one of the great mining success stories in the corporation's modern history (see chapter 16).

Back in the 1950s and 1960s, when Phelps Dodge was the most integrated of America's Big Three copper producers, most of its metal production not sold on the open market went to its own wire, tube, and cable subsidiaries. By comparison, Anaconda consumed only about one-quarter of its mine output. One result was that whenever demand for red metal dropped, Phelps Dodge had a captive market for its copper. By the 1980s, alas, the gears of production no longer meshed so well, and the earnings of several of its fabricating units declined.

As it entered the 1980s, Phelps Dodge began to divest itself of underperforming assets. In 1980, following a three-month study by Douglas

Yearley, then president of Phelps Dodge Sales Company, and to avoid investing additional capital and realize cash, Phelps Dodge sold its 40 percent stake in CONALCO to Swiss Aluminium for $125 million. Nearly two decades later, Yearley, then chairman, recalled that the corporation had a difficult partner in the Swiss aluminum producers: "Never be a minority partner in a large venture if you want to control your destiny." In all, Phelps Dodge basically broke even on its 1960s and 1970s venture into aluminum—and that was welcome news.[18]

Through the early 1980s, Phelps Dodge made the classic moves a business normally does during a major crisis. It reduced the cost of producing copper by reevaluating and revising engineering and mining plans at each mine and processing facility, tightening control on inventories and receivables, and limiting capital expenditures to projects that offered clear and substantial cost savings. It sold off assets and properties not essential to its core copper business in order to raise the cash it needed to contain its debt and interest costs. In 1982 it also skipped the first of several quarterly dividend payments to its thirty-eight thousand shareholders. It reduced the number of employees, both salaried and hourly paid, and it finally halted production altogether. The lengthy shutdown that began in mid-April 1982 hit production workers especially hard, but no one was exempt from painful cuts. On April 1, 1982, in what was certainly no April Fool's joke, company management announced that beginning the following month it would trim salaries for all nonunion personnel by 4 percent up to $40,000 and by 8 percent for any in excess of that

amount and eliminate cost-of-living allowances.

By all those measures and a host of smaller ones, Phelps Dodge achieved major cost reductions and cash savings. Still the corporation lost money until it seemed that there was nothing left to trim but wages and benefits received by its union-represented workers. *Business Week* noted in 1982 that, after contending with the Environmental Protection Agency, "Munroe's next biggest challenge will be to cut labor costs, which now represent more than 40% of production outlays." The magazine predicted that obtaining relief from unions "will be no easy task." That was a classic understatement. Phelps Dodge dealt with some of the toughest, most militant unions in the United States, and trimming even the least of their hard-won gains would likely precipitate a bitter labor-management conflict.[19]

In any public discussion of the ongoing copper crisis, management increasingly took note of the escalating cost of labor during the early 1980s. After all, wages and benefits constituted the largest single operating cost at Phelps Dodge copper mining and processing facilities. The corporation often reminded production workers that between the late 1940s and 1982 the average hourly wage of copper miners in the southwestern United States had increased almost nine times, from slightly less than $1.50 to $13.00 an hour. That was when Chilean mine laborers earned less than one-tenth that amount, yet their products competed head to head with America's in world markets.

It soon became clear that for both sides the flash point was the cost-of-living adjustment (COLA). Unionized workers in the American copper industry first gained COLA during bargaining in 1971. After it went into effect on July 1, 1972, the controversial provision automatically bumped up the basic (non-overtime) wage every three months. Thus in the early 1970s, in addition to the recently acquired encumbrance of long-term debt, Phelps Dodge Corporation shouldered the heavy financial burden of automatic cost-of-living adjustments.

During the runaway inflation of the late 1970s and early 1980s, company managers recognized that the ever-escalating cost of COLA added greatly to the cost of producing copper. Wage rates increased an average of 12 percent a year, while copper prices climbed only an average of 3.5 percent a year. Even worse, when converted into 1982 dollars, copper prices had actually dropped 34 percent since COLA went into effect, but for unions COLA had already become a sacrosanct benefit. Perhaps if the price of copper had risen modestly with the tide of inflation, COLA might have remained a low-profile issue, but in 1982 when the price of copper slumped to a level below what it cost to produce, Phelps Dodge believed that COLA contributed mightily to the economic one-two punch that staggered the corporation. "This kind of relationship simply cannot continue indefinitely," warned Munroe.[20]

One thing that made COLA irksome to Phelps Dodge management was that quarterly adjustments were tied to the Consumer Price Index, two components of which (accounting for more than 25 percent) were medical expenses and home ownership. But a substantial number of Phelps Dodge employees lived in low-rent company housing; the company also provided them with good hospitals and with few exceptions paid the medical bills of employees and their dependents. Yet the program pushed wages higher and higher in every quarter but one. And not just higher: wages escalated at a compounding rate.

To explain to workers Phelps Dodge's growing concern about COLA as well as other matters that contributed to the copper crisis, Munroe, Kinneberg, and Jack Ladd, head of labor relations, visited each major production facility in Arizona and New Mexico during May 1982 when all the mines were shut down. The message was one they would hammer home repeatedly in the year to come. As Kinneberg phrased it: Phelps Dodge wanted to compensate its Western Operations employees well and to "provide the job security we would all like. But we cannot do this if it costs us more to produce our copper than we can sell it for. One of our principal cost problems is that increases in wage rates and COLA have far outpaced increases in our productivity." He warned that an adjustment had to be made, but all attempts to persuade union leaders that COLA presented a problem were unsuccessful. That did not bode well for contract negotiations in 1983.[21]

DURING THE EARLY 1980s, not many things seemed to go right for Phelps Dodge. But punctuating everything and leaving by far the most indelible mark on those who recall the grim times was the bitter strike that erupted in July 1983, mainly over the contentious issue of COLA. Long after many other unhappy events of the early 1980s are forgotten, people will remember the great strike of 1983, often with tears welling up in their eyes.

Chapter Thirteen

HAMMERED *on the*
ANVIL *of* ADVERSITY

E ARLY ON THE morning of August 19, 1983, residents of Clifton, Arizona, awoke to the
sound of helicopters in the dark skies above and of heavy trucks winding down the canyon
road that led into their secluded town. A seemingly endless convoy of eighty-seven military
vehicles, including armored personnel carriers and one tank retriever, and thirty-three Department of
Public Safety vehicles took up positions in the troubled Clifton-Morenci district. The dawning light
revealed state police officers dressed in blue helmets and riot gear and specially trained sharpshooters
silhouetted on nearby hills. Governor Bruce Babbitt went on statewide television to make clear that
he would "do whatever is necessary to prevent violence and uphold the law." He feared that Arizona
tottered on the brink of a great tragedy because of rapidly escalating tensions between labor and
management at Morenci.[1]

As if the ongoing copper crisis had not already pummeled Phelps Dodge enough, in 1983 it
suffered the most bitter strike in its 150-year history, one that a future chairman, Douglas C. Yearley,
labeled "a defining event in the corporation's history. No doubt about that." The 1983 strike also ranks

---

PHOTOGRAPHER MICHAEL GING *captured the angry scene in Morenci on August 9, 1983, that caused Phelps Dodge to agree to
a ten-day cooling off period.*

303

among the biggest and most significant labor disputes in Arizona history. In fact, it is no exaggeration to say that it reshaped the course of labor relations across the entire United States.[2]

The ongoing copper crisis of the 1980s and the 1983 strike were closely related events. Understanding why labor and management ended up confronting one another in a showdown reminiscent of the frontier West requires understanding both the financial troubles of the corporation (discussed in the previous chapter) and several decades of Phelps Dodge labor relations in copper mining and manufacturing. Key individuals are important too. As always, the story of Phelps Dodge Corporation is that of its people, in this case Chairman George Munroe, President Richard Moolick, and the senior vice-president and Western Operations general manager Arthur Kinneberg, three critical players along with several others in the unfolding drama. For them, no less than for Phelps Dodge's production workers, the 1983 strike meant being hammered on the anvil of adversity.

TELEVISION AND NEWSPAPER coverage of events tended to focus public attention on Phelps Dodge spokesmen in Arizona, notably Kinneberg. Rarely did the New York–based leaders Munroe or Moolick appear on the evening news or in print. Throughout the conflict, Moolick remained only a shadowy figure in public consciousness, and that suited him just fine. He much preferred to work behind the scenes. Let protesters march along Park Avenue chanting with their bullhorns, "Hey, hey, ho, ho;

George Munroe has got to go," it was Moolick who probably did more than anyone else to define the company's hard-line position. In fact, adamant that Phelps Dodge must not yield to striking unions on the cost-of-living adjustments, he yet feared at times that even his own chairman might compromise on the issue. "I didn't have any anti-union feeling," recalled Munroe, "and I had a lot of sympathy for the strikers, particularly the Mexican-Americans. The union had done a lot for them." He added, "We had to do what we did, but I didn't relish it." As for his assessment of Moolick: "Dick is very bright . . . and strong minded."[3]

Events of 1983 do not suggest any power struggle among top Phelps Dodge executives, but they do highlight, as perhaps nothing else could, fundamental differences in their management styles and personal outlooks. They offer insight into the human dimension that is a fascinating part of any large corporation's history. Moolick, a navy combat pilot in the Central Pacific during World War II, was ready—even eager—to confront organized labor in a knock-down, drag-out fight, to produce copper in defiance of striking unions, to defend a financially ailing Phelps Dodge. Munroe, also a navy combat veteran of the war in the Pacific, was temperamentally less inclined to confrontation and more to promoting collegiality and building consensus. He had no passion to challenge the unions and expose the company and its employees to the inevitable risks that would follow, but in the circumstances Phelps Dodge faced in 1983, he saw no realistic alternative. So, despite differing temperaments and management

styles, Monroe and Moolick were in agreement on the course Phelps Dodge had to follow.

Kinneberg, a 1945 graduate of the Naval Academy at Annapolis and a solid, handsome man who oversaw Phelps Dodge's Western Operations for the New York–based chairman and president, was effectively caught in the middle. A seasoned executive, he needed to work successfully with both Moolick and Munroe and with top Phelps Dodge personnel in the Southwest, all in addition to responding to spokesmen for organized labor.

Moolick was single-minded in his resolve, so much so that, in early August 1983 when he learned that two Phelps Dodge executives in Arizona agreed to temporarily shut down facilities in strife-torn Morenci and thus appeared to compromise with organized labor, he boarded

a flight from New York swearing to "fire all those bastards." His language was rough, the product of a career path that led to New York's Park Avenue by way of the corporation's big mines in Arizona and New Mexico. After graduation from the University of Arizona's School of Mines, Moolick began his career at Phelps Dodge in 1949 at the Morenci Branch. Starting as a junior geologist, he gradually climbed the management ladder. Following a two-year stint in Douglas as assistant general manager of Western Operations, he went to Phelps Dodge headquarters in New York in 1975 as a vice-president.

Moolick was in the East, but certainly not of it. Even in the heart of Manhattan, he made a point "to saunter" along the city's busy streets "like a good westerner" in conscious defiance of New York's never-stop-moving mentality. That strong streak of stubbornness branded him a tough-minded man in an industry known for its rugged ways. He sincerely believed that only his kind of strong medicine could save Phelps Dodge.[4]

Moolick's concerns for the future of the enterprise were well founded. *Business Week* had observed in mid-1982 that Phelps Dodge Corporation had suffered through a "wave of shutdowns, pay cuts, and firings in the last nine months—all reactions to the continuing losses that are anticipated." It added that the entire copper industry "is in dire straits today, because prices have dropped to their lowest level in real terms since the 1920s. But most of the major producers have found shelter in the arms of rich oil companies and can ride out the current storm protected." Independent Phelps Dodge, it emphasized, "has no security blanket." In Moolick's mind as he flew from New York to Phoenix in August 1983, it was Phelps Dodge against the world—and he was ready to do battle.[5]

Highway to "Heaven in '67"

RELATIONS BETWEEN PHELPS DODGE and its workers had been relatively benign from the end of World War I until the later New Deal years of the 1930s. The corporation's mines, mills, and smelters continued to be nonunionized. Tensions between labor and management surfaced mainly in the fabricating plants of Phelps Dodge Copper Products Corporation, that portion of the enterprise added at the onset of the Great Depression and led for many years by no-nonsense Wylie Brown, a man wholly unafraid to battle tough unions of the greater New York area.

Uncle Sam clamped a lid on labor-management squabbles for the duration of World War II, but the stresses of round-the-clock wartime production when combined with the patriotic desire of Phelps Dodge to avoid a slowdown in production of vital copper and copper products fostered union activity at several of its mine, mill, and smelter facilities. By the time the war ended, Phelps Dodge bargained regularly with many different labor organizations, though the dominant one was the International Union of Mine, Mill and Smelter Workers. That, of course, did not insulate the corporation from the wave of postwar labor unrest that swept through basic industries of the United States,

notably coal and steel, during the second half of the 1940s. Most work stoppages at Phelps Dodge were neither particularly lengthy nor especially bitter, except at the Bayway fabricating facility in northern New Jersey, where disputes during the post–World War II years sometimes flashed into violence. One bloody strike there lasted more than seven months.

The 1950s were generally prosperous years for the copper industry, and Phelps Dodge shared what it regarded as a reasonable portion of its prosperity with its unionized workers, not only in the form of wage increases but also in fringe benefits. In 1956, a year of record earnings, the highlight of the era of good feelings was the three-year agreement Phelps Dodge reached with the Mine-Mill union several months before the contract in place was scheduled to expire. Beneath the surface calm, however, significant rumblings were occurring within organized labor. First of all, at contract time all major United States copper producers dealt not just with one union, but with many. Mine-Mill usually got the headlines. As the self-proclaimed successor to the former Western Federation of Miners, it was militant and highly vocal. Both during and after World War II, Mine-Mill had profited from huge gains in membership.

Despite that, the union was not as all-powerful as either its numbers or combative language suggested. Indeed, at Phelps Dodge several smaller but nonetheless robust craft unions affiliated with the conservative American Federation of Labor held representation rights in job areas relating to construction, maintenance, and operation of large mine equipment. In addition, several separate railway brotherhoods represented employees who hauled the ore by trains in the open-pit mines in Arizona. Mine-Mill leaders recognized that this type of fragmentation diminished the bargaining power of all unions. Accordingly, at Mine-Mill's 1946 annual convention in Denver, union leaders put in place the first part of an ambitious, step-by-step, long-range strategy to negotiate contracts for the entire nonferrous metals industry through a single united labor group, which they hoped the Mine-Mill union would lead.

The road to realization of the union's ambitious 1946 plan, so filled with obstacles in the best of times, grew far more difficult after the Congress of Industrial Organizations (CIO) expelled the Mine-Mill union in February 1950 because of its "consistent, unswerving support of the policies of the Communist Party." To replace the discredited left-wingers, the big labor congress invited the United Steelworkers of America to organize the copper industry. That union, despite its undeniable power, at first attracted few members among mine workers, who remained fiercely loyal to the Mine-Mill union. On the other hand, in the brass mills and other fabricating plants, where the CIO granted the United Auto Workers jurisdiction, raids on Mine-Mill were more successful. With the exception of Kennecott, whose president, Charles R. Cox, had come from the steel business and thus was comfortable dealing with the Steelworkers union, copper management tended to fear United Steelworkers' apparent strength. It thus favored the Mine-Mill union as the lesser of two evils, despite its political tilt to the Left. In any case, through the 1950s, most strikes that resulted from disagreements in the mines were

both peaceable and relatively short, at least in comparison with what happened in 1983.

Before labor-management relations reached that historic impasse, however, several other milestones marked the course of organized labor at Phelps Dodge. In early 1967, the old Mine-Mill union was merged into the United Steelworkers of America. The expanded Steelworkers union was eager to flex its newfound muscle by forging a coalition of twenty-six separate unions that by functioning as a single unit would greatly enhance union power in dealing with the copper companies.

In pursuit of that goal the Steelworkers union early in 1967 assembled the National Nonferrous Industry Coordinating Committee, comprising representatives from all unions involved in that year's bargaining with the major copper producers. The unions hungrily eyed the companies' record profits in 1966, when America's four largest copper producers piled up earnings that exceeded $400 million after taxes, and they formulated their bargaining goals accordingly. Included in those goals was a demand for industrywide bargaining, patterned after the arrangement in the steel industry. The unions defined *industrywide* to mean not only copper mines, smelters, and refineries, but also copper fabricating plants and lead and zinc mines and smelters owned by the copper-producing companies. That may have been one reason industry negotiators labeled labor's demands a quest for "Heaven in '67."

Business Week reported in May of 1967 that "the strongest massed force" of unions that had ever confronted the nonferrous employers was warning that "strikes will occur" if the industry didn't "settle big" in the midyear bargaining. Negotiations stalled as the June 30 contract expiration deadline approached. Workers remained on the job and bargaining continued for two weeks after the contracts expired, but on July 15 some forty-five thousand nonferrous workers from California to New Jersey walked out. For the next eight months, a coalition of twenty-six unions ("a union of unions") shut down most of the nation's copper mines, smelters, and refineries in what was perhaps the longest industrywide strike in United States history and certainly the longest for the copper industry. Yet none of the big copper companies went into the red in 1967, even though their mammoth electric shovels and rotary drills remained silent during almost the entire second half of the year.[6]

Profits at Phelps Dodge plunged 39 percent, from $82 million to $50 million. The lengthy work stoppage idled nearly all of the corporation's mines and related facilities, except for one refinery. There was no violence, however, and at its company stores in Arizona, Phelps Dodge continued to extend credit to striking workers. It even provided them a big Christmas surprise early in December when it announced that it would distribute nearly a million dollars in 1967 accrued vacation pay. The corporation's several fabricating plants, most of them represented by unions other than United Steelworkers, continued business as usual until their contracts expired in early February. As for the ongoing miners' strike, James Beizer, metals analyst for *Iron Age*, an industry trade journal, concluded that Phelps Dodge probably suffered the greatest "shellacking" of the four major producers

because, unlike the others, it had no foreign copper production to offset losses in the United States.[7]

The protracted walkout hit local economies hard, but the nation at large scarcely seemed to notice, much to the disappointment of the union coalition. The main reason the extended shutdown had little significant impact was that manufacturers had earlier stockpiled copper in anticipation of a long strike. Only in mid-March 1968, after President Lyndon Johnson and other federal officials increased pressure on copper companies and striking unions to resolve their eight-month dispute, did Phelps Dodge reach a settlement for all its struck properties. It agreed to raise wages of its Arizona mining employees an average of 5.2 percent a year over the life of the new three-year contract. Given the rate of inflation at the time and the high cost to workers of a strike that had dragged on for so long, the union coalition had not won an impressive victory. One publication concluded that the workers involved had forfeited an average of $5,200 apiece in wage income, a sum that, because of the settlement's modest gains, would take them more than twenty years to recover.

After failing in their grand quest for industrywide negotiations and settlements, unions retreated during the next round in 1971 to much narrower goals limited to copper mining, smelting, and refining. United Steelworkers was used to dealing with the steel industry, but American copper companies were much more subject to the ups and downs of the world commodity market than were steel companies. Further, in the copper industry, unlike the steel industry, each company negotiated with its unions separately because each had substantially different costs and operating circumstances.

Thus, union strategy as it evolved during and after the 1967 bargaining was for the union coalition to hammer out a settlement with the company it regarded as being willing to come closest to meeting labor's terms. Once it had that first agreement in hand, the union coalition would then demand that other copper companies agree to settlements incorporating all the major features of the initial "pattern" settlement. The coalition would strike any holdout company. Displaying little flexibility in tailoring settlements to conditions peculiar to an individual producer, the unions would hold firm until all the companies eventually fell into line.

That was one major cycle of behavior, but there were others that developed in America's copper industry as well. Every three years from 1968 to 1983, negotiations resulted in an impasse followed by a strike. "A strike is never pleasant and I personally wish that we at Phelps Dodge never had to experience such interruptions of our normal routine," lamented Kinneberg in late 1980, a year marred by another long strike at the corporation's Arizona properties.[8]

By then, something of a ritual had developed between labor and management. Union workers treated the work stoppages as extra vacation time; they were confident that their jobs would be waiting for them when negotiators finally resolved the dispute. In fact, the corporation invariably sustained strikers through the lean days of the strike by continuing to extend them credit at Phelps Dodge Mercantile stores and by deferring rent on company-owned housing. During down times, Phelps Dodge

routinely hired outside construction workers to make major repairs to its facilities, and at the El Paso refinery it was often the strikers themselves who did routine maintenance work to ensure that the plant would be ready to resume production once they signed a new contract. After the 1983 strike, one man in El Paso remembered negotiations in years past when the price of copper was good: "You could ask for anything and they would give it to you." At one point the unions gained a holiday celebrating the birthday of Lee Trevino, the acclaimed Texas-born, Mexican American professional golfer, because they already had off all popularly recognized holidays.[9]

Though the basic mode of contract negotiations followed by strikes had not changed since 1967, something that influenced corporate attitudes had. Back when strikers idled America's copper industry for more than eight months—from July 15, 1967, to March 19, 1968, at Phelps Dodge—most of the companies had little or no long-term debt. Phelps Dodge could easily afford to let its copper-producing facilities sit idle and hope that curtailed production would push up the market price for copper. The corporation also continued to pay dividends. But by the late 1970s, America's largest copper producers had heavy debts that required payments, and hence they needed regular cash flow. If workers struck and halted production for long periods of time, the companies still had to pay the interest on their debts. Factor into that simple equation the multimillion-dollar losses that Phelps Dodge suffered in 1982 and early 1983, and it is not surprising that the strike of 1983 did not follow the long-established ritual.

No longer was Phelps Dodge willing or able to wait out a lengthy strike. Yet given the troubled times in America's copper industry and Phelps Dodge's new gospel of efficiency, there was greater need than ever before for speedy and peaceable resolution. At issue for Phelps Dodge management was nothing less than the long-term survival of the corporation—not to mention the thousands of jobs it provided.

Hard Times and Hardening Positions

FROM THE TIME Phelps Dodge temporarily suspended copper production in April 1982 until the end of June 1983, when labor contracts at its Arizona mines and El Paso refinery expired, the corporation's campaign for greater efficiency had resulted in a reduction of more than 10 percent in the cost of producing copper. But Phelps Dodge had to leave untouched the largest single element of its production costs, the wages and benefits of its union-represented employees, because those items were established by the existing labor agreements and could not be changed until those contracts expired. Nonetheless, in their public presentations during the preceding two years, Munroe and his lieutenants had left no doubt that the company was determined to address the issue of wages and benefits in the 1983 bargaining.

The typical Phelps Dodge miner in 1983 earned more than $26,000 a year, received another $10,500 in benefits, and lived in a three-bedroom house rented from the company for as low as $50 a month. For more than half a century, residents of the company's mining towns had regarded Phelps Dodge as something of a Santa Claus. Besides providing them high-paying jobs, it supplied carpenters and painters to maintain their low-rent company houses. Even disregarding the benefit of subsidized housing, a Phelps Dodge miner earned about 50 percent more than the average manufacturing worker in the United States.

Early in 1983, President Moolick met with top officials of other American copper companies to discuss their respective views of the forthcoming negotiations. While all the industry leaders agreed about the need for labor cost relief, there was no consensus on whether they might look to the unions for help. Divisions of opinion ran unusually deep on the issue of negotiating changes in the cost-of-living adjustment (COLA). Moolick declared that future COLA increases must be ended, and he urged copper producers to achieve that goal by standing united on the issue. However, many of his fellow executives were concerned that any attempt to eliminate or modify COLA would surely cause a bruising confrontation with the unions, one that the companies might well lose in the end.

In mid-April 1983, a full ten weeks prior to the scheduled expiration of the labor contracts of the major producers, and before Phelps Dodge had held its first bargaining meeting with the unions, Kennecott and its unions announced that they had reached a settlement. The agreement called for wages to be frozen for the three-year term of the contracts and provided Kennecott a modicum of additional relief in areas where its benefits exceeded those of the other copper companies. Significantly, Kennecott's settlement provided for continuation of COLA unchanged in any respect.

Because of the prevailing ritual of labor relations in nonferrous metals—and because Phelps Dodge did not have access to the money-laden pockets of a major oil company, as Kennecott did with Standard of Ohio, Anaconda with Atlantic Richfield, Cyprus with Standard of Indiana, and Duval Corporation with Pennzoil—Phelps Dodge executives concluded after analyzing the Kennecott settlement that they had only three choices in 1983. First, they could avoid a strike by accepting the pattern settlement, including continuation of COLA. Second, they could decline to duplicate the so-called pattern, take a strike, and shut down mining and related operations until the dispute ended. That was what Phelps Dodge had done previously. From the perspective of Munroe, Moolick, and other top executives, for them to agree to a Kennecott-style settlement meant "slow death" for Phelps Dodge, while to shut down because of a strike meant "rapid death."

A third option was to take a strike and continue to produce copper. Exactly where that road might lead was anybody's guess. There was no modern precedent for it within the copper industry. Recalled Kinneberg: "It was a difficult decision, but one that Phelps Dodge had to make. To close down on the heels of a $78 million loss over the previous 18 months would have been suicidal. It also would have been devastating to agree to another imprudent labor settlement that would have added tens of millions of dollars to Phelps Dodge's cost of producing copper over the next three years."[10]

Prestrike negotiations for Phelps Dodge's Arizona mining operations began in Phoenix in early May. From the outset the union spokesmen took the position that the Kennecott settlement had established the pattern that Phelps Dodge and all of the other companies would be required to duplicate without meaningful variance. For its part, Phelps Dodge repeatedly pointed out to the union negotiators that in order for the company to have a chance of surviving in the intensely competitive world copper industry it needed to contain, or even reduce, the labor component of the cost of producing copper. Jack Ladd, Phelps Dodge's director of labor relations and chief spokesman in the negotiations, said that part of that goal might be accomplished by improving worker efficiency, such as by combining jobs and removing arbitrary barriers to the use of employees. But he made it clear that such measures would not be sufficient. In particular, he stressed that the inexorable increase in wage costs caused by COLA had to be halted. Ladd informed his union counterparts, "This year it's going to be different. We're going to keep working." They dismissed the warning.[11]

On June 13, two and a half weeks before the old contract expired, Ladd suffered a serious accident, and vice president Matthew P. ("Pat") Scanlon took his place at the bargaining table. "Pat had all the 'fun' at Morenci," recalled Ladd, tongue in cheek, of the tense confrontation that erupted there the following August after a month of relative calm. As the deadline approached, Scanlon gave the union negotiating committee Phelps Dodge's "final" economic offer. It called for no wage reductions for employees already on the payroll, and only a few cutbacks in benefits, such as transferring to employees some of the costs of medical care. It also called for rolling existing COLA into wages and making the increase permanent. In return, Phelps Dodge proposed that there be no future cost-of-living adjustments and that a schedule of lower wage rates be applicable to new hires.[12]

The two sides stood miles apart. From the unions' perspective, Phelps Dodge's proposal struck at the hard-won principle of pattern bargaining. As a consequence they gave it little serious consideration, though management repeatedly warned that the corporation intended to work through any strike. Perhaps the unions thought Phelps Dodge was bluffing. Besides, the union coalition was supremely confident: Since its formation in 1967, it had used its hard-line strategy five times without a single defeat. In Phelps Dodge's eyes, on the contrary, the company's offer was a fair one under the circumstances, but union leaders had locked themselves into pattern bargaining; further, labor's behavior seemed to be dictated in part by internal union politics, by a contest for the presidency of the Steelworkers union.

As the end of June approached, three of the major American copper producers—Magma Copper Company, American Smelting and Refining (ASARCO), and Phelps Dodge—were still refusing to match the Kennecott settlement. Magma agreed to meet the pattern on June 29. Late the following day the Federal Mediation Service arranged a meeting between Frank McKee and Robert Petris, representing the union coalition, and John F. Boland, Jr. (a senior partner of the Evans, Kitchel and Jenckes law firm in Phoenix), and Scanlon, for Phelps Dodge. Neither side budged from its prior position.

Once again, Phelps Dodge insisted that it intended to operate its facilities during any strike. That night, just half an hour before the midnight deadline, one more producer, ASARCO, capitulated. Now Phelps Dodge stood alone.

The Great Strike of 1983

When July 1 began, the midnight shift failed to appear, and picket lines went up as some twenty-five hundred members of thirteen separate unions struck Phelps Dodge facilities at El Paso, Morenci, Douglas, Ajo, and Bisbee. Tyrone's employees were on a different contract cycle and thus they remained at work, as did Hidalgo smelter workers, who had never unionized. Coincidentally, that same day the *Wall Street Journal* reported that employees at the United Steelworkers of America's headquarters in Pittsburgh went on strike to demand a 10 percent wage increase.

Moolick was in Phoenix, and after consulting with Kinneberg he ordered immediate implementation of Phelps Dodge's previously formulated plans to work despite the walkout—and despite the unions' reputation for invincibility. At its five struck properties, the corporation put into place the wage-and-benefit package it earlier had offered the union committee, and it continued producing copper as best it could. About fifty union-represented workers crossed picket lines that first day.

In operating its plants in defiance of union strikers, Phelps Dodge employed a tactic used infrequently by the managements of large American companies during the decades after

World War II, although it gained respectability in business circles after President Ronald Reagan replaced unionized air traffic controllers who went on strike in 1981. To say the least, what Phelps Dodge did was highly controversial and generated much criticism, and not from union spokesmen alone. "I was ridiculed by the industry for even trying to buck [the unions]—they said it couldn't be done," recalled Moolick.[13]

One opinion voiced by some outside observers was that the 1983 strike represented a failure of communication, but that claim was groundless because Phelps Dodge repeatedly stated exactly what it intended to do in the event of a strike. Another claim made both during and after the strike was that the corporation's underlying purpose was not to seek a settlement, or even relief from labor costs, but to engage in old-fashioned union busting. There prevailed a "warlike atmosphere," observed Alex Q. Lopez, the chief union negotiator. "Every time the company does something, we suspect there must be a dark and devious reason for it."[14]

Contrary to labor's assertions, neither Moolick nor any other Phelps Dodge leader set out to break the unions, as a close reading of unfolding events will prove. Besides, the nation's copper industry was virtually 100 percent unionized, and every facility of Phelps Dodge was unionized except for the Hidalgo smelter in New Mexico and the Norwich Rod and Wire Plant in Connecticut.

Munroe emphasized back in 1982 that "our battle is not with the principle of unionism" but with socialized foreign competition subsidized by international lending agencies. For most people that was a fairly esoteric distinction. It was hard for Phelps Dodge's cold economic logic to counter fiery gut-level passions stirred by the unions' campaign of intimidation and misinformation. "I went to each of the operations in the West last spring," Munroe later recalled in the *Arizona Republic*. "I held open meetings there with our employees. I tried to explain why we had to take certain steps." But he came away disappointed. "The turnouts were smaller than I had hoped for. One reason for that was the unions. They told the workers that the union leaders would go to the meetings and listen for them."[15]

Midway through the first half of 1983, about when Munroe sought to explain Phelps Dodge's position to its employees, Moolick instructed Kinneberg to have the managers of branch operations prepare to keep producing copper should their unionized workers walk out. With that goal in mind they carefully estimated their available workforces and worked out production schedules, they trained supervisors and other nonunion personnel to operate production equipment, and they made security arrangements to control entry to the plants. They also prepared for the grim possibility of union violence to lives and property, and they had applications for temporary restraining orders drafted in case the pickets got out of hand and threatened to destroy civil order.[16]

Back in late 1980, Kinneberg had stated that "the 'union' and the 'company' are not objects—they are people." Certainly that was never more vividly illustrated than in the grim confrontation of 1983, when "the company" was personified by Munroe, Moolick, and Kinneberg, but not just those three alone. There was, for example, Pat

Scanlon, who was in Palo Alto, California, attending his daughter's graduation from Stanford when he received a telephone call from Kinneberg, who said, "Ladd just fell off his windmill and broke his back. You've got to take over negotiations." Though he had not been actively involved at the bargaining table for more than ten years, when he was chief negotiator for Western Operations, Scanlon caught a flight to Phoenix the next day. A corporation vice-president at that time, he took over until Ladd returned to work in early August in a wheelchair.[17]

Moolick's counterpart in the Steelworkers union was Frank S. McKee, who had served as chief negotiator for the labor coalition since 1974. When it came to personifying such words as *tough*, *abrasive*, *irascible*, and *confrontational*, McKee and Moolick were more alike than either man would admit, both having been early graduates of the school of hard knocks. Moolick recalled three meetings with McKee: "One was in Phoenix at the insistence of federal mediators. In that meeting we discussed golf, because Frank had recently been on the famous Saint Andrews links in Scotland." They held two more meetings in secrecy at a small French restaurant on 57th Street in New York City. "They were arranged by a prominent New York City lawyer—and guess what we discussed—golf." Moolick recalled that "I never even asked him what his handicap was." All three face-to-face encounters were unproductive because neither man budged, but added Moolick with no hint of guile, "I liked the guy."[18]

McKee's union activities dated from 1940 when he was hired at the same Bethlehem Steel plant in Seattle where his father had worked for thirty years. In time the younger McKee became director of District 38 of United Steelworkers on the Pacific Slope. Still later he advanced to secretary-treasurer of the national organization, while continuing to serve as chief negotiator in copper. As a bargainer, McKee early won a reputation as intelligent, resolute, and unyielding, three traits that produced handsome returns for unions during the first years of his tenure but resulted in disaster in 1983.

Of course McKee and his associates were fully aware of the economic distress of the copper industry, for they had watched a long procession of layoffs and closures during the preceding two years. But they saw absolutely no benefit to their members from union concessions to the copper employers. They felt that any special favors for struggling companies would only undermine the entire wage structure of the industry without benefit to overall employment. Why sacrifice a uniform system of wages and benefits that had been built up brick by brick through a series of hard strikes over a long period of time? To Scanlon, on the other hand, the union's bargaining position was primarily political, while Phelps Dodge's was primarily economic. He believed it would have been impossible for McKee to allow Phelps Dodge a more favorable settlement than Kennecott had signed earlier because McKee was in effect campaigning as a tough-as-nails candidate to succeed Lloyd McBride as head of the Steelworkers union.

When the strike began, Phelps Dodge continued to produce copper by scheduling two twelve-hour shifts per day to operate the mines and plants around the clock, seven days a week. The corporation initially cobbled together crews composed of supervisors and other nonunion employees from the struck locations, and union-represented employees who either had been on layoff or who refused to go on strike. Management personnel recruited from Phelps Dodge's nonmining operations and from its headquarters in New York augmented those crews. Secretaries drove huge haulage trucks at Morenci, and New York executives stored their stylish suits and donned work clothes suitable to the heat and grit of the smelters and the El Paso refinery. Two women from the refinery office worked on the loading dock stacking heavy copper cathodes aboard boxcars. Despite the activity in the plant, union spokesmen claimed that the refinery was not actually producing any copper but rather was running the same loaded train around and around to create the illusion of production.

One Park Avenue executive who helped maintain production at Morenci was young Steve Whisler, then age twenty-eight, who earlier had come east to Phelps Dodge headquarters from Western Nuclear in Denver. At the start of 1983 he was serving as assistant corporate secretary and legal counsel to the exploration division. In response to Moolick's call he volunteered for duty in Morenci, a place he had never seen before, but where he remained from early July until the end of September. For twelve hours each day he did a variety of jobs, including changing tires on haulage trucks and pulling hot slabs of copper off the anode wheel. That was a tense time, and having to run the picket-line gauntlet at each shift change "wasn't a very pleasant experience," recalled the future president of Phelps Dodge Corporation with a generous measure of

understatement. Another New Yorker who went west to help produce copper was Renny D. Warren, the corporation's treasurer, who worked in the El Paso refinery as a "hot sheet" man.[19]

Also coming temporarily to the Southwest was Jack Coulter, a Phelps Dodge vice-president. He was a former special agent for the Federal Bureau of Investigation who had learned labor relations as an attorney for Phelps Dodge Copper Products Company. Coulter was a ubiquitous presence whom some Phelps Dodge officials in Western Operations feared as "Moolick's Mole," a man sent out from New York to spy on them. Although Coulter did report directly and usually daily to Moolick, he was one of numerous executives who headed west to help during a very difficult time.

In the early days of the strike, Phelps Dodge sought repeatedly to persuade its regular employees to come back to work. During the first five weeks of the strike it sent out sixteen appeals, some of them written by Chairman Munroe himself. Union leaders were just as assiduous in trying to rally their members to stand firm. "Picket Lines Solid as PD's Back to Work Fizzles" read one headline in the *Phelps Dodge Workers Strike Bulletin 1983* at Morenci for July 6. "Eastern copper barons are once again trying to force their will on working people." McKee claimed that Phelps Dodge had hired "1983-Pinkertons" in an attempt to "bust the unions and impose its will on the workers." The same issue of the union paper contained a sidebar headlined "66th Anniversary of Bisbee Deportation." It included the words, "Historians agree that the Bisbee deportation that affected innocent men stemmed from the demand of the copper companies for greater profits." The *Strike Bulletin* also called the Bisbee deportation "clearly a lesson that must not be forgotten." But just as clearly the union had forgotten its highly charged World War I context.[20]

The *Strike Bulletin* for July 18, 1983, featured a poem attributed to the popular American author Jack London. Its opening words were, "After God had finished the rattlesnake, the toad, the vampire, He had some awful substance left with which He made a scab." When Phelps Dodge offered a $25,000 reward for information on damage caused at the Mercantile in Ajo, each company poster was matched by an anonymous reprint of the full London diatribe under a banner headline that purported to offer "A $25,000.00 Reward for the Disappearance of a Scab." Police investigated but never made any arrests.[21]

The unions' heavy-handed tactics and uncompromising stance alienated some of their own members and split the ranks of strikers. Eventually, about two-fifths of the union members returned to work. The bitter tug-of-war was especially hard on the many Hispanic employees of Phelps Dodge. In remote mining towns in Arizona, the meaning of union was associated with family heritage, ethnic pride, and years of struggle. Some unions, although not all, had fought

---

ONE OF A SERIES *of advertisements that Phelps Dodge published in newspapers throughout Arizona to help the public better understand the facts underlying the 1983 strike and the package of wages and benefits the corporation offered its workers. Phelps Dodge believed that the advertisements had an important positive influence on public opinion.*

various forms of racial and ethnic discrimination since their inception, and that was a primary reason Mexican Americans in mining had signed up. Many were very loyal to organized labor because they believed their unions had made them first-class citizens. The 1983 strike thus sharply divided the mining and smelting communities involved, often pitting brother against brother and neighbor against neighbor.

Many second- and third-generation Phelps Dodge employees had never worked for anyone else, and some of them considered subsidized housing, medical care, and numerous other fringe benefits to be entitlements. Without question, employment in a Phelps Dodge town meant more than a job. "It was a way of life," emphasized Leonard R. Judd, a onetime member of the hod carriers union, who served as president and chief operating officer of Phelps Dodge Corporation from 1989 to 1991. "If you lose a job in Morenci, you lose a company house, you will move, your whole lifestyle changes. A move to the city meant no more hunting and fishing."[22]

"The unions led them into an unrealistic position that was damaging to people once they were not working for Phelps Dodge," added Munroe. Of course, unions will dispute any assertion by management to that effect, but there was simply no arguing the fact that some of the workers who eventually crossed picket lines had formerly been loyal union members who concluded that labor intransigence would cost them their jobs. That was one reason tensions mounted with each passing day. Said one worker who crossed the picket line to feed his three children, "I'm happy—except at 6 o'clock when they yell at me." And yell the pickets did, lining

both sides of the highway as one caravan of cars headed in the plant gates as another one drove out. Their message was always the same: "Scab, scab, —— you, scabby-boy! You're a ——ing scab!" With stony faces staring straight ahead, non-striking workers had to listen as a protester publicly announced their names with a bullhorn. Deputy sheriffs and Department of Public Safety officers silently watched the twice-daily ordeal from positions nearby.[23]

In response to the ongoing harassment, Phelps Dodge issued "Procedures for Crossing a Picket Line" that advised non-striking employees to approach and cross cautiously in an automobile, at a slow rate of speed, with windows rolled up and doors locked, and to avoid any conversation with pickets. Above all, "Do not carry anything in the vehicle which may be perceived to be a weapon."[24]

Phelps Dodge obtained court orders, which were routinely ignored, limiting the number of pickets outside each of its Arizona facilities. Further, the state highway department built a berm along roads leading into the Morenci complex in an unsuccessful attempt to prevent anyone from parking near Phelps Dodge property. Greenlee County supervisors imposed a 9 P.M.–to–6 A.M. curfew, but members of the Clifton City Council, which was pro-striker, instructed city police not to enforce it. At first most pickets seemed satisfied simply to vent their rage on nonstriking workers by denouncing them as scabs or by making obscene gestures, even the women and children. One striker's wife displayed her babe in arms with its middle fingers specially taped to form an obscene gesture. Moolick assigned men to preserve everything on video, as had been company policy from the time the new technology came into vogue in the early 1970s. Phelps Dodge security people didn't carry firearms, only cameras, which in later legal proceedings proved far more effective than any weapons.

Along with mass picketing came growing resort to illegal activity by hard-core strikers, their families, and outside agitators. Since it was against the law for strikers to threaten publicly those who chose to work (although many of them did so anyway), apparently some of them did it anonymously over the phone. Six thugs with baseball bats beat in one worker's car and broke his house windows. A striker surreptitiously scattered roofing nails along the road taken by buses that ferried employees to work at Morenci. Sheriff's deputies, after picking up the nails, went to the local hardware store to ask the proprietor if he remembered anyone making a large purchase of roofing nails. "Yes, just yesterday," he replied. Did he know the purchaser? "No, I don't. But wait—he paid for them with his credit card!" At about that time Governor Bruce Babbitt ordered extra Department of Public Safety officers to the Clifton area to maintain order, though strikers claimed that the officers acted like Phelps Dodge guards and seemed more interested in protecting strikebreakers than in keeping the peace.[25]

At the El Paso refinery, city police maintained order and prevented most picket-line violence. Early public support for strikers diminished after local newspapers reported how much more strikers earned than the typical worker in that border metropolis and after a feature story detailed how strikers in mid-August viciously attacked a police dog named JoJo with steel rebar, two-by-fours, and baseball bats. So severely injured was the six-year-old German shepherd that he was never able to return to work. One striker who beat JoJo was fined $200, the maximum penalty for injury to a police dog.

In remote Ajo on the night of July 27 someone fired a gun at the home of Keith Tallant. The .22-caliber bullet passed through two walls and a fragment lodged in the brain of Tallant's sleeping three-year-old daughter, Chandra. Her father had been among the first workers to cross the picket line, and thereafter his family suffered incessant public harassment and threatening phone calls. The unions denied all responsibility, but in the wake of Chandra's shooting, Arizona's major daily newspapers published vehement antiunion editorials, and most of them blamed the tragedy on the inhumanity of the strikers. Phelps Dodge offered a $100,000 reward for the apprehension and conviction of the person or persons who shot Chandra Tallant.

There was growing concern that violence would continue to escalate in the isolated mining communities where there was no substantial law force apart from local sheriff's deputies and scattered officers of Arizona's Department of Public Safety. Said one replacement worker, "It's hard to sleep—I wake up at every noise. I have to leave my wife there; I taught her to shoot the gun, and that's all I can do." No one was quite certain how the strike would end.[26]

The shooting of little Chandra Tallant gave Governor Babbitt the rationale he needed to step up civil protection in strife-torn areas. After visiting the convalescing child at her hospital bed in Phoenix, he promised her mother that he would take action to prevent additional injuries or death, a point he made clear also to AFL-CIO officials who later questioned his response. Chandra survived, although her 25-year-old father died a short time later in an automobile accident when his car mysteriously ran off the road. Even as Tallant's coffin left the church, a noisy group of strikers kept yelling "scab!" Police investigators concluded that the young man was not the victim of foul play.

From the beginning of the strike, people on both sides of the dispute wondered whether Phelps Dodge's salaried personnel, its assorted supervisors, engineers, geologists, metallurgists, clerks, and typists, working together with a small but ever-growing cadre of union employees, could actually produce significant quantities of copper. The company reported that Ajo output for the month of July was 10 percent above the normal, nonstrike budgeted goal; Morenci mine production was 88 percent of capacity; and combined production in July for all Phelps Dodge copper smelters was 72 percent of projection.

The unions charged that Phelps Dodge was running Morenci without proper maintenance, but the corporation claimed it could train anybody to do two-thirds of the jobs in an open-pit mine in two weeks. What it could not do was keep its workers from growing tired. By early August the demanding production schedule left its Morenci force of some nine hundred men and women close to exhaustion. That was one reason Phelps Dodge stepped up its appeals to striking workers to return to their jobs.

Simon Peru, a shovel-and-drill repairman and crane operator at Morenci, stayed out five weeks. It was only when union officials advised him that he was ineligible to receive strike benefits because he had some money in the bank that he got angry and returned to work. He concluded that the Machinists union had been quick to take his dues but slow to provide benefits, even a few dollars to pay his utility bills. He joined the rising chorus of those who concluded that "we were wrong" to strike.[27]

On August 5, about five weeks after the strike began, Phelps Dodge announced that it intended to hire new employees to fill out its complement of workers at the struck facilities. The decision to replace permanently all strikers who chose not to return to work proved to be a major turning point in the dispute. The potential for a truly bloody confrontation, which had been present from the start of the walkout, now began to escalate ominously. Pickets used bullhorns to shout their threats and obscenities. There were open threats to kill nonstrikers and their families, to seize the plants, and to evict or kill those working inside. At the same time it appeared increasingly obvious to Phelps Dodge that sheriff's deputies and local courts were either unable or unwilling to maintain civil order in Morenci, Douglas, and Ajo.

Stepping Back from the Precipice of Violence

INTIMIDATION AND ACTS of violence increased alarmingly in the Morenci area. An automobile caravan of more than a hundred militants threw rocks through the windows of non-striking workers' homes in the adjoining town of Clifton. On Monday, August 8, the Greenlee County sheriff, Robert Gomez, asked Governor Babbitt for a contingent of National Guard troops because of the mounting commotion and highly vocal threats of violence outside the entrance to the Morenci mine and plant. Later that afternoon, the intimations of disaster caused Babbitt to fly to Clifton. He landed in a thunderstorm both literally and figuratively. In the county courthouse he shuttled back and forth between labor representatives in one room and management in another. His message to each side was to take one step back from the "precipice of violence."

After several hours of discussion, Scanlon for Phelps Dodge agreed to a ten-day freeze on hiring replacement workers; the union leaders committed to attempting to control the mobs at the mine and plant gates and to secure free and safe access to persons working therein; and both sides promised to return to the bargaining table under the auspices of the Federal Mediation Service. Early the following morning Babbitt flew back to Phoenix, where later that day he reported the substance of the agreement to television viewers. The governor stressed to listeners that Phelps Dodge was "a company that by any measure is up against the wall," and he deplored the "continuing violence and intimidation" by strikers in Morenci.[28]

On Tuesday morning, August 9, almost as Governor Babbitt spoke, Scanlon, the Morenci Branch manager, John Bolles, and a few other Phelps Dodge representatives were in the administration building a quarter mile from the main plant gates at Morenci. From inside the building they could hear the roar of an angry

"OH, WE MAY BUST UP YER HOUSE A BIT, SHOOT YER KIDS, FIREBOMB YER CAR, SPIT IN YER FACE, MAYBE EVEN KILL YA OR SOMETHIN'---BUT STAND IN THE WAY OF YER RIGHT TO WORK? NEVER!"

crowd composed of as many as two thousand strikers and supporters who had massed outside the gates. Many carried baseball bats, clubs, chains, and similar weapons. Guns could be seen in a number of the vehicles parked alongside U.S. Route 666. At approximately 9:30 A.M. the top Department of Public Safety officers on site came to the administration building and advised the company representatives that an assault by the mob appeared imminent. Major Ernie Johnson, in command of the force of approximately 130 Department of Public Safety officers guarding the company's property, said that he doubted that his men could repulse a charge by the armed mob. "We'll take a lot of them with us," he said, "but eventually they'll get us and they'll get through."[29]

At about 10:00 A.M., international and local

The 1983 strike was perceived by many nonunion observers as anything but a David and Goliath contest between powerful Phelps Dodge and weak organized labor, as this Steve Benson cartoon illustrates. The image from Phoenix's Arizona Republic *captures well the problems replacement workers faced when they crossed union picket lines. In all, from the end of June 1983 until the end of April 1984, Phelps Dodge's "Log of Incidents Reported" detailed more than a thousand cases of harassment and vandalism in the Morenci area alone.*

union representatives arrived to warn Scanlon, Bolles, and the others that they had lost control of their own people, some of whom had remained up all night drinking. The union leaders stated that they spoke not on behalf of the unions but only as bearers of a message from the mob leaders. The message was that if Phelps Dodge didn't shut its Morenci facilities down by that noon for a period of ten days, "they will go into the plant, mine, and administration building and drag everyone out and if that means killing them, they will kill them." Armed strikers in automobiles were keeping in radio contact with mob leaders, the spokesmen warned; if the mob stormed the gates the automobile squads would go to the homes of employees who were working and kill their spouses and children. People on both sides talked of a "blood bath."[30]

Scanlon, who graduated from high school in the old coal camp of Dawson, where his father was office manager of the local Phelps Dodge Mercantile, and who attended Harvard University on a scholarship, later recalled that moment with unusual clarity—and he chose his words carefully: "I was scared shitless." He knew all too well that the mob was perfectly capable of carrying out its death threats, that many lives might be lost that day.[31]

Neither Scanlon nor Bolles had any previous experience with a situation this desperate, and they did not want to risk causing the loss of lives. Scanlon sought to reach Moolick in New York by telephone but was unable to do so. He then called Kinneberg in Phoenix, who at that moment was on his way to a meeting with Governor Babbitt, who had returned earlier that morning from his mediation efforts in Clifton.

Kinneberg, like Moolick, had worked up the ranks of the mining industry and along the way gained considerable practical experience. Starting in 1948 at Kennecott's open-pit copper mine in McGill, Nevada, he moved on to Chile, where he spent seven years working in smelter operations. As an executive at Phelps Dodge, he was tough, as was the popular reputation of mining men, but without being abrasive. That was one asset in his favor when he met with Governor Babbitt. But before they could talk together, Kinneberg received an urgent message to return immediately to his office.

Scanlon and Bolles were on the phone from Morenci, and both men urged a shutdown of operations there at noon—or people would be killed. Kinneberg could tell over the phone that tensions were rapidly rising in Morenci, a fact Jack Coulter listening in on the speaker phone in Phoenix confirmed. Kinneberg asked for a few minutes' time. With Moolick still out of his office, apparently Kinneberg and Scanlon reached Munroe in New York at nearly the same moment, about 10:30 A.M. Arizona time. When Scanlon described the circumstances to him, Munroe said, "You're there, you can best judge the situation, you make the decision." To Kinneberg a few moments later, Munroe responded to the idea of a shutdown in the face of threatened mob violence with the words, "You've got to do it."[32]

Scanlon, with support from Kinneberg, hastily typed a memo stating that the facility would be shut down for a ten-day "cooling-off period." At approximately fifteen minutes before noon the union representatives left the administration building with copies of the memo to be distributed to the mob and read to them over a bullhorn. Workers inside began shutting down the plant. Strikers thought they had won a big victory.[33]

When Moolick learned of the ten-day suspension of operations at Morenci he was enraged, and the next morning when he boarded a plane for Arizona he fully intended to fire Scanlon, Bolles, and even Kinneberg "if he stuck his head around the door." The latter threat was probably pure bluster, but it showed how angry Moolick was. Scanlon's hands visibly shook and his voice quavered when he met him at Phoenix's Sky Harbor Airport, as both men recollected. "He was convinced that people would have been killed without the shutdown," observed Moolick. "I told him he was frightened by Steelworker rhetoric." It wasn't the rhetoric that scared Scanlon, he later recalled, but what he knew had nearly happened—or what might yet happen—in Morenci. He was scared too for managers who would bear the brunt of Moolick's wrath.[34]

The last thing Scanlon said to Moolick before the Phelps Dodge president and the vice-president Jack Coulter boarded a chartered plane for the hour's flight from Phoenix to Clifton was, "Dick, please do not jump to any conclusions until you talk to the people who were actually there. You can't know how bad it was unless you were actually there."[35]

Moolick did not fire Scanlon or anyone else that Wednesday. He had changed his mind about halfway to Arizona, concluding to himself, "That's stupid. It doesn't make sense." Later he admitted that Scanlon did the right thing: "I was lucky I wasn't in" when he called New York. Scanlon "is a very bright guy." In retrospect, mused Moolick, "it worked out beautifully." The

unions during the cooling-off period redirected their anger from Phelps Dodge to Governor Babbitt.[36]

That same day, August 10, when the Phelps Dodge president reached Clifton, Sheriff Gomez, an old acquaintance, picked him up at the airport. "Nobody will ever shut down Morenci again unless it's me," declared Moolick, a point he reemphasized when he met with his Morenci staff a short time later. "For the most part they were a strong crowd, although some were obviously frightened. We had no sooner started our meeting than we received a bomb threat—a bomb set to go off in fifteen minutes. I told the group we had twelve minutes to talk before we had to vacate the building."[37]

Phelps Dodge pledged to compensate all Morenci employees who were unable to work because of the ten-day shutdown. But Moolick, meeting with Scanlon and the governor later that day, declared that when the cooling-off period ended, operations would resume. Scanlon, still distraught by the near bloodshed in Morenci, had tears in his eyes as he pleaded with Babbitt, "Governor, you've got to do something. Declare martial law, send in the National Guard. Because if you don't, a lot of people are going to get killed over there." Babbitt refused to say at the time what, if anything, he would do when the shutdown ended.[38]

On August 15, Phelps Dodge mailed eviction notices to Morenci strikers fired for misconduct on the picket lines and gave them ten days to vacate company-owned housing. Two hundred replacement workers marched on the Arizona Capitol that same day to urge the governor to send National Guard troops to protect them after

STOP UNION VIOLENCE

Rabbit·Babbit HOPS FOR UNION Thugs could BUY

PROTECT Our FAMILIES!!

HOW MANY MORE Kids Will be SHOT

⟳ DISPLAYING SIGNS *that speak to the tenseness of the times, families of Phelps Dodge workers petitioned Arizona's Governor Bruce Babbitt to provide them protection from "union thugs."*

work resumed. A nonunion boilermaker from the Morenci plant dramatized their plea by dumping baseball bats, a logging chain, and a sledge hammer on the capitol floor, all items pickets had supposedly used to threaten replacement workers. At the same time, Phelps Dodge again made it clear to Babbitt that it intended to resume operations as soon as the truce expired.

ON AUGUST 17, as the end of the ten-day cooling-off period neared and as pressure mounted for him to maintain order in Clifton and Morenci, Arizona's governor activated seven National Guard units to provide logistical and

tactical support to state police and local law enforcement officials in the troubled area. The massive show of military muscle two days later was almost without precedent in the state's history. In fact, except for a strike in 1916 when Governor George W. P. Hunt dispatched 450 National Guard troops to Clifton, it was the largest deployment of state military personnel in all Arizona history. In total, some 325 national guardsmen traveled to the Clifton-Morenci area to back up 425 Department of Public Safety officers, a number that included over one-half of Arizona's state police force. Some Clifton residents felt as if their town had been invaded. One news commentator likened the scene to the war in Vietnam but, referring to the badly divided community, observed that Americans used to be on the same side.

From that day on, the unions directed a large part of their anger toward the Department of Public Safety and Governor Babbitt. Because he appeared to the unions to have sided with Phelps Dodge, they denounced the once-rising star in the Democratic party as "Governor Scabbitt," a term of opprobrium apparently first coined by the head union negotiator McKee. Union spokesmen claimed that Phelps Dodge had "just knighted another Arizona governor to their order of the copper collar." But when they accused Babbitt of being in the company's back pocket, he retorted, "I'm in the back pocket of the American judicial system." Actually, on several occasions Babbitt sharply criticized Phelps Dodge management for its uncompromising stand during negotiations.[39]

This much seems clear: the ten-day cooling-off period ending with the governor's show of force turned the tide in favor of Phelps Dodge. "It swayed public opinion back to the side of the company," recalled James L. Madson, general superintendent, operations, at Morenci during the strike. Thus protected by Arizona national guardsmen and Department of Public Safety officers, Phelps Dodge reopened the main gates at Morenci shortly before 6 A.M. on August 20 and resumed production as promised and without incident. Smoke rising from the tall stack signified to all that work was under way again.[40]

Phelps Dodge estimated that 35 percent of its normal workforce returned. Immediately after the resumption of operations the corporation again began hiring additional replacement workers. It gave each new employee a written statement, signed by a Phelps Dodge representative, affirming that he or she was a permanent replacement and would not be laid off in order to make a place for a striker who wished to return to work. By this and other means, Phelps Dodge quickly bulked up its workforce to the desired level.

At the time it should have been clear to all but the most diehard union members that they had lost the struggle. Officially the strike lasted

---

⟨ PAINFUL MEMORIES ⟩

IT WAS ROUGH," *recalled Simon Peru during an interview early in 1997. Family members permanently separated with an exchange of insults, and high school classes split into contending factions as children of strikers cursed the children of non-striking workers as "scab daughters," or worse. Of the infamous 1983 Morenci strike, Peru's wife, Velia, remembered, "I dreaded going to the store. People would follow you around and cuss at you." Union officials supposedly stalked former members who dared to cross the picket line, and workers' wives were nervous whenever their husbands left the house. "I don't wish this on anybody," Velia stressed.*

More than one worker packed his most precious belongings in a suitcase and left it standing just inside the front door for a fast getaway in case someone firebombed his home. James Madson, former Phelps Dodge Mining Company executive vice-president, sent his family out of town during a couple of especially tense times. His father-in-law, a strong union man who retired before the 1983 strike, "still will not talk to me or my wife." Or so it remained until a family crisis changed things in 1998, but the two men never discussed the strike.

(Based on interviews with Simon and Velia Peru, Safford, Arizona, January 5, 1997; and James L. Madson, Phoenix, Arizona, January 2, 1997.)

---

twenty months, but in the eyes of Phelps Dodge management the walkout was effectively over by the end of August 1983. That was when Arizona guardsmen, who had set up a tent city near the struck plant in Morenci, considered the area calm enough for them to move out and return home.

Management and labor held occasional but always fruitless talks during the remainder of 1983, but by the time the dispute dragged into the next year it consisted mainly of a legacy of bitterness and sporadic violence. In the end, the amazing thing was that no one was killed in any of the several outbursts of violence that attended the 1983 strike and its aftermath. Phelps Dodge assumed the cost of repairing striker damage to employees' cars. As for the unions, said one former member who decided to return to work in defiance of labor leadership, "They're obsolete."[41]

Within a year after the strike began, Phelps Dodge's Arizona labor force was composed of twenty-two hundred hourly paid workers, of whom 58 percent had been employees before the strike and 42 percent were new hires. The corporation placed all strikers, except for the 208 it discharged for strike-related misconduct, on preferential rehiring lists in accordance with the rules of the National Labor Relations Board, to be offered reemployment as vacancies occurred. Any striker who was rehired reacquired his or her prestrike seniority. But many would never again work for Phelps Dodge.

Picket lines remained in place, for the walkout officially continued for two more years, but after July 1, 1984 (copper production having returned to normal levels), non-striking workers began circulating petitions calling for decertification of all unions at Phelps Dodge's Arizona and Texas properties. The unions lost every one of the elections that eventually took place. The result was the largest decertification of labor unions in United States history (thirty-five locals of thirteen international unions in Arizona, New Mexico, and Texas). At its copper facilities across the Southwest, Phelps Dodge became a nonunion company for the first time since World War II.

Upset by the losses, United Steelworkers urged other unions to continue the fight against Phelps Dodge by withdrawing pension funds from banks that had backed the company. Another way the union coalition sought to undermine Phelps Dodge was by badgering it on environmental issues. A prime example of that tactic took place at the Douglas Reduction Works. At a public hearing held in 1982, the Steelworkers union had voiced strong support for granting legislative waivers to extend the life of the smelter, but organized labor changed its mind after the strike. Seeking to cause economic harm to the corporation and its non-union workers at Douglas, the United Steelworkers joined environmentalists to pressure the Environmental Protection Agency to close the smelter earlier than mandated.

In retrospect, note the authors of *Capital and Labor in American Copper*, various factors explain the unions' failure. Once government intervention had restored order and thus ended most illegal activities, it became obvious the strike lacked popular backing, and labor's cause could not be sustained by peaceful persuasion alone. Only recently, many Phelps Dodge employees had been on layoff for months because of low copper prices. Having been recalled to work just weeks before the strike began, few of them, apart from union militants, had any real interest in being idle once more.

Coalition leaders expected that their past history of lengthy but ultimately successful strikes would be repeated. In 1983 they gambled and lost everything—and not just at Phelps Dodge. The nation's other big copper producers took note of labor's weakness. They realized that if Phelps Dodge could break a strike in prounion Clifton and Morenci other companies could break one anywhere, and that realization further weakened organized labor in the copper industry. That was an ominous trend for unions, which claimed thirty thousand other dues-paying copper workers. In fact, Arizona has experienced no major copper strikes since 1983. Over in New Mexico, the labor contracts at Tyrone were slated for renewal in 1984. In the wake of the 1983 strike, Phelps Dodge abolished cost-of-living adjustments and adopted a two-tier wage system at Tyrone, and unions there had no will to fight back.

In mid-1984 the Steelworkers union held a presidential election to pick a successor to the late Lloyd McBride. Frank McKee waged a bitter contest with Lynn Williams, but when all the votes were counted, Williams emerged the winner. McKee stepped down from his union offices, including chairmanship of the copper coalition, the following November. To succeed him, Williams selected Edgar H. Ball, Jr., the international secretary of the union, who had

originally achieved prominence in 1967 as the union's negotiator at the Phelps Dodge telephone cable plant in Fordyce, Arkansas.

Long retired from Phelps Dodge, Moolick fifteen years after the big strike was chairman of the National Mining Hall of Fame and Museum, which he helped launch in Leadville, Colorado. During our visit together in the fall of 1997 he was by turns amiable and combative on the subject of organized labor and Phelps Dodge. "Time was, big unions were considered invincible," he reflected. "We demonstrated that nobody was invincible." Or indispensable. Moolick himself "chose" to retire at the end of 1984. The main reasons for his abruptly stepping down as president of Phelps Dodge are part of the corporation-saving drama discussed at length in the following chapter.[42]

Suffice it to say here that Moolick became a controversial figure around the corporation, and there are many different opinions of his role in the strike. "I made Phelps Dodge lean and mean, having to break the unions to accomplish this, and getting very little help from New York in getting the job done," he now claims. Many former colleagues wince at such tough talk, which a few dismiss as just another "Moolickism." They much prefer to speak of all the teamwork that contributed to Phelps Dodge's successes back in the dark days of 1983 and 1984. Yet, looking back from his own unique perspective, the current corporation president, Steve Whisler, who participated in the daily struggle at Morenci during the summer of 1983, gives Moolick his due. "Dick provided the backbone that was necessary in those kinds of situations."

L. William Seidman, once Moolick's fellow vice-chairman, describes the fate of his colleague as "a human tragedy because he had saved the company, in my view." The leadership controversy surrounding Moolick will probably never be resolved.[43]

WHEN THE WALKOUT began on July 1, 1983, some financial analysts doubted that Phelps Dodge could survive—at least not for long—because of financial troubles that began with the onset of the copper crisis. By late August, rumors circulated up and down Wall Street that the corporation was edging toward bankruptcy. That scenario is a bit too dramatic, claims the retired chairman George Munroe, who notes with more than a hint of pride that Phelps Dodge rallied to confound all the hand-wringing prognosticators. True, it contained its labor costs by winning the bitter strike, but that is only one part of the full story of how Phelps Dodge battled back from the brink.

Chapter Fourteen

"FIX IT, SELL IT, *or* SHOOT IT"

I N 1981 AND 1982, well before the worst of the copper crisis, Phelps Dodge relocated some of its Arizona office personnel to Phoenix. Three and one-half floors of leased space in a twin-towered, twenty-story skyscraper provided a much more visible and centralized location for Western Operations than had the buildings in Douglas, where it had been headquartered since 1916. Further, the move to the state's main metropolis offered good airline connections to the rest of the United States and an opportunity for Phelps Dodge to respond positively to Arizona's rapidly changing political and demographic landscape.

As population increasingly gravitated to Phoenix and Tucson after World War II, the Phelps Dodge mine, mill, and smelter facilities located in remote and sparsely populated counties lost political influence, though not until after the Supreme Court's *Baker v. Carr* ("one man, one vote") decision in 1962 began shifting power to the cities was that fact fully clear. Environmental issues grew so politicized during the 1970s that Phelps Dodge realized it needed a bigger presence in Phoenix and Maricopa County, location of nearly 60 percent of Arizona's population in 1980 and center of the

---

PHELPS DODGE CORPORATION'S *high-rise headquarters in Phoenix. In the early 1980s the building was intended to house only its western regional offices, but in 1987 the main administrative offices were relocated to Phoenix after 150 years in New York City. A wide-angle lens tilted back gives the building a soaring appearance.*

state's government. At times it seemed that the senior vice-president and general manager John Lentz spent half his time commuting between Douglas and Phoenix to attend to political matters.

From the beginning the Phoenix offices contained a "Board of Directors' Meeting Room," although in the early 1980s no one in Arizona could expect that 2600 North Central would become the corporation's home address after a century and a half in New York City. But with relocation of Phelps Dodge headquarters in 1987, the copper-trimmed high-rise soaring above Phoenix, a metropolis named for the bird in Greek mythology that rose from its ashes, seemed to symbolize the return of the company to robust health after a near-death experience in the early part of the decade.

"Simply Going through a Meat Grinder"

T HE YEARS 1983 and 1984 saw the American economy rebound from recession, but for the nation's copper producers, the times remained exceedingly tough. Phelps Dodge suffered a net loss of $63.5 million in 1983—only a slight improvement over the flood of red ink the previous year—and reported a net loss of $5 million for the first quarter of 1984 as copper prices dipped even lower than a year earlier, from 73.9 cents to 64.8 cents a pound on the New York Commodity Exchange. One thing that kept the 1984 shortfall from being even worse was $25 million that Phelps Dodge gained from the Washington Public Power Supply System early

that year to settle a dispute over a terminated uranium contract.

Approximately every fifty years since its founding, Phelps Dodge had undergone a period of fundamental structural change. Each time it emerged a stronger, more energized company. The most astonishing restructuring and turnaround since Anson Phelps launched the company 150 years earlier took place during the first half of the 1980s. The difficult work of reinventing Phelps Dodge began in earnest in mid-1983 when executives began pruning underperforming subsidiaries among its several fabricating businesses, some of which it had acquired back in the early 1930s.

The work of negotiating asset sales became the responsibility of G. Robert Durham, at that time a senior vice-president. He worked closely with the New York law firm of Debevoise, Plimpton, Lyons and Gates to dispose of the corporation's interest in CONALCO, the aluminum firm, in 1980. That turned out to be only a warm-up for the sale three years later of several major businesses in its copper-fabricating sector. During the twelve-month interval between its 1983 and 1984 annual meetings, the corporation gained some $144 million from sales of Phelps Dodge Brass Company, Phelps Dodge Cable and Wire Company, Phelps Dodge Communications Company, and Phelps Dodge Solar Enterprises (the latter for slightly more than $1 million). Since those operations were making little or no profit at the time, their sale had no significant impact on corporate earnings, but gone from the familiar landscapes of production were numerous fabricating facilities, including big telephone cable plants located in

Elizabethtown, Kentucky, and Fordyce, Arkansas; antenna and broadcast products plants in New Jersey and Arizona; a high-voltage research laboratory in Yonkers; and a small laboratory in Elmsford, New York, devoted to fiber-optic telecommunications research. For a time, all that remained of Phelps Dodge Industries were the International and Magnet Wire subsidiaries. That was because Magnet Wire was a leader in its industry, a good copper user, and most important of all, a solid money-maker.

The corporation still had many good assets, but it could not continue indefinitely to sell them off one after another just to fund current operations. That would be like burning the superstructure of a steamship one board at a time to maintain fire in the boiler and just enough pressure to keep the vessel idling along. "We'd bring in this $60 million or that $100 million," Durham later recalled, but red ink continued to stain the corporation's ledgers. Phelps Dodge also raised $72.5 million through sale of 2.5 million shares of stock in a public offering, and some of that infusion of new money went toward reducing Phelps Dodge's burden of debt. But it grew increasingly clear that neither stock sales nor the restructuring done to date would address the fundamental problem the corporation faced during the early 1980s.[1]

Primarily, that was the continuing low price of copper. The most aggravating thing, according to the chairman, George B. Munroe, was not the flux and flow of copper prices, to which Phelps Dodge had long ago learned to adjust. "The problem has been the ignoring of market forces during recessionary periods by the relative newcomers to our industry—the socialized mines of the less-developed countries overseas—and their abuse of our markets to unload their excess production."[2]

That was why in January 1984, Phelps Dodge joined with other domestic copper producers to seek temporary tariff protection from foreign imports. The United States Tariff Commission strongly recommended relief for the industry, but President Ronald Reagan rejected the proposal. The corporation also lobbied Congress and federal policy makers to advise them that loans by the World Bank and the International Monetary Fund to copper-producing nations in South America and Africa inflicted great damage on Phelps Dodge and the domestic industry.

Meanwhile, on July 25 the corporation reported a net loss of $25.7 million for the first half of 1984. If that ominous trend continued for the rest of the year, it would represent at best only a slight improvement over the multimillion-dollar losses of the previous two years. At the time the corporation continued to derive nearly 90 percent of its annual revenues from copper and copper products, a far higher percentage than any of its major American competitors, and though world copper inventories declined rapidly during 1984, the all-important price of copper still showed no sign of a sustained upturn.

It gradually grew evident that a fundamental change had overtaken the industry: once familiar cycles were no longer predictable, and once familiar indicators of future trends had become obscured. For example, consumers resumed purchases of big-ticket items that utilized significant amounts of copper, and that should have been good news for Phelps Dodge, but in 1984 something unusual happened to the price of copper. Typically it followed the general upswing in business activity, but this time it remained low despite renewed prosperity and near record world consumption of copper. Could anyone say with real confidence when the price of copper would bounce back? In the long run it was certain to rise. The question was, how much longer could Phelps Dodge wait? During 1984 the corporation lost money virtually every business day.

Munroe himself may have been confident that Phelps Dodge was not on the brink of bankruptcy, but many of his lieutenants were not so certain. On one occasion Edson L. Foster, vice-president and former corporation controller, confided to Richard Pendleton, Jr., a senior vice-president, that at the present rate he felt certain the corporation would go under. Such dire forebodings filtered through the ranks of employees. "I remember every day coming to work and wondering if that was the day it was going to fall," reflected Richard Rice, then a top-level engineer at Western Operations.[3]

"It was a very gloomy period," recalled Durham. It got even gloomier when Douglas Yearley, shortly before he became part of the senior management team, presented the results of his careful analysis of copper prices. Yearley, then a vice-president, at his own expense and on his own initiative (but with Munroe's blessing) had rented an apartment in Manhattan during the summer of 1984 and spent his after-work hours immersed in study. Only rarely did he go home to New Jersey for the weekends. What inspired such intense commitment? "Self preservation," he now states without a moment's hesitation. "Twenty-five years with a corporation that's about to disappear. My instincts were telling me that it was bad, but I didn't have hard data."[4]

Personal experience and the long hours he spent in study at the New York Public Library shaped what Yearley wrote. "Beyond 1985, the outlook for the copper price is murky and troubled. It is hard to be optimistic," he observed in the confidential report he gave to Munroe in August. Phelps Dodge "must attempt to further reduce both variable operating costs and fixed costs to lower the break-even point. All costs not essential to copper must be cut. All assets not essential to copper and not producing cash flow must be sold with the funds generated going to reduce debt."[5]

Yearley's dismal forecast, correct as it turned out, projected that for three more years the price of copper would average sixty-five cents per pound, and that was at a time when Phelps Dodge's cost of production averaged seventy cents and more. Munroe, who registered no outward sign of emotion about the findings of the report, said to Yearley a short time later, "I want you to present this to the board" at its September meeting. Later Yearley recalled thinking that somebody had "died in the boardroom when I finished." According to his arithmetic, Phelps Dodge would lose money for every pound of copper it sold. Unless something was done, the corporation would simply bleed to death. Executives used many different metaphors, including a fatal hemorrhage and a sinking ship, to dramatize the peril.[6]

Given that observers both inside and outside the corporation thought Phelps Dodge stood at the edge of bankruptcy, it is vitally important to understand Munroe's perception of unfolding events—and especially the reasons for his long-range optimism. Although hindsight suggests that the trend in copper prices was down, "in living through the period, it was not that simple," asserts Munroe. "The large loss in 1982 was a shock, but the price increased and we essentially broke even for the first half of 1983." Phelps Dodge common stock held steady at around $28 a share. "We have a chance to be in the black in 1983," he reported to the *Wall Street Journal* in midyear. Unfortunately, "the second half of 1983 was heavily impacted by both the strike disruptions and declining prices," he later recalled. He nonetheless derived some comfort from the fact that at the end of the year, despite the large losses reported in 1982 and 1983, the corporation's long-term debt was 10 percent lower than it had been at the end of 1980.[7]

Munroe was a very thoughtful though somewhat reserved leader. He worked hard and often remained busy at his desk until late in the evening, and he was confident that an upturn was imminent—at least until he saw Yearley's report in August 1984, which Munroe later described as "awfully good." And awfully frightening, he might have added. Until then he had been certain that Phelps Dodge had not yet approached the edge of the abyss. He expected that sales of its underperforming assets and new money raised by stock offerings would help sustain the corporation until the price of copper finally rebounded. He also hoped for some form of tariff relief from Uncle Sam. "However, by August it was clear we could wait no longer."[8]

In fairness to Munroe, it should be noted that each of the half dozen men who chaired Phelps Dodge during the twentieth century faced problems unique to his time; yet not one of his predecessors confronted more fundamental structural changes in the world copper industry. Most notably these were the rise of environmentalism and the nationalization of foreign copper mines, which together during the late 1970s and early 1980s had a "devastating effect on the whole industry," recalled Munroe. Yet, in terms of what had always happened in the past, he had every right to believe that copper prices would improve at almost any time. "Prices continued low in 1984, but worldwide inventories of copper were declining at a record pace," and that again gave him cause for hope.[9]

Some of his more pessimistic lieutenants—those who had access only to the corporation's published reports and not to intimate details of its bank relationships—thought the chairman required a serious crisis to show him why the future survival of Phelps Dodge seemed so doubtful to them. Certainly during the summer of 1984 they believed the crisis was at hand. Until then, it seemed to some of the more impatient executives almost as if "an element of paralysis" had set in. "We were simply going through a meat grinder," recalled Durham. Munroe may not have agreed that the corporation was "hovering on the edge of bankruptcy," but he did later describe the situation he and Phelps Dodge faced as "damned uncomfortable." Was the corporation in truly serious trouble? "You bet!" responded Munroe without a moment's hesitation.[10]

What to do? As recently as February 1984 copper had sold for sixty-five cents per pound, but by midyear it had dropped nearly five cents more on the New York Commodity Exchange. Copper priced at approximately sixty cents per pound, adjusted for inflation, meant that it neared its lowest levels of the twentieth century. Even if Phelps Dodge shut down every mine and processing facility to await a price rebound, it would still lose money because of the cost of maintaining a closed plant. Also it would have had to buy copper in the open market to fulfill its contracts. Indeed, what to do? "Easy decisions get made fast," observed Munroe. "Things that you'd really rather not do, but are forced to do, come later."[11]

"The Famous Five Plan"

MUNROE AND DURHAM became close. Unlike most Phelps Dodge employees, both resided in Manhattan and had no long commutes. They could thus linger in the office to unwind with after-hours chats. Durham recalled that by August 1984 even Munroe was beset by momentary doubts and that the chairman, gentle and relentlessly composed, once confided that he could not bear to think that he might be standing at the helm if the corporation slid into receivership. One day when Munroe seemed uncharacteristically glum, Durham emphasized that there was still a chance to save Phelps Dodge if "total objectivity" prevailed in the painful decisions that needed to be made. Durham told Munroe that he believed a committee of four or five executives should study the corporation in detail and recommend ways to turn things around (something Durham himself had done six years earlier at Phelps Dodge International). It was Munroe alone who had to decide how to proceed.

Just a year or so earlier, the chairman had set up two temporary committees, one headed by Richard Pendleton, Jr., and another by Frederick DeTurk, president of Phelps Dodge Industries, and challenged each to develop a list of suggestions detailing how the corporation ought to proceed. Pendleton's group huddled for three or four days on a Chesapeake Bay plantation, while DeTurk's group met in a hotel in suburban Westchester County. It was probably the first time that middle-level executives from both the mining and the manufacturing sides of the corporation had joined together to talk strategy. Each team produced a written report, but many of the suggestions—including one calling for forming strategic alliances with Japanese companies—seemed too drastic at the time. Munroe quietly set them aside. But that was

then. Now he was ready to take the next step.

Munroe concluded that the best way—indeed, the only way—to stanch the continuing hemorrhage of red ink was to undertake drastic self-medication that went far beyond anything done to date. That was why on August 23, 1984, even before Yearley shared his gloomy forecast with the members of the board of directors, but after Munroe saw it, the chairman sent a "highly confidential" memo to his top managers explaining the plight of the corporation and designating five senior executives, Durham, Yearley, Richard Pendleton, Edson Foster, and Leonard Judd, "as a committee to consider and recommend to the Senior Management Committee for its review one or more plans for reorganizing and downsizing the company to permit it to remain viable if copper prices continue near present levels for an extended further period." It was more of a challenge than a command.

Munroe's memo emphasized that members of the special committee "should assume that the copper price realized by Phelps Dodge is 65 cents (in 1984 dollars) through 1987." Emphasizing the breadth of the committee's mandate, the memo continued: "Consideration should be given to discontinuing any operation that generates a cash drain. Administrative functions should be tailored to fit the minimum needs of the reorganized company, without regard to present office locations or staffing, other than leases or other existing obligations and provision for fair treatment of employees affected by downsizing." Potentially ominous for the corporation's future as an independent producer of copper was Munroe's request that "the value of the company's large tax loss carry forward to a possible merger partner should also be considered." He assigned Durham to "coordinate times and places of meetings of the committee."

Munroe's charge required the Committee of Five to have a report ready for consideration by October 2. He fully understood that drastic measures had to be taken sooner rather than later. The Gang of Five, as these vice-presidents were sometimes called, had to work quickly to develop a strategic plan to save Phelps Dodge. About low copper prices they could do nothing, but they could suggest ways to reinvent Phelps Dodge as a low-cost copper producer.

Their closed-door meetings began after Labor Day 1984. Through most of the next month, committee members huddled in secret in the corporation's boardroom on the sixteenth floor of the Colgate Palmolive Building on Park Avenue. Each session typically began at eight in the morning and sometimes lasted far into the night. Noon lunches often consisted of sandwiches and provided a touch of austerity that spoke to the seriousness of the search for ways to save Phelps Dodge. One of the five members, Judd, spent nearly every evening at his computer carefully examining statistics for each mine to calculate which could turn a profit.

All through the summer of 1984 (and well before formation of the Committee of Five), Judd had periodically called Ramiro Peru in Phoenix, where he was supervisor, Financial Analysis, to request that he "analyze Morenci's costs, or take a look at Douglas." Eventually the two men conferred by phone almost every day as Peru supplied Judd with reams of data for him to analyze at home on his computer after work. "It

was all very hush hush," recalled Judd, who felt under great pressure to reduce all costs associated with the mines, mills, and smelters of Western Operations. His analysis, incidentally, went on independently of the copper-price study that Yearley was doing at much the same time. On one occasion when Peru flew from Phoenix to New York, he remained up virtually all night "crunching" numbers for a meeting the next morning. It was demanding work, recalled Peru, later a Phelps Dodge senior vice-president and chief financial officer, but "I learned a ton about the company." In an informal sense, Judd also served as his business mentor.[12]

Durham recalled the dynamic that developed in the Committee of Five meetings. Foster understood finances well and knew what it would take to turn Phelps Dodge around. He did much of the number analysis. Having formerly worked for Anaconda, he had as an added incentive the specter of that celebrated corporate collapse. Pendleton, a Yale-educated lawyer, did most of the administrative writing. Yearley "was so valuable because he knew the copper industry. He was 'Mr. Phelps Dodge' to the merchants and manufacturing people." Judd, who was the only one of the five committee members to have a master's degree in business administration, was "a perfect combination of a mining guy and a good businessman." His careful study of mining properties was a real contribution. "We had to get the mines competitive," Durham emphasized as he explained Judd's crucial role. "If we didn't do that, we were still just 'bandaiding' the company." Durham, a "great catalyst" in Judd's words, later described the atmosphere within the group as a "good collegial environment." That

WHERE PHELPS DODGE MEETS THE MARKETPLACE

Phelps Dodge power cable and building wire photographed at
Phelps Dodge Corporation's open-pit copper mine, Morenci, Arizona.
At Morenci, up to 300,000 tons of material are moved daily
by 9- to 22-cubic yard shovels into trains and 100-ton trucks.

You're looking at one of the great resources for America's copper power cable.

You may never have seen a high-voltage power cable like the one on the large reel. But electrical contractors for new industrial plants have. And so have engineers at your electric utility.

It's called extruded solid dielectric cable—and has up to seven pounds of copper per foot. This type cable can be made flame retardant—meeting 70,000 and 210,000 BTU flame tests.

Our Phelps Dodge Cable & Wire Company produces wire and cable like this from 600 volts through 138,000 volts. And oil-impregnated paper cable through 550,000 volts.

And it all starts here. Each year, our U.S. mining operations produce on an average about 600 million pounds of copper—about 20% of all U.S. production. And our U.S. manufacturing divisions transform about a billion pounds into a wide variety of products—from copper rod to condenser tube, from magnet wire to broadcast antennas.

We go to the source. So that people who depend on copper can depend on us.
Phelps Dodge
300 Park Avenue,
New York, New York 10022.

**phelps
dodge**

The copper people from Phelps Dodge.

꙾ OVER THE YEARS, *Phelps Dodge advertising has ranged from newspaper advertisements and posters to products catalogs and brochures. The Morenci mine during the railroad era formed the backdrop for this 1970s advertisement for copper power cable.*

Phelps Dodge solar absorber plates photographed at
Phelps Dodge Corporation's open-pit copper mine, Morenci,
Arizona. Phelps Dodge first invested in these mountains in 1881,
and will still be mining here well into the next century.

Deep in this copper mine is the key to harnessing the sun's energy more efficiently.

You can probably guess what these are just by their shape: solar absorber plates. When the sun's rays strike their surfaces, they heat circulating fluids that will, in turn, heat your household water or your home.

Simple? Not entirely. Under extreme conditions, temperatures within a collector can soar to 400°F during the day, then drop to freezing at night. It takes a tough plate to withstand this kind of stress. And using all copper, our Solar Enterprises unit builds solar absorber plates that meet the test.

No other practical material can match copper's heat conductivity. That's one reason our U.S. mining operations produce on average about 600 million pounds of it each year—about 20% of all U.S. production. And our U.S. manufacturing divisions transform about a billion pounds of copper into everything from copper rod to condenser tube, from magnet wire to mobile antennas.

If you depend on copper, you can depend on us.
Phelps Dodge, 300 Park Avenue,
New York, New York 10022

phelps dodge

The copper people from Phelps Dodge.

The Morenci Mine *and Phelps Dodge Solar Enterprises addressed issues important to Americans during the energy-conscious 1970s.*

Condenser tubes

as advertised by the Los Angeles Tube Mill of Phelps Dodge Brass Company. Financed by the federal government, the West Coast fabricating plant of Phelps Dodge Copper Products Corporation launched production in 1942. It was sold in 1983.

Phelps Dodge copper and copper alloy condenser tubes. For better performance... a better source.

Phelps Dodge *Communications Company manufactured vehicular antenna systems in addition to FM broadcast antennas, marine antennas, and many other related products. It was one of several properties that Phelps Dodge sold during 1983 as part of a corporate restructuring of assets.*

Vehicular Antenna Systems

POLICE

phelps dodge Communications Company

San Francisco...by Willy Mucha, Chevalier de L'Ordre des Arts et Lettres

San Francisco... jewel of the West.
Gateway to the Pacific. Grandeur built on a proud heritage. A Phelps Dodge city
where it's all happening.

Look closely, all round the Bay, you'll find Phelps Dodge everywhere. Our copper pipe and tubing transporting gas and water...our coaxial cable carrying radio and TV signals ...our copper-nickel condenser tubes helping to generate steam at local power plants...our high-voltage cables distributing electricity downtown and to outlying towns. Telephone wire and cable, communications antennas, magnet wire, bronze valves and fittings, copper mill products... wherever you look, you'll find Phelps Dodge products at work.

While you can't always see us, our experience and imagination are at work everywhere across the land.

phelps dodge
Copper, Aluminum Alloy Products

THIS BROCHURE COVER *advertises submarine cable made by ICONEL, Phelps Dodge Venezuela, during the 1990s.*

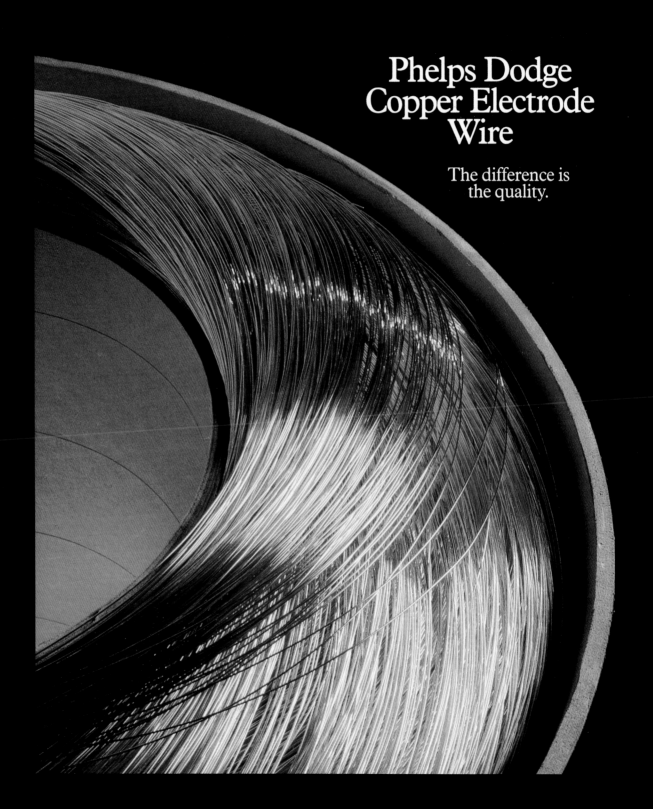

A BROCHURE COVER *for Phelps Dodge copper electrode wire, a product the Campbell Soup Company and other food processors use to seal their cans. For photographers the never-ending challenge was to portray cable and wire products in dramatic new ways.*

Phelps Dodge Copper Electrode Wire

The difference is the quality.

Raven®

Carbon Blacks

A Family Of Quality Carbon Blacks For All Applications

Columbian Chemicals offers a complete line of carbon blacks for superior performance in coatings, printing inks and plastic applications. From the jettest of mass color and blueness in tone to the strongest in tint, Columbian spans the spectrum.

Many of our blacks are specifically designed for maximum system compatibility to assist dispersion and assure stable systems. Selective grades are also available with balanced properties to meet electrical conductivity, which is key to performance.

For more information and assistance in your selection, please call
1 (800) 235-4003 (Outside of Georgia)
or
1 (404) 951-5707

or write to:
Columbian Chemicals Company
1600 Parkwood Circle
Suite 400
Atlanta, Georgia 30339

Columbian
Chemicals
Company

Circle No. 56 on Reader Service Card.

As Advertised In... **Modern Paint and Coatings**

was especially important to Munroe, who recalled that "I left off the committee anyone who should have been left off as unable to work collegially."[13]

Phelps Dodge had little maneuvering room. Those five people had a mandate to propose anything they wanted. "There were no sacred cows," Durham recollected. The committee mantra, Yearley stated, was "fix it, sell it, or shoot it, in that order." Members asked, "What is underperforming and what can we do about it?" When Judd was later asked whether he found it hard to sleep at night after each day's intense work, he smiled and quipped, "All during this time I slept like a baby: I woke up every two hours crying." Turning serious, he quickly added, "It was hell."[14]

After weighing every conceivable alternative, members of the Committee of Five drew up a plan that enunciated three basic goals. First and foremost was to reduce the cost of producing copper to less than sixty cents per pound (the prudent planning price based on the corporation's view of the market at that time). Second was to reduce the burden of long-term debt. The third and perhaps most radical goal was for Phelps Dodge to expand into noncopper businesses to help it better weather low points in the copper cycle. That meant a historic departure from business as usual. Production of copper would remain central to everything it did, but henceforth the corporation would diversify into enterprises on which variations in the price of red metal had no impact.

The Committee of Five's strategic plan suggested numerous ways that Phelps Dodge could trim costs. Because the company's mining and processing operations were losing money, it suggested operating the Morenci and Tyrone facilities on a more efficient continuous basis, together with the Douglas and Hidalgo smelters, the refinery, rod mills, and Phelps Dodge Magnet Wire, and it urged closing the old, inefficient smelter at Morenci, as well as the Ajo mine and mill, which had closed the previous August for an indefinite period. The New Cornelia Branch, which at the time was the company's high-cost producer, permanently shut its smelter at the same time. Changes occurring at Ajo that summer, or anticipated in the strategic plan, were necessary but painful. Jack Ladd, Phelps Dodge's director of labor relations, found it heartrending to see the vice-president and general manager Arthur Kinneberg have to go to Ajo in 1984 to tell employees who had faithfully braved the harassment of pickets that they were out of work after the smelter poured its last copper anode on August 12.

Further, the plan called for immediate sale of Western Nuclear, Phelps Dodge Fuel Development Corporation, the South African mine at Black Mountain, the Australian mine at Woodlawn, the Far East properties of Phelps Dodge International, and the Laurel Hill Refinery, which had closed earlier in the year. It proposed to sell as much as a 40 percent stake of the Morenci operation, a radical suggestion that evoked gasps among some people. Sell its crown jewel? But that was not all. The plan also suggested selling a minority position in Phelps Dodge Magnet Wire and for disposing of Phelps Dodge's 16 percent interest in Southern Peru Copper "at the earliest date when timing would indicate proper values could be realized." Finally, among the other properties Phelps Dodge considered for sale were the Mercantile, the Dawson Ranch, various townsite facilities, and its interest in the Apache Powder Company, a manufacturer of explosives.

Committee members suggested consolidating a downsized Western Operations staff in the vacant Phelps Dodge Mercantile building in Tyrone, "which can be converted for that use with relatively modest expenditures for partitions, etc.," and they urged moving corporate headquarters on Park Avenue to a less expensive location within New York City. "With lean staffs at all locations, direct communications between the operations, the Western offices and the corporate office should improve." Neither proposal was ever implemented, however. In retrospect, notes Judd, the proposed move of Western Operations to the Silver City area was "too severe," though the later move of corporate headquarters to 2600 North Central Avenue in Phoenix saved Phelps Dodge an estimated $3.5 million a year. Some people fretted that relocation from New York to Arizona might cost the corporation access to important financial circles, but that never happened. The relocation had a far greater impact on personnel. In the end, Phelps Dodge offered only about fifteen employees the chance to make the transfer, and almost all chose to do so.[15]

Members of the Committee of Five explained their restructuring plan to Munroe on September 27, only shortly after the chairman returned to New York from the annual American Mining Congress meeting in Phoenix. The "hunker down" plan, as Judd liked to characterize it, elicited no visible show of emotion from Munroe. The committee of vice-presidents could only guess what he might do next. As for President Richard T. Moolick, he was one day behind Munroe in flying back from Phoenix—although the timing of his return to New York would not have made any difference in what happened a few days later.[16]

"Bull" Durham Takes Charge

DURING SEPTEMBER 1984, as members of the Committee of Five met behind closed doors to deliberate what measures to forward to Munroe and the Phelps Dodge directors, the corporation learned that the Reagan administration rejected American copper's request for quotas or tariffs on imports. It received more bad news the following month. As Phelps Dodge closed its books on 1984's third quarter, the largest independent copper producer in the United States faced the inevitability of having to report multimillion-dollar losses for the third consecutive year. That was a stretch of hard times unmatched even during the Great Depression. Did the corporation have a future, at least as an independent producer without recourse to the deep and money-filled pockets of an oil company? A year earlier an analyst for Salomon Brothers investment firm had claimed that "independent copper producers were going the way of dinosaurs and flying reptiles."[17]

It was in their meeting of October 3, 1984, against this backdrop of bad news, that the corporation's directors approved the harsh assets-restructuring plan proposed by the Committee of

Five. Board members placed only two constraints on the proposed cost-reduction projects. Each one had to be sound for long-term management of the resource, and each capital project for which funding was sought had to demonstrate both a substantial internal rate of return and a prompt payback.

A short time later, Munroe entrusted implementation of the Committee of Five Plan to Durham, who in January 1985 at age fifty-five became the corporation's new president and chief operating officer. "I guess I'm considered a kind of turn-around guy," he later reminisced. Another veteran member of the Committee of Five, 45-year-old Len Judd, took charge of Western Operations as senior vice-president after Kinneberg retired and Arthur Himebaugh replaced him as general manager. "With a new management team in place, Phelps Dodge is clearing the decks to mount with the beginning of the new year an all-out, no-holds-barred final drive to reduce production costs to an absolute minimum," observed Bill Epler, editor of the mining publication *Pay Dirt*. Emphasized Munroe, "Survival is the name of the game—and we intend to survive."[18]

One management casualty of those tense weeks was President Moolick. Not being named to the Gang of Five, he later said with acerbity, did not trouble him at the time. However, he regarded the finished document with unprintable contempt and scorned certain sections as wholly unworkable. In anticipation of his dissenting reaction, some committee members grew concerned that their capable but occasionally irascible president would be forced to carry out a restructuring plan in which he had no personal interest or that he would actively oppose.

That was why, when they handed Munroe the finished version of their plan, they added a sealed envelope containing a "highly confidential" memorandum dealing with future leadership. "We ducked this issue in our report, because we did not want it to distract from or impair speedy implementation of its other components. But we did not feel we could fairly duck it entirely; hence, this separate memorandum to you." In the one-page document committee members carefully broached the idea of early retirement for both Moolick and Kinneberg effective January 1, 1985. They reasoned that a person not already on the verge of retirement must have "long-term responsibility" for carrying out their restructuring plan. However, they were concerned that neither Moolick nor Kinneberg "could sit on the sidelines, remaining as senior management, and give the new man the free hand he must have to move quickly. We concede that mistakes may be made without their overview, but feel that the risks are much greater with three men, each with strong views, trying to run one job." There followed the leadership shuffle that William Seidman later called an unprecedented "internal takeover." But "nothing could be more natural," Munroe insisted, than for the authors of the Committee of Five Plan to implement it.[19]

Following the directors' discussion and approval of the asset-restructuring plan, it was Munroe's unenviable task to relay word of the change of presidents to Moolick, who he knew would be enraged. Moolick recalled, "When George came into my office on October 3rd about 2:30 P.M. I was not aware of any meeting. So, when George told me that the board wanted me to move to vice chairman [a resurrected position the short-fused Moolick regarded as largely honorific], I told him, 'I quit.' His face turned red, and he finally asked me, 'When?' I looked at my watch and said, 'Well, it's 2:32.'"[20]

Moolick finally agreed to continue through January 1, 1985, as special consultant, though he vented his frustration at the directors' November 1984 meeting. Kinneberg, the other executive mentioned in the committee's secret memo, was happy to retire. "I didn't have anything to quarrel with," he later recalled. "I'd about had enough." Having wrestled with numerous and seemingly intractable problems in recent years, Kinneberg was ready for a well-deserved rest. But Moolick left Phelps Dodge after more than thirty-five years of service a bitter man: he believed that members of the Gang of Five claimed credit for reforms he had already suggested or initiated.[21]

Ironically, only two months after the Committee of Five presented its report to the Phelps Dodge board, *Business Week* pronounced the corporation dead—or nearly so. The magazine printed an obituary of sorts in its December 17 issue headlined "The Death of Mining." No question mark followed the statement, and the magazine's cover illustration consisted of an ingot of metal with handles added to make it look like a closed coffin. "Analysts contend that Phelps Dodge Corp., the second-largest U.S. copper producer, is especially vulnerable because it relies on copper for almost all its revenues." It was difficult to see how Phelps Dodge could survive, lamented an analyst for the brokerage firm Smith Barney, Harris,

Upham and Company who was quoted in *Business Week*.

"The Death of Mining" appraisal was premature, and as it turned out, Phelps Dodge management had the last laugh. Looking back, Munroe called the article silly, but to his employees at the time it was certainly frightening. "I must admit that the facts in 1984 supported *Business Week*'s gloomy conclusion about Phelps Dodge," added Yearley at the publication's "President's Forum" in 1993, a far more prosperous time for copper and Phelps Dodge than in early 1985, when the corporation unexpectedly rallied to confound the analysts. It was a far different enterprise by the late 1980s too.

Another major change of leadership came in early 1987 when chairman Munroe at age sixty-five followed his lieutenants Kinneberg and Moolick into retirement after nearly thirty years of service at Phelps Dodge. The most recent decade had been especially trying for him, fraught as it was with daunting challenges posed by environmental and labor issues, serious inflation, and subsidized competition from newly nationalized foreign copper companies. Yet under Munroe, Phelps Dodge had moved ahead, most notably by risking its resources to build a clean-burning but enormously expensive flash smelter at Hidalgo and by committing to the solution extraction–electrowinning process it first implemented at Tyrone in 1983–84 to produce low-cost copper.

Kinneberg, who maintained a "great deal of respect" for Munroe over the years, recalled that if the chairman "had any faults it was because he wanted to be very sure" before he committed the corporation to various expensive or technologically advanced projects, "and he did move fast on some of those things. He knew what he was doing." Like no Phelps Dodge chairman since the 1920s, Munroe left the company for an active retirement. He served on the boards of directors of several large corporations, including the New York Times Company, as chairman of the board of trustees of Dartmouth College, and as chairman of the finance committee of the Metropolitan Museum of Art.[22]

Durham, the man who succeeded Moolick as president in 1985 and Munroe as chairman in early 1987, was the person who most visibly led the corporation's turnaround. He was always quick, however, to acknowledge the valuable spadework done earlier by his colleagues. "When we put our plan together in the fall of 1984, we were able to build on much good cost-reduction work already done by my predecessor, Richard T. Moolick, and by senior vice-president Arthur H. Kinneberg—both now retired—particularly in the stabilization of labor costs and development of a highly motivated and productive workforce," Durham reported to the American Mining Congress on October 8, 1986.[23]

There were others too who deserved credit. Patrick J. Ryan, a senior vice-president who had charge of exploration and international mining operations, received a "nice basket of projects to fix," as he recalled with his usual understatement. Every one of them was a money loser; still, closing companies was difficult for the amiable South African because of the jobs it cost employees. Ryan, however, had little choice but to shut down Western Nuclear because the Phelps Dodge uranium company possessed only

low-grade, high-cost mines; and he sold its wholly owned subsidiary, Phelps Dodge Zeolite Corporation (formed just two years earlier) to Tenneco, Inc., in 1985. He likewise sold all assets of Phelps Dodge Fuel Development Corporation and closed down the recently formed (1981) Small Mines Division. In total, all funds that Ryan and his colleagues generated by sale of a mixed group of mining assets added up to approximately $90 million, which Phelps Dodge Corporation reinvested in productive capital programs or used for debt reduction.

The man who decided what steps to take next was the stocky, personable Durham. He was often known to his colleagues simply as "Bull"—a nickname he acquired as a teenager from the Bull Durham brand of tobacco once sold in cloth pouches, which was an apt description of his determination to shove aside hallowed traditions at Phelps Dodge and bulldoze a new path to profitability. If the strategic plan failed, the corporation would probably go bankrupt. Fortunately, added his colleague Ryan, "'Bull' Durham was a very strong-minded guy."[24]

Durham was the latest in a line of Phelps Dodge leaders who hailed from America's heartland. Robert Page came from Ohio, Munroe and Durham were both from Illinois, as was Durham's successor, Yearley. Durham was quick to emphasize that he was from southern Illinois while the others were from the greater Chicago area—and his roots were distinctly blue collar. He was the son and grandson of coal miners. In the Durham home in the gritty town of West Frankfort a family hero was John L. Lewis, the legendary leader of the United Mine Workers Union. The Durhams were familiar with tough times, strikes, layoffs, and work-related casualties, and for such reasons the father urged his sons to get out of the mines. Born in 1929, Durham graduated from Purdue University in neighboring Indiana in 1951 where he studied economics and "the business side" of labor law. It was there that he acquired his nickname. "I've known people who think I have no other name."[25]

Durham came to Phelps Dodge twice. The first was back in 1967 when the copper producer acquired the Ashby Corporation, where Durham was president, as part of its entry into the business of aluminum fabrication. Building on previous expertise, he helped develop Phelps Dodge Aluminum Products Corporation and served as its executive vice-president until that short-lived subsidiary was merged with Consolidated Aluminum Corporation (CONALCO) in 1971. A short time later, Durham became president and chief executive officer of CONALCO, in which Phelps Dodge retained a large minority stake.

Between 1967 and 1973 everything seemed to go right for the rising young business executive, but then, without warning, Durham's world collapsed. Physicians told him he had cancer, and their prognosis was "not all that good initially." In July 1973 surgeons removed a kidney. Durham retired from CONALCO while still in his early forties and sought to enjoy life while he still could. "When you go through a major operation you start seeing things differently."[26]

Amazingly, Durham beat the odds and recovered fully. He was financially well off, "never intending to work again," but later in the 1970s he received a discreet inquiry from Ted

Michaelsen, the suave president of Phelps Dodge Industries, asking if he was bored. Durham wasn't, but he agreed to return to Phelps Dodge briefly as an outside consultant to suggest ways to turn around its financially ailing subsidiary, Phelps Dodge International Corporation. In short order, Durham presented his strategic plan to Michaelsen, who invited him to implement it in 1978 as chairman of International. Subsequently, Durham was elected a vice-president of Phelps Dodge Corporation in 1980 and a senior vice-president two years later. The events of late 1984 and early 1985, and most notably the need to reorganize Phelps Dodge, offered a variation on his previous assignment at Phelps Dodge International.

Overall, the news for 1985 was good. Despite no real movement in the price of copper, Phelps Dodge Corporation's losses dropped to $7 million in January, down from $10 million per month in late 1984. The earlier pruning of underperforming assets and the recently approved Committee of Five Plan seemed to be working. The following month the company appeared to have earned a million dollars' profit, its first in many months. *Appeared* was indeed the operative word because before Durham had time to celebrate, William Tubman, now retired vice-president and corporate secretary, recalled how he delivered information that spoiled everything. It was Tubman's unenviable job to explain to Durham that Phelps Dodge had just settled a lawsuit out of court for a million dollars. There was no profit after all. "'Bull' got so mad I thought he was going to slug me," Tubman later recalled with a touch of nervousness in his laughter. "I remember the incident," confirms Durham a bit sheepishly. "But that's life."[27]

In the end, there was news worth celebrating. When Phelps Dodge returned to profitability during the first quarter of 1985, it was the only major copper producer in the United States to do so. Things had improved enough that in April 1985 the corporation raised $50 million by selling 2.5 million shares of common stock. Further, during all four quarters the corporation earned $29.5 million and enjoyed its first profitable year since 1981. The positive financial tidings of 1985 probably resulted more from changes made in 1983 and 1984 than from the Committee of Five Plan—the "flow of red ink was basically on its way out before the committee met," Munroe still insists—but their taking the long-range view ensured that the recovery would continue well past 1985 and even with no significant increase in the price of copper.[28]

The 1984 write-offs, which in addition to sale of assets included shutting down high-cost production at inefficient operations, reduced Phelps Dodge's costs by eliminating depreciation previously charged, and lowering costs gave an immediate and significant boost to earnings. By 1986 the corporation had whittled down its long-term debt to $423 million, the lowest level in a dozen years. Likewise, it downsized its work-force by slightly more than half, to sixty-four hundred people, mainly through selling off underperforming assets. Phelps Dodge common stock recovered from a low of $12.88 a share in 1984 to reach $20.75 in the fall of 1985. Incredibly, all of the rebound occurred without any boost from the price of copper, which continued to languish until 1987. Then, finally, during a single month the price rose from just over sixty-one cents a pound to more than sixty-nine cents.

That was very good news for Phelps Dodge, which through its aggressive campaign to fix, sell, or shut down underperforming properties had repositioned itself to become the nation's premier low-cost copper producer. It was a positive initial step toward realizing a major financial payoff from the corporate strategy first enunciated in the Committee of Five Plan. When the dust of the copper crisis of the 1980s finally settled, it was Phelps Dodge alone that stood tall.

~~~ *Chapter Fifteen* ~~~

TURNAROUND TIME

T HE TURNAROUND HAS been gritty, not glamorous," observed *Fortune* magazine of Phelps Dodge Corporation in October 1985. Regardless of how outsiders characterized it, managers who oversaw the toilsome process of recovery usually credit employees "in the trenches" for showing "real passion" in their commitment to the Committee of Five Plan. Leonard R. Judd, a member of that committee who returned to Arizona to lead a critical part of the turnaround, recalls how the employees of Western Operations battled hard to save their corporation and return it to profitability. "Our plan was very complicated," Judd now explains with a twinkle in his eyes and a hint of irony in his voice: "We took bad mines and shut them down, and we took good mines and expanded them."[1]

In addition, after Judd, a senior vice-president, took charge of Western Operations, he shrank the size of the Phoenix office staff from around 170 to 63 members. Nothing that saved money was too small to be overlooked, and that included getting rid of all decorative plants in the Phoenix offices, along with the person who watered them.

Judd recalls that the key to successfully "reinventing the West" was speedy implementation of the Committee of Five Plan. Everyone was "running and gunning," as he describes the furious pace of

---

~~~ CHIEF ARCHITECTS OF *Phelps Dodge's astonishing turnaround were (left to right) Patrick J. Ryan, J. Steven Whisler, G. Robert Durham, Douglas C. Yearley, and Leonard R. Judd (see color image following page 208).*

345

activity that developed early in 1985 after the board of directors formally approved the radical restructuring program. Keeping the momentum up were Judd's lieutenants—a coterie of diligent individuals including Arthur E. Himebaugh, general manager of Western Operations; Dushan P. ("Duke") Milovich, general superintendent of the Tyrone Branch; Richard Rice, manager of engineering services; Carl Forstrom, manager at Morenci; and Nicholas S. Balich, controller, Western Operations—who together with their employees tackled all kinds of demanding work assignments with "enthusiasm and energy," recalled Ramiro Peru, because of their newfound sense of purpose. He explained that, until the team of westerners had the Committee of Five Plan in hand and a distinct sense of direction and purpose, morale in the mines, mills, and smelters "was horrible." Judd, who clearly does not share the former chairman George Munroe's rosy assessment that a corporate recovery was already observable by late 1984 or early 1985, insists that "this company didn't turn around until we put that plan in place."[2]

No matter how the about-face may be described, or what or who was responsible, the resurgence of Phelps Dodge during the second half of the 1980s and early 1990s was one of the most dramatic chapters in its long history. It is a story of how, in response to fundamental and long-term changes in the American copper industry, Phelps Dodge aggressively implemented cost-cutting technology and successfully diversified into businesses outside its familiar terrain of metal mining.

The initial payoff for the difficult work of revitalization came during five fabulous years from 1988 to 1993, when shareholders who had faith enough in Phelps Dodge to reinvest their quarterly dividends enjoyed annual increases averaging a whopping 33 percent of the total value of their investment. In 1995, in fact, Phelps Dodge ranked a very respectable twelfth among Fortune 500 enterprises providing the highest total return to investors between 1984 and 1994, just ahead of Coca-Cola and a little behind Walt Disney. This was fitting recognition of how restructuring helped Phelps Dodge thrive once again following the corporate cliff-hanger caused by the copper crisis of the early 1980s. "In most cases, it's been more successful than our wildest dreams," observed the chairman Douglas C. Yearley. "We feel good about the copper industry."[3]

Seeking Salvation through New Technology

USING THE BASIC blueprint provided by the Committee of Five Plan in late 1984, President and later Chairman G. Robert Durham maneuvered the company through a period of cost cutting even more drastic than in 1982, all the while continuing the creative methods of financing new technologies begun back in 1983, when Phelps Dodge launched construction of its first solution extraction–electrowinning (SX/EW) plant. Given the corporation's financial plight at the time, the decision to embrace the new technology was an act of courage by the then chairman, George Munroe, who displayed similar

THE EVOLUTION OF *copper production continues as the solution extraction–electrowinning process creates its own landscape of production. Looking very much like the familiar sprinklers used to irrigate lawns and golf courses, these at Morenci use water and highly diluted acid to work SX/EW's magic.*

The "Magic" of SX/EW

ANY SUFFICIENTLY ADVANCED *technology is indistinguishable from magic," the scientist and author Arthur C. Clark once wrote. In that sense, the term* magic *is certainly appropriate to describe solution extraction–electrowinning technology. It was unquestionably different from the traditional way of processing copper ore, which was mainly a "blacksmith" business that required crushing, grinding, and heating. For one thing, SX/EW was a much quieter process. A plant was relatively inexpensive to build compared to a smelter, and the new technology required only a third of the labor needed for precipitate or leach copper and about a quarter of the energy needed to smelt and refine it.*

The pyrometallurgical method of smelting used fire. It was successful at removing up to 98 percent of copper from ore, although traditional smelters had been widely criticized since the late 1960s for releasing sulfur dioxide into the air and creating significant sulfuric acid disposal problems. Hydrometallurgical methods, using liquid processes, created virtually no harmful gas emissions or toxic waste. Some people might even pay a backhanded compliment to the Environmental Protection Agency for creating the conditions that made the new process attractive.

To work its magic, SX/EW technology utilized three proven processes to recover copper: dump leaching, solution extraction, and electrowinning. Without going into basic chemistry and physics, you could liken these three to square dancers handing off copper ions to one another in progressively greater concentrations as red metal moved from low-grade ore to two-hundred-pound, 99.99 percent pure cathodes, all from stockpiles once considered only waste. Further, the new technology addressed the problem of too much copper concentrate going to the company's aging smelters and their need for costly environmental retrofitting. Costs climbed somewhat higher when the corporation had to "mine for leach" after it had used up all its old stockpiles, but the process was still a bargain.

For a variety of reasons (depending on the ore type and grade), in the late 1990s Phelps Dodge continues to smelt copper concentrates not easily amenable to SX/EW processing, and it ships heavy anodes to its El Paso electrolytic refinery for further processing. There a different type of technological magic takes place. During electrolytic refining, insoluble impurities from anodes settle to the bottom of the tanks in the form of a fine black mud, also known as slimes. From this impurity the El Paso refinery recovers gold, silver, selenium, and tellurium at its slimes treatment plant. Income received by Phelps Dodge Sales Company from those precious metals constitutes a bonus that effectively reduces costs of copper production by traditional smelting and refining methods.

audacity when he committed Phelps Dodge to flash smelting at Hidalgo. It could be claimed, in fact, that implementation of flash smelting in the 1970s and of SX/EW in the 1980s represents two of the most significant technological advances at Phelps Dodge since World War II.

Many people will credit the innovative SX/EW technology with having saved the firm during its darkest hour. Reality is a bit more complex. Phelps Dodge, true to its conservative character, was actually among the last big American copper companies to conclude that SX/EW was a good deal—and that was fortunate for several reasons. The process initially had numerous problems to overcome. Its roots extended far back to Uncle Sam's supersecret Manhattan Project, which sought an efficient way to extract a uranium isotope it needed to build atomic bombs during World War II. Following various improvements, the world's first commercial SX/EW plant for copper began working production miracles in March 1968 near Miami, Arizona, at the Bluebird Mine of the Ranchers Exploration and Development Company. At first, though, the selectivity of the reagents for copper was very low.

Phelps Dodge embraced SX/EW technology only after its own research department had studied it thoroughly and only after the technicians and scientists had solved many of its early problems. "Phelps Dodge seldom, if ever, claims 'a first' on anything," noted Richard Rice, the corporation's retired vice-president for engineering. "Our conservatism makes us do that." However, after technologies have proven themselves, "we'll do them better than anyone else." That was certainly the case with SX/EW, in which Phelps Dodge became a world leader after its first such plant (and also the first of its kind in New Mexico) commenced production at Tyrone in April 1984 as the Burro Chief Copper Company.[4]

The main problem in the early 1980s was to interest members of senior management and the Phelps Dodge board of directors in an exotic process that to many of them seemed more like science fiction than fact, recalled Timothy R. Snider, the young chemist who headed the analytical laboratory at Tyrone. Later, after he became president of Phelps Dodge Mining Company in 1998, Snider recalled that as long as the corporation made money selling copper as usual he did not gain anyone's attention. The attitude changed after the onset of the copper crisis in 1981, and especially after goading from Len Judd, the first top-ranked executive to realize the unbelievable cost-saving potential of SX/EW.

In old western movies, when prospectors found gold they shouted "Eureka!" Though the metal this time was copper, initial SX/EW success at Tyrone was a "Eureka moment" for Phelps Dodge. After members of the board of directors gave their approval in 1982 to proceed with a full-scale plant, the corporation still had to raise the needed capital; and therein lay the reason for setting up the Burro Chief Copper Company as a subsidiary operation. Len Judd successfully coaxed the prime contractor for the facility, Brown and Root, Inc., into financing $7 million of the project by forming Burro Chief to protect its investment should cash-starved Phelps Dodge run into further difficulty. Phelps Dodge later bought out Brown and Root's stake, but the Burro Chief Copper Company continues to operate SX/EW facilities.

Brown and Root broke ground in March 1983. A little over a year later, on April 6, 1984, the $35 million plant extracted its first copper from solution and in just three days more had plated the metal onto starter sheets in the tank house. The initial shipment of cathodes left for Phelps Dodge's continuous-cast rod mill in El Paso on April 17. At Tyrone, Phelps Dodge had accumulated numerous dumps that never before had been leached—ore of the type that lent itself easily to the new SX/EW method—and that was a main reason its early cathodes cost less than thirty cents a pound to produce. The cost of actually mining leach material had already been charged against prior copper production. For the same reason, Morenci profited when the SX/EW technology was implemented there too. Together, they helped Phelps Dodge to achieve its goal of becoming a world leader in low-cost copper production.

In sum, SX/EW did not save Phelps Dodge, but the technology resulted in bonanza years once the price of copper improved and money at last flowed in. For about four years in the mid-1980s, a special Phelps Dodge crew worked with Brown and Root to do nothing but build SX/EW plants, and all those facilities originally did was to leach valuable copper from old "waste" dumps. Within a decade, Phelps Dodge produced almost half its copper by SX/EW technology, and that was the single most important reason for its production costs dropping from approximately eighty cents per pound to about fifty cents. Ironically, if Phelps Dodge had rushed to embrace the new SX/EW technology in the 1970s, it would not have had the stockpiles of oxide ore available in the mid-1980s when it needed them most.

A New Day at the Chino Mines Company

BETWEEN 1981 AND 1987, Phelps Dodge dramatically reshuffled its landscape of production in Arizona and New Mexico. In 1982 the company relocated its western regional headquarters from Douglas to Phoenix; two years later it closed its New Cornelia Mine, then its Ajo smelter in 1985, eliminating nearly six hundred jobs. At that time the town of Ajo lost its hospital and nearly half its population. Phelps Dodge also closed its Morenci smelter in January 1985, idling about 450 workers, some of them permanently. The shutdown was necessary because the corporation had spent $91 million in an ultimately unsuccessful effort to comply with government sulfur dioxide emission standards.

Next, in January 1987 Phelps Dodge shut down its aging Douglas smelter, once again because of the impossibly high cost of complying with government emission mandates. The community thus lost a $9 million annual payroll, by far the largest source of income in the border trading center of fourteen thousand people. After early 1987, the corporation operated no smelters at all in Arizona and only one major mine in the state, the open-pit complex at Morenci. There, in general accordance with the 1984 Committee of Five Plan, the work of mining and milling continued around the clock, every day of the year, and was aided by computers that ran equipment

⌁ Historic Chino ⌁

YEARS BEFORE PHELPS DODGE *produced low-cost copper at Chino using the solution extraction–electrowinning method, the process of technology transfer had once linked the Chino and Bisbee operations. That was back in 1917, when Phelps Dodge launched its first open-pit mine. To provide guidance as it moved through unfamiliar technological terrain toward successful operation of its Sacramento Pit, Phelps Dodge hired an adviser from the Chino Copper Company.*

Chino's history actually dates back to some of the oldest copper mines in North America. In the early 1800s, the mine on the Santa Rita del Cobre Grant served as the principal source of copper for the Mexican mint in Mexico City. One estimate suggested that Santa Rita produced as much as forty-one million pounds of copper prior to 1845, the year northern Michigan began to yield large quantities of red metal. The New Mexico deposit produced another 124 million pounds of copper between 1845 and 1910, the year Chino became the first open-pit facility in the territory.

The Santa Rita Mining Company, as it was named in the late nineteenth century, actually ran out of high-grade ore in the early 1900s. A young mining engineer named John Murchison Sully, a recent graduate of the Massachusetts Institute of Technology, spent almost a year evaluating the New Mexico property for General Electric, directing a drilling program and delineating a large deposit of low-grade copper amenable to open-pit mining methods then being perfected at Bingham Canyon in Utah. But General Electric, which already owned a productive copper mine, was unimpressed by any ore deposit that assayed only 2.73 percent copper.

Others, however, did have the necessary faith. In June 1909 the Chino Copper Company was born; Sully was its general manager, and Daniel C. Jackling its vice-president and managing director. Its first steam shovel bit into the ground in September 1910, a date that places Chino among the American pioneers of open-pit technology. In fact, only famous Bingham in 1907; Ely, Nevada, in 1908; and Miami, Arizona, in 1910 antedated it. That same year, 1910, Chino Copper erected a concentrator ten miles south of the mine at a Santa Fe Railway siding known as Hurley, where it had good water and a place to dispose of tailings.

From 1914 until the end of World War II, the big Santa Rita mine accounted for more than 90 percent of all copper mined in New Mexico. As for ownership, in 1924 the Chino Copper Company and Ray Consolidated Copper Company merged. Two years later, they joined Nevada Consolidated Company to form a mammoth outfit called Nevada Consolidated, which the newly formed Kennecott Copper Corporation acquired in 1933.

and kept track of every shovel, locomotive, and truck in the mine. In its quest for efficiency, Morenci switched to trucks, instead of trains, to haul rock or muck out of its pits.

During the same time, big changes came to Phelps Dodge operations located eighty miles east of Morenci and over the mountains in New Mexico. In 1986 when Standard Oil of Ohio announced its willingness to sell various Arizona and New Mexico properties belonging to its Kennecott subsidiary, Phelps Dodge paid approximately $88 million for a two-thirds interest in the Chino Mines Company complex located southeast of Silver City. Besides an immense open-pit copper mine, the property included a newly renovated smelter, which along with the Hidalgo smelter gave Phelps Dodge two reduction works in New Mexico to offset loss of production caused by the imminent closing of its Douglas Reduction Works. The corporation subsequently inaugurated an SX/EW facility at Chino on October 31, 1988, shortly after it introduced the new technology at Morenci.

Seemingly secure in its status as one of the Big Three copper producers in the United States, Kennecott enjoyed many good years until the 1960s and 1970s. That was when it experienced one financial setback after another. It unsuccessfully battled an antitrust suit all the way to the United States Supreme Court, waged another lengthy court fight to recover $68 million for assets expropriated by Chile, and spent millions more to implement pollution controls. A series of proxy fights further drained Kennecott's corporate revenues and left it strapped for cash.

At its Chino facility, the decade of the 1980s

brought monumental changes not seen since the open-pit mine complex began almost seventy years earlier. First, Standard Oil of Ohio, or SOHIO, paid $1.8 billion to acquire struggling Kennecott, which issued its final annual report in 1980. Prior to that date, Kennecott had made its most recent major expenditure on the Chino smelter during 1973–74, when it built a $30 million acid plant to capture about 60 percent of the sulfur dioxide gases produced during the smelting process. To help raise funds to further modernize the Chino complex—its smelter at Hurley, constructed in 1939 and unable to meet environmental regulations, operated only under a special permit—Kennecott sold a one-third interest in the facility to the Mitsubishi Corporation of Japan in 1981.

Mitsubishi agreed to pay about $116 million for its one-third ownership and one-third of all costs of modernization. In all, the two partners spent about $405 million to upgrade mine, concentrator, and smelter facilities at Chino, and that included adding a state-of-the-art flash smelter. Directors of SOHIO gave their approval on September 24, 1982, for Chino to launch a $128 million smelter modification program that included a Canadian INCO flash furnace, new-fashioned technology that increased the smelter's capacity and met New Mexico's environmental standards by recovering at least 90 percent of all sulfur entering the plant. But profits remained elusive, and even after Kennecott spent bundles of money and lowered production costs, copper prices stayed depressed. That was frustrating for SOHIO, which had problems of its own with declining oil prices.

A shake-up at Standard Oil of Ohio made

AN AERIAL VIEW *of the Chino Mines Company property, part of the New Mexico operation in which Phelps Dodge acquired a two-thirds interest in 1986. Back in 1914 the assets of the Chino Copper Company included 16,700 acres of land, 10 steam shovels, 21 locomotives, a fleet of 124 ore cars of various sizes, more than 20 miles of standard gauge track, a machine shop, mill, company homes, and boarding- and rooming houses.*

British Petroleum, formerly a 49 percent owner, the new operator of SOHIO and its Kennecott properties. In March 1985, Phelps Dodge's first break-even month for its Western Operations, Len Judd happened to notice in the *Wall Street Journal* that British Petroleum had grown irritated with SOHIO and Kennecott. Only four years earlier, marriages between oil and copper had been the rage—so much so that by mid-1981 oil companies controlled more than half of all American copper producers—but by the mid-1980s the romance had faded, and several celebrated industrial divorces were in the works. The independent copper producer had not been such a dinosaur after all. Both Judd and President Durham simultaneously recognized a great opportunity for Phelps Dodge to pick up valuable new properties. Durham told Judd that "we need to go to Cleveland" and talk to Standard Oil about Kennecott acquisitions.[5]

On the basis of a handshake at SOHIO headquarters, Judd thought that Phelps Dodge had acquired both Chino and Ray, a large open-pit mine complex in Arizona, but over the weekend British Petroleum agreed to sell Ray to ASARCO. It was all part of the intense horse trading, but Judd was himself certainly no amateur when it came to the art of deal making. Chino, for example, had a refining contract with ASARCO that proved a real blessing to Phelps Dodge because ASARCO charged only three cents per pound to refine ore that cost Phelps Dodge an additional two cents to refine. Judd, however, feigned opposition to the ASARCO refining contract and thereby forced Kennecott to throw in its Lone Star ore body at Safford ostensibly to make the deal palatable to Phelps Dodge.

On September 11, SOHIO announced the sale of Kennecott's two-thirds ownership of the Chino Mines Company to Phelps Dodge for approximately $88 million, a sum far less than Kennecott had recently spent to modernize the smelter alone. All in all, it was a smart move by Judd, Durham, and Chairman George Munroe. Acquisition of Chino was an obvious bargain, but as Bill Epler, longtime editor of *Pay Dirt*, noted, "there aren't many live prospects out there these days willing or able to buy a large copper operation." The new smelter cost less than half of what Phelps Dodge once anticipated spending to bring its old Morenci reduction works into compliance with government air quality standards. The Chino Mines acquisition was thus both a good fit and a good bargain for Phelps Dodge.

Continued Epler, "Now that Phelps Dodge has added the union contracts at Chino to those at Tyrone, observers are wondering if the Steelworkers union will continue to push its corporate campaign to hurt the company in every way possible in retaliation for the 1983 strike that the unions lost. After all, the continued viability of Phelps Dodge is vital to a large and growing number of union members. Some believe the campaign will now be allowed to quietly expire." For its part, Phelps Dodge agreed to honor the four-year union contracts Kennecott had recently negotiated at Chino, but as Judd warned the new Phelps Dodge employees in January 1987, "We're not a sugar daddy. If this mine is a mistake, I'm history and you're out of a job."[6]

When Phelps Dodge took over responsibility for Chino Mines in January 1987, the acquisition increased the corporation's overall copper production by about a third. Phelps Dodge personnel also rapidly "PD-ized" Chino to make it more modern and efficient. A few months later, on May 6, 1987, Phelps Dodge announced plans to spend $55 million to construct a new solution extraction–electrowinning plant at Chino capable of producing at least 90 million pounds of copper a year. The SX/EW facility—operated like the Tyrone facility as part of the Burro Chief Copper Company—went on line in October 1988 with a daily capacity of about 120 tons of copper cathodes. Today Phelps Dodge's nonoperating partner at Chino is Heisei Minerals Corporation, jointly owned by Mitsubishi Material Corporation and Mitsubishi Corporation, the Japanese business empire founded in the early 1870s as a shipping company.

Acquisition of Chino was significant to Phelps Dodge in two major respects. First, it reaffirmed management's commitment to the copper business at a time when it was easy to doubt the future profitability of red metal. At the time of the acquisition, the price of copper still hovered around sixty-one cents a pound, and Chino, despite its recent modernization, struggled to break even. Second, the Chino purchase reflected the growing confidence among Phelps Dodge managers and engineers that they could produce low-cost copper. Southwestern New Mexico soon recorded an all-time high in red metal production. Never before had so much copper poured from the mines of Grant County. And there was still more to come. In 1998 Phelps Dodge added another significant copper property in the area when it purchased Cobre Mining Company for approximately $115 million and thus gained the Continental property, which

comprised an open-pit copper mine, two underground copper mines, two concentrators, and the surrounding eleven thousand acres of land, all located conveniently close to the Chino Mines complex. Though it temporarily closed Cobre later that year because of low copper prices, it stood ready to expand both facilities to develop their synergy at some future date.

More Than Copper

PERHAPS THE MOST radical part of the Committee of Five Plan that plotted a corporate turnaround in the mid-1980s was its call for diversification outside copper. For the previous century, copper and Phelps Dodge had been virtually synonymous, even if the corporation had in recent years ventured into aluminum, uranium, lead, fluorspar, and other minerals. As part of the new strategy, Phelps Dodge intended to diversify enough to end an almost total dependency on copper and its unpredictable boom-and-bust cycles. It sought to acquire basic industries that would continue to earn profits during down times for copper and thus enable the parent corporation at least to cover its dividend.

The problem for Phelps Dodge was that its previous ventures into aluminum, uranium, and oil and gas had left some industry analysts and stockholders skeptical about all its diversification projects. At the corporation's 1985 annual meeting, one shareholder pointedly asked the new president, G. Robert Durham, "How can you assure stockholders that you will be more successful this time around?" His response was,

"We had some bad luck. We had an aluminum venture; we fortunately made a proper and valid sale of that. We also took a move into uranium, and if you could have predicted the downturn in the uranium market future five, seven years ago, you are smarter than we are. We have had some bad luck. But we have done our best to clean up those disappointments."

Ms. Travis: "Mr. Durham, is that bad luck or poor management?"

Mr. Durham: "That is a judgmental factor."[7]

In the mid-1980s, Phelps Dodge finally enjoyed a backhanded stroke of good luck. That was after the corporation grew interested in acquiring the Bath Iron Works in Maine, a diversification prospect that greatly appealed to both Durham and Munroe because the firm was America's number one builder of frigates and destroyers for the navy. A long-established enterprise, it had a big backlog of orders that looked like money in the bank. In fact, all during the cold war it had been a highly profitable enterprise, and as an added attraction, the defense-spending cycle typically offset the copper cycle. With those advantages in mind Phelps Dodge bid $500 million for the Bath Iron Works, but came in second. "Thank God we lost it," Douglas Yearley emphasized during an interview in 1997. It turned out that the workforce was angry; further, the end of the cold war and consequently a sharp downturn in defense spending were near at hand. "But for the luck of the draw, our diversification could have been a disaster." Phelps Dodge directors rejected another acquisition prospect, a conglomerate called Standex Corporation, as "too far outside our scope," recalled Durham.[8]

Among other options, Phelps Dodge considered gypsum companies, but Munroe, who served on the board of Manville Corporation, worried about asbestos litigation, and since some major gypsum producers had also once mined asbestos, he warned against acquisition. That proved excellent advice because asbestos litigation soon beset two major American producers and dragged down their earnings.

Another idea was to merge Phelps Dodge and the Cabot Corporation, a big American producer of carbon black. The two enterprises were then roughly the same size, but nothing came of the idea. Nothing, that is, except that as a result of the Cabot contact, Phelps Dodge management grew intrigued by the carbon black industry, and by Columbian Chemicals Company in particular. On November 4, 1986, Phelps Dodge announced that it intended to purchase Columbian. Just one day later, Phelps Dodge again made news when it announced plans to relocate its headquarters from New York to Phoenix. Things were moving fast.

Why Columbian Chemicals Company? Because it offered Phelps Dodge a compatibly sized original manufacturing business that generated good cash flow, ranked among the leaders in its industry, had strong growth potential, and had able management already in place. At the time Phelps Dodge could not spare top executives of its own to turn around another company. Nor, added Durham, did it care to venture into making any consumer goods. On all counts, Columbian Chemicals Company has measured up well. Today, like Phelps Dodge Magnet Wire, it is a process-oriented business and makes a high-quality industrial product. In

On the east bank *of the Ohio River near Moundsville, West Virginia, the Marshall plant of Columbian Chemicals Company produces valuable carbon black.*

Columbian's case that product is carbon black, a substance derived from petroleum-based feedstocks that it carefully decomposes under controlled conditions to form various types of special-use carbon particles. Carbon black is essentially pure carbon, like diamonds, but has a different molecular structure.

Most important, Columbian's special carbon blacks greatly improve the tread wear and durability of so-called rubber tires, so-called because they actually contain a substantial amount of carbon black. With its large and active surface area, molecular carbon black bonds easily with molecules of rubber to increase resistance to tearing and abrasion and also to increase the stiffness of rubber compounds. That is why carbon black is essential to all types of mechanical rubber goods, such as gaskets, hoses, conveyor belts, and moving sidewalks. In ancient times, Chinese makers of stick inks burned oils such as tung and sesame and used feathers to collect the carbon residues. Carbon black is still favored for high-quality printing inks as well as for modern electrostatic toners, plastic films, telephone cable jackets, and insulation for high-voltage electrical cables. One of Columbian's customers is Scotland Yard, which uses a specialized form of carbon black for fingerprinting.

From 1986 to 1990, Phelps Dodge invested more than $20 million per year to bring Columbian's manufacturing plants to a level competitive with those of any carbon black producer in the world. Within five years it had earned back its money: "There aren't too many investments where you get payback in less than five years," declared Yearley enthusiastically.[9]

Columbian Chemicals Company was clearly on the move, adding a manufacturing property here and removing another there to strengthen the enterprise. At the end of 1998 it manufactured carbon blacks in twelve plants in nine countries and employed well over a thousand people worldwide. That was when, following acquisition of the carbon black business of Brazil's Copebras, S.A., for $220 million (and the world's largest carbon black plant), Columbian proudly ranked second only to Cabot Corporation in world production. With its major presence in North and South America and Europe, Columbian became a truly global supplier.

After a decade as part of Phelps Dodge, how has Columbian changed? For one thing, emphasis on safety in Columbian's several plants increased noticeably. It also discovered

COLUMBIAN CHEMICALS COMPANY HAS a lengthy history of its own—one very nearly as old as that of Phelps Dodge itself. It traces its roots back to the mid-nineteenth-century days of lamp black, which artisans made by slowly burning creosote and other oils in cast steel pans by incomplete combustion. They allowed thick smoke to settle out of the air in a confined environment and collected the soot or carbon to make black pigments used mainly for printing inks and to a lesser extent for paints and lacquers.

Prospectors searching for oil beneath the hills of West Virginia and western Pennsylvania discovered that natural gas could be burned under controlled conditions to produce a black pigment, too. The substance known first as gas black and later as carbon black is about as fine as cigarette smoke. It is part of a family of products that includes graphite, acetylene black, vegetable black, and bone black, the latter made by charring animal bones in ovens or retorts and grinding them to pigment fineness so that they can be used for tints, artists' colors, and artificial leather coatings.

Each of the carbon family siblings possessed distinct characteristics, such as the natural lubrication that made graphite suitable for common "lead" pencils. Among the most recently discovered forms of carbon are fullerenes, named for R. Buckminster Fuller, inventor of geodesic domes, "bucky balls," and Phelps Dodge's own Dymaxion Bathroom of the 1930s (see chapter 8). A bit of early doggerel used to explain members of the carbon family to children contained these words: "Carbon diamond is arranged in a regular array while carbon soot is put together in a helter skelter way."

Further, carbon black, depending on its application, could be engineered to form a very active particle or a rather inert one. The rougher its edges, the more eager it was to bond with other materials. Depending on how it is made, it can be shaped like bunches of grapes or be strung out like beads on a string. "It is incredibly complex material," observes Rodney Taylor, vice-president for technology at Columbian and a fuel scientist who clearly relishes the challenge of digging deeper into the mysteries of carbon black.

Columbian Chemicals today offers more than eighty carbon black grades to satisfy the exacting requirements of users in a variety of different industries. It manufactures each one to meet a specific application, whether that be the finest high-color black for an automobile finish or an inexpensive low-color black for painting iron castings.

At the close of the twentieth century, carbon black in all forms is highly engineered smoke. Like that of magnet wire and numerous

considerable common ground with its parent corporation. Both enterprises maintain a strong work ethic, both utilize the talents of many engineers, and both are technology driven. And, stresses Columbian's President John T. Walsh, "we must religiously pay attention to our costs, and we have been through a series of tough times." One important and valuable difference: Columbian Chemicals Company and its carbon black have much flatter business cycles than copper. Good times or bad, people still buy tires. In fact, by the late 1990s carbon black sales accounted for such a rapidly growing percentage of Phelps Dodge earnings during a time of low *other products we seldom notice though they are vital to modern life, the evolution of carbon black relates to the main currents of American history in surprising ways. Consider automobile tires, for example, which are today the biggest users of carbon black. In 1895 the Michelin Company in France introduced the world's first tires with compressed air for use on a sputtering new invention called the automobile. Pneumatic tires lasted a mere twenty-five hundred miles before they needed to be replaced, if they did not suffer a blowout first. Their natural rubber simply wasn't strong enough. Happily for modern motorists, a chemist discovered by accident in 1904 that when carbon black was mixed with natural rubber it became a significantly more durable product. After that lucky discovery, drivers could go as far as twenty thousand miles on one set of tires, and much farther today. Without carbon black, tires acted like giant erasers and rapidly left all their rubber on the road surface. Carbon black can be engineered to reduce heat buildup in fast-running tires.*

The importance of that discovery was not fully appreciated until automobiles began to multiply rapidly after 1910 and extended tread wear grew important to motorists. Introduction of reinforcing cords into tire carcasses was important too, but without carbon black, the cords would easily have outlasted the rubber. Today, the average automobile tire includes about five pounds of carbon black.

The Columbian Carbon Company, formed by a combination of several different manufacturers in 1921, was long allied with the Binney and Smith Company, a firm that specialized in pigments and whose Crayola crayons have become synonymous with American childhood. In 1955 the two firms went their separate ways. Columbian maintained a far lower profile than the Crayola maker, although by the early 1960s it had emerged as a major international producer and distributor of carbon black. From 1962 to 1979 it was part of the Cities Service Company. On its own once again, it adopted the name Columbian Chemicals Company in 1980 to reflect its broad scope of operations. Phelps Dodge bought Columbian Chemicals in December 1986. It was President G. Robert Durham's ideal "smokestack industry," although it was no longer one that released black clouds of smoke or soot into the atmosphere.

(Based on a telephone interview with Rodney Taylor, February 11, 1999, and "The Columbian Chemicals Story," unpublished ms., ca. 1997.)

copper prices that Columbian was the single brightest star within the corporation's universe. That development affirmed the wisdom of the diversification strategy developed by the Committee of Five back in 1984.[10]

A New Manufacturing Landscape

DURING 1987, Phelps Dodge Corporation assimilated both Columbian Chemicals and Chino Mines companies quickly and smoothly into its family of operations. The two new acquisitions meant that Phelps Dodge moved into 1988 about a third larger than it had been in 1986. But Phelps Dodge was far from finished refashioning its manufacturing landscape. In March 1988 it obtained Accuride Corporation for $272 million in cash and assumed debt. By gaining control of North America's largest producer of wheels and rims for medium and heavy trucks, truck trailers, and buses, Phelps Dodge further confirmed that it intended to lessen its dependence on copper.

Accuride, which originated in 1905 as Firestone Steel Products, the first subsidiary operation of the Firestone Tire and Rubber Company in Akron, Ohio, took its present name in 1986 when it went independent. Phelps Dodge held the Henderson, Kentucky–based enterprise for a decade until selling 90 percent of its interest for $480 million effective January 1, 1998 (and the remainder later in the year). Proceeds from the sale enabled Phelps Dodge to buy the Cobre Mining Company in New Mexico and the Copebras carbon black plant in Brazil.

Back in the late 1980s, both Accuride and Columbian Chemicals had enlarged the holdings of Phelps Dodge Industries, which as recently as the mid-1980s included only Phelps Dodge Magnet Wire and Phelps Dodge International. Another addition before the end of the 1980s was Hudson International Conductors, a family-owned and -operated specialty wire manufacturer until Phelps Dodge bought it in November 1989. The purchase price was approximately $60 million in cash and assumed debt; annual sales were around $85 million.

In 1910 the Hudson firm (dating from 1902) pioneered a process for plating silver on copper wire for applications in the electrical industry, and it subsequently developed special alloys and complex constructions for stranded conductors. During the years that followed, the company specialized in high-strength, high-reliability, and carefully engineered wire. Especially from the 1960s through the mid-1980s, Hudson prospered along with America's expanding aerospace industry, and its biggest single market became commercial aircraft. With the end of the cold war it continued to produce fine wire to meet aerospace requirements for guided missiles, satellites, and even the Hubble space telescope. In addition, its high-temperature nickel-plated conductors for the automotive and appliance markets turned up in fuel system sensor wiring, waffle irons, and hair dryers; its silver-plated copper wire was used to make computers and printer cables.

In May 1996, Phelps Dodge Corporation acquired yet another family-owned specialty products firm, Nesor Alloy Corporation. The New Jersey–based enterprise was a leading manufacturer of tin-, nickel-, and silver-plated wires, and one of Hudson International's competitors. Phelps Dodge united the firms as Phelps Dodge High Performance Conductors, an operating division of Phelps Dodge Industries. Also part of the union was Phelps Dodge's venerable Bayway facility, transferred in 1990 from Phelps Dodge Magnet Wire to Hudson International, which had for years used an alloy that Bayway supplied. Incidentally, another specialty product from Bayway was trolley wire, and one of its customers was Amtrak, which extended it over electrified parts of its tracks serving New York's Pennsylvania Station.

In all, by the mid-1990s an enlarged and revitalized Phelps Dodge Industries accounted for approximately a third of the corporation's annual income and about 49 percent of its assets. From 1988 through 1997, the operating cash flow of Phelps Dodge Industries averaged almost $150 million annually, and in 1995 it topped $243 million. The strategy to diversify out of copper as outlined by the Committee of Five back in 1984 paid off most brilliantly during 1998, when, as a result of another downturn in the price of copper, annual earnings generated by Phelps Dodge Industries for the first time surpassed those from Phelps Dodge Mining Company.

The Decade's Winners and Losers

THE REST OF the story for America's Big Three copper producers in the 1980s was one of triumph or tragedy—a time of rebirth or restructuring or shaking out—depending on your perspective. For Phelps Dodge the period was unquestionably one of newfound success and ascendancy, but what happened to its two major competitors was sad and provided a "but for the grace of God go I" perspective on the whole survival struggle. It clearly illustrated the profound and lasting structural changes that overtook America's copper industry.

Phelps Dodge for most of the early twentieth century ranked well behind Anaconda and Kennecott, although by the time of World War I it had pulled ahead of Kennecott in its copper-producing capacity. During the difficult years of the 1920s it dropped back again. Alas for Anaconda, its greatest properties were located in

Chile where government expropriation in the early 1970s hurt it and Kennecott, but the former far more than the latter. After that, Phelps Dodge steadily closed the gap to become America's number one producer.

Standard Oil of Ohio acquired Kennecott for $1.8 billion in 1981. The timing couldn't have been worse. SOHIO suffered $59 million in losses during the first six weeks of ownership! Kennecott then reported losses of $189 million in 1982 and $91 million in 1983, much as Phelps Dodge did at that same time. During those same years Kennecott laid off three thousand employees and temporarily suspended its Chino and Ray operations. By the late 1990s, Kennecott's only remaining large copper asset was its venerable Bingham Canyon property, which it closed in 1985, reopened in 1988, and then sold along with what remained of the company to RTZ (Rio Tinto Zinc), the giant British mining company, in 1989.

An even more poignant example of what might have happened to Phelps Dodge during the grim days of the early 1980s was Anaconda, which had grown to become the largest nonferrous mining company in the world, and second only to United States Steel as a metal producer. It seemed powerful, but it was also vulnerable, as Chile's takeover of its copper properties in 1971 illustrated. The loss of the huge Chuquicamata mine staggered Anaconda: Chilean expropriation cost it approximately $450 million; Kennecott lost $150 million. Worse still, Anaconda had borrowed heavily from New York lenders to expand its holdings in Chile and had counted on its Chilean profits to repay those loans. To pay off its bankers, the company shed assets, including its forest products division that owned and managed more than six hundred thousand acres of prime timberland in western Montana. In nearby Butte, Anaconda had labored to save its onetime crown jewel and even talked of restoring the "Richest Hill on Earth" to its former grandeur and profitability.

Robert O. Anderson, chairman of Atlantic Richfield Company, in the seventies wanted to acquire Phelps Dodge. Turned down earlier by Chairman Robert Page, in late 1976 he closed a deal with Anaconda, and by January 10, 1977, the copper company became an operating subsidiary of Atlantic Richfield (ARCO). At the time ARCO was awash in cash because its Alaska pipeline had just begun production, and the steady flow of black gold from the North Slope gave it plenty of money to invest. Wanting to diversify, it thought the world of rising commodity prices would surely buoy copper as well as oil. ARCO's goal was to restore the Anaconda Company to its former greatness.

For ARCO, as for SOHIO, the venture into copper turned out to be a disaster. A loss of $332 million in 1982 hit Anaconda hard, and it never was able to trim production costs that averaged nearly twice the market price for copper. A few months later the company shut its Berkeley Pit in Butte and allowed water to flood its underground mines. Anaconda was so ill that the best ARCO could do was liquidate the business permanently after a century of operations. By 1986 the oil company had exited hard-rock mining and sold off Anaconda's remaining assets. Once the industry leader, Anaconda ceased to exist (though ARCO had to retain its environmental liabilities).

The marriage of oil and copper during the 1980s proved a debacle, one of many during that decade of wrenching shake-outs for American copper producers. By 1990, the nation's four biggest survivors were Phelps Dodge Corporation, Magma Copper Company, ASARCO, and Cyprus Minerals, all of which made giant strides to modernize their operations. At the time, Phelps Dodge alone produced more than 40 percent of all copper mined in the United States.

The Road Ahead

CHAIRMAN DURHAM IMPLEMENTED yet another mandate of the Committee of Five Plan when he relocated Phelps Dodge's headquarters—and not merely to another address in New York City. A great believer in the Sunbelt, he chose Phoenix. The move consolidated top management in a single location and was logical because recent pruning of New York–area manufacturing plants had shifted the corporation's center of gravity noticeably west toward its mining company properties.

In December 1987, Durham announced that, because of Phelps Dodge's strong financial recovery, he had concluded the 1984 restructuring program and was no longer willing to sell any remaining properties identified in that plan, among them the corporation's share of the Black Mountain Mineral Development Company in South Africa and Southern Peru Copper. Now Durham wanted to streamline Phelps Dodge itself. In September 1988 he divided it into two operating divisions, each with distinct goals and personalities. One was Phelps Dodge Mining Company, headed by Len Judd, forty-eight, and the other was Phelps Dodge Industries, headed by Doug Yearley, fifty-two. Both men reported to Durham.

One of Durham's great strengths was his ability to fix problems. He seemed bored with everyday management: "If it got quiet, 'Bull' would create a project. We tried not to let it get too quiet," recalled Yearley. Nonetheless, Durham retired early on June 30, 1989, at age sixty. Whether he realized it then, in his future were still more fix-it projects for other corporations. "When I was elected president in 1984 and chairman in 1987," Durham said of his impending retirement, "I had an understanding with my family and with my colleagues at Phelps Dodge that my term in office was not to be determined by the mandatory retirement age but rather by the earlier reaching of goals we had set for the company." He noted that those goals had been reached. Offering a retrospective look at what Bull Durham accomplished between 1985 and 1989, a former senior vice-president, Richard Pendleton, Jr., observed in a 1998 letter to Richard T. Moolick, "I have to tell you that (love him or hate him) Bull did an excellent job of implementing that business plan." Durham was, according to William G. Siedenburg, the vice-president of research at Smith Barney, Harris, Upham and Company, the "mainspring" of the turnaround.[11]

After Durham took early retirement, Yearley became chairman and chief executive officer and Judd became president and chief operating officer. At the time Judd continued to serve as president of Phelps Dodge Mining Company. His professional training, both in terms of formal education in mining and business administration and hands-on experience, typified what it took to manage the corporation's technologically and financially sophisticated mining operations.

Judd had a remarkable career path. A native of Butte, Montana, and son of a diamond driller who did exploratory work for the mines, he originally attended the Montana School of Mines in his hometown before completing a degree in metallurgical engineering at the University of Arizona. Work in thirteen different jobs after he joined Phelps Dodge the first time in 1963 gained him valuable firsthand knowledge of the corporation's mine, mill, and smelter operations in the Southwest. In addition, he attended school on the side to earn a master's of business administration in 1968, a professional degree few people in Western Operations had at the time. After a four-year stint at the Cities Service Company, he returned to Phelps Dodge in 1973 to oversee the new smelter complex at Hidalgo, and from there he continued up the career ladder.

Yearley possessed a similar wealth of training and experience on the manufacturing and marketing side. Between them, Yearley and Judd knew the corporate landscape intimately. Further, they were both veterans of the Committee of Five (like Durham) who would be running Phelps Dodge in general accordance with the strategic plan they laid out in 1984. Judd, however, continued only until 1991, when he reluctantly chose to retire at age fifty-two because he developed debilitating, even life-threatening, cluster headaches. The malady began in 1983, shortly after he moved to New York, and disappeared six months after he retired, but Judd declined to return formally to the world of corporate business. Choosing to retire as president of Phelps Dodge at the peak of his career was for Judd "the most difficult thing I've ever done in my life."[12]

Yearley, on the other hand, remained at the helm of Phelps Dodge until the start of the new millennium, and for a time he added the title of president to that of chairman and chief executive officer. Beginning in 1989 his strategy was to move Phelps Dodge as fast as prudently possible beyond mere corporate survival and commit it once again to a course leading to financial strength and profitable growth. "It's not going to be a revolution," he prophesied at the start of the 1990s. "We're not going from being a copper company to making cigarettes or perfume. But we do intend to build on our strengths and, hopefully, grow and stabilize earnings."[13]

Copper remained Phelps Dodge's core business, and the corporation publicly renewed its commitment to red metal by investing more than $275 million, a sum that included the purchase of the Chino Mines Company, during just the first three years of the Committee of Five Plan. The commitment continued after Yearley took the corporate reins in 1989. During the next four years, Phelps Dodge strengthened its existing

PHELPS DODGE CORPORATION
EARNINGS REBOUND

Year	Net Income (Loss) (in millions of dollars)	Average Copper Price per pound (COMEX)
1980	91.3	$1.01
1981	69.3	0.84
1982	(74.3)	0.73
1983	(63.5)	0.77
1984	(267.8)	0.61
1985	29.5	0.61
1986	61.4	0.62
1987	205.7	0.78
1988	420.2	1.15
1989	267.0	1.25

operations by investing $125 million in SX/EW technology, $250 million more on other projects intended to boost productivity and cut costs at its mining operations, and $125 million to upgrade its manufacturing plants. In addition it launched construction of the Candelaria mine complex in Chile.

All in all, Phelps Dodge, which had been labeled one of the aging behemoths of American industry in the early part of the decade, bounded into the 1990s with the vigor of a colt. In fact, as early as the second half of the 1980s, the corporation posted record copper production for five consecutive years. Chino in 1989 had the best year in its eighty-year history. The result was that when Durham retired, the corporation's balance sheet was far stronger than it had been in a decade, and the future looked bright. The restructuring has been hailed as one of the 1980s top ten corporate success stories.

The biggest financial payoff from the Committee of Five's strategy came initially in 1987 and 1988 when a dramatic climb in copper prices combined with the corporation's ability to produce low-cost copper to make Phelps Dodge quite flush for the first time since 1980. "No one breathed quite normally until copper prices jumped in 1987," quipped Yearley. That year the company had record earnings, and 1987 was only the first of several phenomenally good years. By the end of 1988, long-term debt had dropped to

slightly less than $450 million from more than $600 million when the strategic plan went into effect in 1985.[14]

It was reward time. too for Phelps Dodge stockholders. The corporation radiated newfound optimism when it resumed its dividend in September 1987, and it increased its payout by an impressive 33 percent the following year. Stockholders received an extraordinary dividend of $10 per share in 1989. Those dividends were paid almost entirely out of cash flow generated during those years. Greatly enhanced stock performance reflected the positive changes. Common stock that hit rock bottom in 1984 when it sold for $12.88 a share rebounded to a high of $62.75 by the time Durham left. After Yearley succeeded Durham, the price climbed from slightly more than $31 following a two-for-one split in 1992 to nearly $90 by 1997, before declining copper prices pushed it back down into the lower $50s range in late 1998 and early 1999. That was still about eight times higher than the 1984 low.

Finally, among the developments that forecast a bright future for Phelps Dodge was an exciting discovery its exploration personnel made in the desert of northern Chile. During the 1980s and 1990s, when the corporation not only restructured and turned itself around financially, it also reached out with increasing confidence to distant parts of the globe, and not just in mining but also in manufacturing.

Chapter Sixteen

FROM COPPER KINGDOM
to GLOBAL ENTERPRISE

ALONG THE ROAD back from the brink in 1984 are several outstanding landmarks in Phelps Dodge history. One is solution extraction–electrowinning (SX/EW) technology that the corporation first implemented at Tyrone, and later at Morenci and Chino, to produce low-cost copper. Another is the general make-over of the landscape of production at Morenci after Phelps Dodge had sold a 15 percent stake in the property to the Sumitomo Corporation. Buoyed by a growing sense of confidence, Phelps Dodge dramatically expanded its role in the international arena, particularly after its discovery of a world-class ore body in Chile.

Make-over at Morenci

AT MORENCI, THE internationalization of Phelps Dodge was subtle. It began after President G. Robert Durham and Leonard R. Judd, a senior vice-president, traveled to Japan in late October

~ AERIAL VIEW OF THE *landscape of production at Morenci in 1947. In the right foreground is the concentrator; the smelter stands in the center. After the reduction works shut down in late 1984, the smelter sat idle for the next ten years. Demolition began in August 1995, at which time workers recovered another four million pounds of copper from its maze of ducts, bins, and hoppers.*

1984 in search of investment capital to implement yet another portion of the Committee of Five Plan adopted only days earlier. Initially the corporation had placed as much as 40 percent of its Morenci complex on the auction block, but after lengthy negotiations it instead sold a 15 percent share to Sumitomo Metal Mining Company, Ltd., and Sumitomo Corporation for $75 million in February 1986. It never did sell the other 25 percent, and Phelps Dodge executives now consider that a stroke of good luck.

Certainly the Japanese "made a great buy," reflected Durham, but the company had to have the money. "What was left to sell? Only a small piece of one of Phelps Dodge's mines." In return, the stake supplied Sumitomo, a major world copper smelter operator, with a portion of Phelps Dodge's concentrate requirements. The deal thus gained the cash-strapped company much-needed capital, and the 15 percent of Morenci concentrates it dispatched to Sumitomo was the portion Phelps Dodge lacked smelter capacity for in 1986. Judd called it an "unbelievable blessing" for both parties.[1]

About a year of negotiations went into forging their partnership, but one may wonder what brought the two old-line enterprises together in the first place? They actually had much in common. Like Phelps Dodge, Sumitomo originated as a trading company (some three centuries earlier) and became a major force in the global copper industry. Its Sumitomo Metal Mining Company ranked among Japan's major mining and smelting enterprises; its Sumitomo Electric Industries was Japan's largest maker of electric wires and cables. In short, because their lines of business were so similar, the American and Japanese companies were not strangers to one another.

Judd had used consultants from Sumitomo to help him solve problems at the Hidalgo smelter. Even earlier, Phelps Dodge and Sumitomo had been wire-and-cable manufacturing partners in three countries, including Thailand. In the late 1950s, Sumitomo Electric sent engineers to Phelps Dodge's fabricating plant in Yonkers, New York, to learn how to make power cables. Because of their years of collaboration, "Sumitomo was the logical choice," emphasized Durham. It was a sign too that any past distinctions between its domestic and international operations came to mean less and less at Phelps Dodge during the 1980s.[2]

To part with even 15 percent of the corporation's crown jewel was a tough decision. When executives broached the idea of selling Sumitomo a stake in Morenci, some Phelps Dodge board members thought management had been "smoking pot," recalled Douglas Yearley. They regarded that mine as "our mother lode." Later, after its deal was in place, management advised concerned observers that day-to-day operations at Phelps Dodge Morenci, Inc., would not change substantially from those at the familiar Morenci Branch. In many ways that was true, but money from Sumitomo did pay for improvements that increased the efficiency of the sprawling complex and altered its landscape of production.[3]

Over the years, Phelps Dodge had produced copper at Morenci in various forms: initially as loaflike ingots that came from smelters located in both Clifton and Morenci, then as flat anodes from the big Morenci smelter that operated from 1942 through 1984, and since 1987 as sheets of SX/EW-produced cathodes. Phelps Dodge's initial investment of $90 million in SX/EW technology enabled it to recover copper from dumps that contained material too low grade to run through Morenci's mills. As elsewhere, SX/EW technology dropped production costs for electrowon copper to approximately thirty cents a pound. After adding a second huge SX/EW plant in 1995, Phelps Dodge Morenci became the world's largest user of that advanced form of hydrometallurgical processing. During 1996 its SX/EW facilities yielded more than 525 million pounds of copper, or about half of all red metal that originated at Morenci.

In its quest for greater efficiency, Phelps Dodge switched from train to truck mining at Morenci during the 1980s. The open-pit mine was originally designed for rail haulage and had been operated that way since the early 1940s, but over the years truck technology had steadily improved. By the early 1980s, some haulage trucks began to work in the pit. Their stepped-up efficiency impressed corporation executives, who committed Phelps Dodge to converting the mine as promptly as practicable from trains to trucks by purchasing a fleet of twenty-nine 170-ton haul trucks at more than three-quarters of a million dollars apiece. In May 1986 all train haulage ended in the pit, and just ahead were other major changes for ore transport out of the mine.

Back in 1980, Phelps Dodge halted production at its Metcalf mine and supplied ore to both the Morenci and Metcalf mills from its Morenci mine alone. Production from the Metcalf mine resumed in 1989, as the Clay ore body in the original Morenci pit showed signs of soon being mined out. That same year Phelps Dodge spent $48 million to install an in-pit crushing and conveying system that reduced truck haulage and fully eliminated the need for trains to take ore to the concentrators. After that, trains hauled only carloads of copper concentrates and bundles of SX/EW cathodes from the Morenci complex downhill to a railroad connection in Clifton.

With the new system in place, haulage trucks made relatively short shuttles between the electric shovels and one of the two in-pit crushers or the low-grade ore stockpiles, where SX/EW technology began working its magic on copper. A conveyor belt several miles long extended from the crushers uphill to feed the mills. Where it dipped downhill, the weight of its ore load generated electricity to offset partially the cost of running the uphill portions. By the mid-1990s, Morenci used its long conveyor belt and fleet of haulage trucks—some of them carried as much as 320 tons—to move about 750,000 tons of rock per day. Louis D. Ricketts, who had installed two conveyor belts at Morenci back in the late 1890s (possibly the first ones to be used in the Southwest), would no doubt have admired such efficiency.

In the late 1990s, Phelps Dodge Morenci employed approximately two thousand people. In total, Phelps Dodge Mining Company employed some sixty-five hundred people in five countries. The most important location outside the United States was Chile, where the alliance first forged with Sumitomo in arid southeastern Arizona flowered anew in the lonely, dry-as-dust Atacama Desert. That was where Phelps Dodge unearthed a major new source of copper. Making negotiations between the two partners easy were the alliance they had forged at Morenci and their growing understanding of one another.

The former chairman George Munroe recalled that the chief executive officer of the Sumitomo Bank had once visited Phelps Dodge headquarters while he toured the United States. Noticing the oversized portraits of the founders and former chairmen hanging in the corporation's New York boardroom, he remarked through an interpreter that Phelps Dodge was the first company he had visited in the United States where he felt a sense of history akin to that in Japanese firms. "This was comforting," thought Munroe.[4]

After its Morenci investment, Sumitomo gained enough confidence in Phelps Dodge to spend another $40 million to acquire a 20 percent stake in the massive ore body that a Phelps Dodge exploration team discovered in northern Chile. As Compañía Contractual Minera Candelaria, the newly developed mine and mill complex became Phelps Dodge's biggest and most productive venture outside North America. For that reason, Candelaria helped raise the curtain on a new phase of international mining enterprise—much as acquisition of Bisbee's Copper Queen Mine had resulted in major new opportunities in the Southwest a century earlier. In addition, discovery of Candelaria highlighted the growing sophistication of Phelps Dodge exploration personnel.

Adventures in Exploration

THE WORK OF exploration was never for the faint of heart. It often meant living in isolated camps, where the main compensations were everyday camaraderie and the special sense of excitement that came with a major discovery. It meant curiosity strong enough to drive searchers despite fears of encountering poisonous snakes and torment from blood-sucking leeches as they tramped through steamy jungles—travails that make for great stories in retrospect.

One of the best raconteurs is Patrick J. Ryan, the jovial South African–born geologist who headed Phelps Dodge's exploration activities for many years during the 1980s and 1990s. He fondly recalled a time during the early 1980s when he investigated old underground workings in the eastern Transvaal of South Africa accompanied by a local guide and a young lawyer and other personnel from Phelps Dodge's headquarters in New York. The mine was dark and infested with bats (some perhaps rabid), and because the roof of the adit was low and uneven, the group sometimes had to walk bent over. Adding to their sense of foreboding was a warning from locals that the place was home to a black mamba, a South African snake so deadly that if it bites a person above the belt there is no effective antidote for its lethal venom. The serpent is commonly nicknamed the Two Step because that is supposedly how far a person can run before being felled by its poisonous bite.

Sure enough, as they slowly proceeded deep into the mine, their guide suddenly yelled out "Inyoka!" the Zulu word for snake. Just ahead in the glare of his light lay a black mamba. Team members nervously debated among themselves what to do. Because everyone wore respirators, muffled speech added to the otherworldly nature of their quest. Meanwhile, their adolescent guide impulsively picked up a loose piece of timber and swung at the reptile, which slithered into the dark, perhaps toward them.

They hesitated a few moments until someone yelled, "There it is!" True or not, that was enough to spook the whole group. Everyone turned and ran toward the distant entrance. Almost immediately the lawyer's hard hat slammed into a low section of roof, and the impact knocked him back onto Ryan, who was himself beating a hasty retreat. Both men tumbled to the dark floor. Whereupon, Exploration's legal counsel and the corporation's future president, Steve Whisler, did the perfectly human thing: he jumped to his hands and knees and crawled off at a record-breaking pace.[5]

"The lifeblood of a mining company is its exploration," Ryan stated with great passion during an interview in early 1996. "That is because each mine has a finite life. You must either replace exhausted ore bodies with new ones, or buy new ones. Good mines are not normally for sale. Without exploration the company is going to die." He noted that "exploration is really the cheapest way to find new mines and grow the company."[6]

Unlike exploration in the days of James Douglas, modern exploration is not an individual effort. Whenever Ryan was asked to discuss the company's activity over the years, he invariably credited teamwork. During an interview back in 1978 he noted, "Of course, most of the detailed work of piecing together the whole jigsaw puzzle of the ore body was done by my staff and these chaps deserve a tremendous amount of credit for all the work that was done in a very thorough and detailed fashion." A typical team draws together geologists, geophysicists, geochemists, mining engineers, metallurgists, accountants, and many other specialists. Like the famous detective Sherlock Holmes, they use an array of tools to obtain the clues they use to build up a picture piece by piece. This evidence will help Phelps Dodge decide whether to take the next step and actually develop a mine.[7]

While engineers and geologists investigate what lies underground, other specialists investigate the availability of a workforce, water resources, electric power, transportation and communications, and a local infrastructure to support the mine complex. Financial analysts study sources of funding and calculate the impact of interest payments on the proposed mine's operating costs and cash flows. The process illustrates how brains as well as brawn are required for success in modern metal mining. At Morenci today, computer programs will help determine the characteristics of an ore body, but the familiar rock pick and compass have by no means been consigned to the shelves of a museum. "There is still a place for the prospector," explained Robert Jenkins, who is manager of international exploration for Phelps Dodge Exploration Corporation. "You've got to get out into the field and beat on rocks. You still must walk the terrain."[8]

The first step toward locating a promising ore body is geologic mapping, which may mean walking over outcrops and noting rock types, the abundance and type of ore minerals, and the thickness and orientation of veins and faults. At Morenci that type of basic information is often compiled with the aid of maps dating as far back as 1905. Historic maps of surface and underground geology sometimes offer the only easy way to learn about areas that are currently covered by low-grade ore stockpiles or otherwise inaccessible. Next, workers drill holes in a grid pattern to obtain samples from the mineralized area. Specialists study the data in order to fashion a three-dimensional model of the deposit, for which they will use computer modeling to predict ore grades between the drill holes. Modern exploration involves a lot of drilling and geophysical surveying because today's searcher for copper will not likely delineate a large ore body using surface evidence alone.

The exploration team forms the advance guard of any mining company. Mining by nature is temporary, and team members might be described as "the clouds" of the corporate landscape, muses S. David Colton, onetime counsel to Exploration and now a Phelps Dodge Corporation senior vice-president and general counsel: "They often come and go and don't give you anything. In fact, you're regularly confronted with disappointment," a sobering fact given a positive twist by A. L. ("John") Lawrence, president of Phelps Dodge Exploration Corporation, who observed in 1998 that "in a global sense exploration is a losing game, but there are players who are consistently successful, consistently adding value to the company, and we're one of them." Successful exploration, furthermore, requires both sustained effort and a

corporate commitment that continues through both good times and bad.[9]

The processes of exploration can take up to six years, as was the case for Candelaria, and sometimes longer. And remember, notes Colton, that "mines are very much like children: each one is different; each one is unique." For that reason, modern Phelps Dodge mines overseas, no less than the historic ones closer to home in Bisbee and Morenci, come with their own adventure stories.[10]

From South Africa to Chile

AT ONE TIME in the not-too-distant past, Phelps Dodge had boasted of its all-American status, though during the late 1950s it reached out cautiously from its established bailiwick in the Southwestern borderlands to invest in the Southern Peru Copper Corporation. In the early 1970s it reached partway around the world to probe for mineral deposits in more distant Australia and South Africa.

There was a pattern to early overseas exploration activity. In the late 1960s and early 1970s, when foreign governments expropriated American and European-owned mines, Phelps Dodge preferred to play it safe. As it surveyed mining opportunities outside North America, apart from its relatively modest stake in Peru, it came to believe that only Australia and South Africa offered reasonable assurance that if it located a promising mineral deposit it would be allowed to develop it.

About thirty miles from the Australian

～ S. DAVID COLTON *became vice-president and general counsel of Phelps Dodge Corporation in 1998, at which time he joined the Senior Management Team. He was elected a senior vice-president in late 1999. Eleven years earlier he had joined Phelps Dodge Mining Company as exploration counsel. Colton later served as vice-president of Phelps Dodge Exploration Corporation. He holds a B.A. degree in economics and a J.D. from Brigham Young University.*

capital of Canberra it became a partner in the Woodlawn Project, a $90 million open-pit mine that after commencing full production in 1979 yielded mainly zinc and some copper. Phelps Dodge earned $2.8 million from the joint venture in 1980, but Woodlawn proved a disappointment, and the corporation sold its one-third interest in 1985 when it pared away underperforming properties to raise money needed for other projects. Also as part of its 1985 restructuring, Phelps Dodge sold its 49 percent stake in the once-promising Cayeli copper-zinc deposit in Turkey.

It was in South Africa that Phelps Dodge made a major find—and not just in terms of mineral deposits. There it gained Patrick J. Ryan, a man in the mold of the elder James Douglas and Louis S. Ricketts, who combined rigorous scholarship with a passion for field research. From the day Ryan was born on a gold mine in South Africa in 1937, until long after he retired from Phelps Dodge as a senior vice-president in 1995, mining formed the focal point of his life. In fact, both his father and grandfather had long been active in South African gold mining. Young Pat launched his own career in mining some eight thousand feet below the surface of the Witwatersrand where he operated a jackhammer. He also attended the South African School of Mines, from which he graduated with a diploma in mining; he later earned a degree in mining geology and master's and doctor's degrees in geology from the University of the Witwatersrand. For Ryan the search for good ore bodies was more than an intellectual pursuit; it was his personal passion. He loved reading,

particularly anything to do with mining, and even when ostensibly on vacation he and his wife spent considerable time visiting mines.

Ryan, after working initially for Anglo American Corporation of South Africa, joined Phelps Dodge's South African operations in October 1970 while they were still in the beginning stages. In short order he discovered a large tonnage of fluorspar in the western Transvaal, on a property he recommended that Phelps Dodge purchase in 1971. Until sold in 1999, Phelps Dodge Mining (Pty.) Ltd., was a wholly owned subsidiary operating the Witkop mine and mill to produce acid-grade fluorspar concentrates, an important mineral that the chemical industry uses to make refrigerants.

Witkop was a good beginning. In 1971 Ryan, Phelps Dodge's newly appointed exploration manager for southern Africa, played a major role in the discovery of the Black Mountain mine in far northwest Cape Province, about 180 miles from Cape Town and at the edge of the Kalahari Desert. Phelps Dodge spent about $22 million on the initial development of Black Mountain, but because it was nervous about investing any more money in South Africa it sought a partner willing to supply the necessary capital to bring the mine into production, about $220 million. That was why in 1977, following discussions with several mining companies, Phelps Dodge selected Gold Fields of South Africa Ltd. as its joint venture partner and retained a 49 percent interest in Black Mountain Mineral Development Company (Pty.), Ltd. The mine came into full production in late 1979. Though Black Mountain never was the lucrative property Phelps Dodge originally hoped it would become, by the mid-1980s the underground mine in South Africa had matured into one of the world's leading lead producers.

The Black Mountain discovery helped make Ryan's reputation at Phelps Dodge. Back in the late 1920s, miners had sunk a shaft on Black Mountain (Swartberg) very close to the site where Ryan and his crew found a rich deposit, but those early seekers came away empty-handed. At least four more times between 1932 and 1963 mining geologists examined the area only to miss its mineral treasure. The well-concealed ore body did not provide clues that fit the searchers' preconceived notions of where to look. The story continues in 1969 with William K. (Bill) Brown, at that time Phelps Dodge's vice-president for exploration, who toured the region and concluded that the company was justified in exploring there for copper and other minerals.

Progress remained halting and uncertain until after Phelps Dodge hired Ryan. He was undeterred by geologists' lukewarm reports. After carefully studying an aerial photograph of the Swartberg range together with a plan that showed surface mineralization, he investigated the area in person toward the end of March 1971. He saw initially the shaft sunk on Black Mountain in 1929 and various copper-stained outcrops. It was nothing very exciting. However, as Ryan picked his way across the black magnetite scree that gives the promontory its name, he recognized what geologists call *gossan*, a Cornish word designating the color of earth—red, brownish, yellow—formed by decomposition of an outcrop of iron compounds and often a good indicator of copper and other metals. Earlier

searchers had apparently overlooked or misread that weathered clue.

After some delicate negotiations about drilling rights with the property's owners, who had begun to look to Rio Tinto Zinc to search beneath their land, Ryan dispatched a diamond-drill operator named Danie Oberholzer to Black Mountain on June 21, 1971. A week later Oberholzer called him back to report, "We've had wonderful rains" in an area that averages only about two inches a year. The news was code for a big find. The driller, following Ryan's directions, had bored at a spot where deep underground he pierced a large mineral seam and thus proved the Black Mountain area rich in lead, zinc, copper, and silver. Later, exploration crews made an equally promising find three miles away at Broken Hill.[11]

Enthusiasts soon hailed the remote area as South Africa's newest treasure-house, perhaps even a second Witwatersrand. It was not—at least not for Phelps Dodge, which finally disposed of its 44.6 percent stake in November 1998—but during the first flush of excitement back in 1971, no one knew how Black Mountain might turn out. Unquestionably it proved a valuable experience for Ryan, whose South African discoveries earned him considerable respect from top Phelps Dodge management.

He certainly "has the touch" when it comes to finding ore bodies, emphasized G. Robert Durham, his onetime boss. In fact, it was Ryan and his enthusiastic team of geologists and other specialists who not only located the Black Mountain deposits in South Africa but also the Candelaria deposit in Chile, which turned out to be the bonanza everyone hoped it would become. "It is very rare that you get a Candelaria," explained Dave Colton.[12]

Phelps Dodge's involvement in Chile dated from 1967 and its investment in COCESA, a wire and cable joint venture. The Marxist government of Salvador Allende expropriated the property in 1973, but following the coup that quickly brought new procapitalist leadership to power, COCESA returned to Phelps Dodge. As its understanding of Chile grew, Phelps Dodge carefully risked investing in the country's mines as well. The payoff came when its exploration team discovered the fabulous Candelaria ore body.

Candelaria is located near Tierra Amarilla, an outpost on the southern fringe of the Atacama Desert in northern Chile. The mine site lies about twelve miles south of the regional center of Copiapó, five hundred miles north of metropolitan Santiago, and about fifty miles inland from the Pacific Ocean. Some places in the Atacama Desert have rarely recorded measurable rainfall, and there are areas where its tan-brown landscape extends to the horizon unbroken by wisps of vegetation. The Sonoran and Chihuahuan deserts, location of Phelps Dodge's major mining operations in the United States, are lush by comparison. Beneath the arid ground of the Atacama Desert, however, lie rich deposits of copper.

Candelaria is a modern chapter in a lengthy and ongoing saga of mining in an area where the hunt for copper and other metals was already under way by the early nineteenth century. By the late 1830s, Chile's numerous mines and smelters allowed it to export some 26 million pounds of metallic copper a year, but because of all the new mines that came into production in the United States in the late nineteenth century, Chile's share of world output amounted to only 5 percent in 1900. By contrast, at the beginning of the next millennium it would account for a whopping 25 percent. The United States, which accounted for between 55 and 60 percent of world production in 1900, had dropped to a mere 14 percent by 1986, about the time that Phelps Dodge stepped up its exploration for more copper in Chile.

Helping give Chile's copper output a big boost in the early twentieth century was the Chuquicamata mine, which almost a hundred years later still ranked among the largest and most productive copper properties on earth. When mining began in 1915, Chuquicamata stood among the world's pioneer open-pit copper producers. The Guggenheims sold the property to the Anaconda Company during the 1920s for a total of $140 million.

In the mid-1960s, *Forbes* magazine contrasted Anaconda and Phelps Dodge as major copper producers. At the time Phelps Dodge was still fundamentally a domestic corporation (and thankfully so during the expropriations of the 1960s and 1970s). It continued to derive all of its 262,400-ton annual production and most of its $327 million in sales and $40 million in profits from operations located within the United States. As for Anaconda, it derived only 22 percent of its copper production and only 12 percent of its total profits from within the United States. *Forbes* noted that Phelps Dodge's fortunes, which outside the United States at the time consisted mainly of a modest 16 percent interest in the Southern Peru Copper Company (later reduced to 13.9 percent), did not hinge, as Anaconda's might, on the outcome of an election in Chile. Few words were ever more prophetic.[13]

By the end of the 1950s the Yankee mining firms in Chile, notably Anaconda and Kennecott, found themselves targets of growing public criticism. Government officials grumbled that the dividends they returned to American investors should be used to fund the expansion of mines in Chile, but the two giants were afraid to do so because of the political storm they saw brewing on the horizon. It remained business as usual for Anaconda's big Chuquicamata mine until 1967, the year the Chilean government shocked the industry by taking a 51 percent stake in Kennecott's El Teniente copper mine; and two years later, in 1969 and 1970, it gained a 51 percent interest in Anaconda's Chuquicamata and El Salvador properties. Allende expropriated the remainder in 1971, with no compensation to the American company, and Anaconda never recovered from the financial disaster.

Then as now, copper was a high-profile industry in Chile. All of the top dozen Fortune 500 firms combined never bulked as large in the United States economy as did just Anaconda and Kennecott in Chile's national economy, and their mine properties proved too tempting. Originally, five separate state-owned companies ran the expropriated mines and ancillary facilities. Following the overthrow of Allende and his Marxists in 1973, the military junta agreed to compensate investor-owned copper companies for their losses, and in that way it encouraged the return of foreign investors to Chile. Even so, Chile's government retained ownership of the five nationalized copper companies, which on April 1, 1976, it merged with the Corporación del Cobre de Chile, or CODELCO, the government-owned enterprise that overnight became the world's largest copper producer. Following the troubled early 1980s, Chileans became responsible copper producers too.

One American copper giant that took newfound interest in Chile's mines in the early 1980s was Phelps Dodge. Uncharacteristically, the traditionally conservative corporation was well ahead of the pack in seeking to do business in Chile after Allende. It did hope to insulate itself from any resurgence of anti-Americanism after it discovered a major ore body by inviting its Japanese partner at Morenci, Sumitomo, to take a 20 percent stake in the new Candelaria mine and mill complex and also by putting together an international group of lenders. Today the sprawling Candelaria complex combines sophisticated engineering design with the technical skills of a highly trained, multidisciplinary workforce to yield copper concentrates at attractive production costs.

Viva Candelaria!

THE STORY OF Candelaria, as is always true at Phelps Dodge, is really that of individuals working to make things happen. It begins in the late 1970s when, at the urging of William K.

Brown, vice-president for exploration, Phelps Dodge pursued some promising leads in the Atacama Desert. Unfortunately, when Brown left a short time later he took most of the corporation's team of mining personnel in Chile with him. That forced his successor, Robert L. Swain, to start anew.

In 1982 Swain and Bert Renzetti, his exploration manager for Chile, recommended that Phelps Dodge acquire control of recently privatized mine property that subsequently became Compañía Contractual Minera Ojos del Salado. Named for a towering volcanic peak in the nearby Andes and located south of the city of Copiapó, Ojos del Salado owned and operated several small underground copper-gold mines and a mill capable of processing eight hundred tons of ore per day. President Richard Moolick, himself a trained geologist, flew to Chile in June 1982 to investigate Ojos del Salado, which he agreed could profitably mine enough 2 percent copper ore to fund an aggressive exploration program that Phelps Dodge planned for Central and South America. But he was so fearful that the Phelps Dodge directors might turn down the acquisition that he never did show them any photos of the mill "because of its ramshackle appearance." They nonetheless gave their approval, and in that way Phelps Dodge Corporation, through its recently formed Small Mines Division, gained a 64 percent stake in Ojos on January 1983 for $1.6 million.[14]

Neither its newly acquired majority interest in Ojos del Salado in Chile nor a stake in Turkey was "expected to make a big contribution to overall revenues or earnings," reported the *Wall Street Journal* in mid-1983. The oracle of Wall Street was right about the Turkey mine but dead wrong about Ojos del Salado, over which Phelps Dodge gained full ownership in 1985. Thirteen years later, in late 1998, slumping copper prices forced Ojos to close down at least temporarily, but until then, two expansion and modernization programs enabled it to generate large enough profits to fund Phelps Dodge exploration activity in Chile and Brazil. That was not even the best part—nor does it explain why Swain, Renzetti, and other Phelps Dodge exploration personnel involved in its acquisition had a right to brag that it opened the door to Candelaria. As Len Judd reflected in 1998, "We'd never have gotten Candelaria without Moolick. He got us into Ojos."[15]

In the early 1980s at the urging of Chairman George Munroe, Pat Ryan relocated from South Africa to Phoenix where he became a senior vice-president in charge of exploration and international mining operations. In that capacity he worked persistently to extend the life of Ojos del Salado at a time when Phelps Dodge cut back everywhere because of the copper crisis. "Please bear with me because we can turn this around," Ryan once begged his skeptical colleagues, who in 1985 pressured him to get rid of the Ojos property. Instead, he arranged to send surplus mine equipment from moribund Western Nuclear to upgrade operations in Chile, which in turn necessitated more ore to feed the Ojos mill. That was how the Phelps Dodge exploration crew came to discover the world-class copper deposit located across the Copiapó River from the mill and about three miles away.[16]

Ojos del Salado leased there what were known as the Lar claims and mined about eighteen hundred tons of oxide copper-gold ore from a small outcropping. It dropped its option on the property in 1984 because of poor metallurgical recovery from the oxide ore at the surface. But it remained clear to Phelps Dodge managers that they still needed to find additional ore if they hoped to keep the Ojos del Salado concentrator running. That was why Phelps Dodge geologists recommended further study of the Lar property, and the corporation optioned it again in late 1985.

Once more it almost gave up its option on the Lar claims, and it even had a letter to that effect ready to mail when Ryan intercepted it at the last minute. He had discovered a clause in the contract that permitted Ojos del Salado personnel to mine the Lar claims to keep their concentrator running while paying out a small royalty. Meanwhile, the Phelps Dodge exploration team carried out a percussion drilling program in search of sulfide ore in an area south of the Lar mine.

One of the drills found enough promising ore to whet everyone's appetite. Using technology that runs electricity through the ground (induced polarization surveys) to reveal what lay hidden beneath the surface, the exploration team delineated a "terrific anomaly," in Ryan's words. Percussion drilling had pierced the topmost portion, but because of inrushing water they had probed no farther. However, as a result of the new information, Phelps Dodge in February 1987 bored a diamond drill hole close to the initial percussion hole. It intersected more than 165 feet of mineralization averaging almost 2 percent

copper, and that proved only a promising beginning. Meanwhile, almost as a bonus, the team found enough additional ore close to the Ojos mine to keep its concentrator running for more than ten years.

As for the copper deposits that formed the basis for the new Candelaria mine (named by Phelps Dodge for Chile's patron saint of miners), more months of exploration followed, both on the ground and in the air. Exploration crews made magnetic surveys while flying across the whole area. That remarkable technology utilizing "stingers" and "birds" dated back to World War II when the highly sensitive instruments were first developed to locate enemy submarines hidden beneath the surface of the ocean. Out of the complicated maze of cross sections emerged geologic and ore body models based on data from more than three hundred test holes and almost seven thousand feet of underground workings completed between 1987 and 1990. In all, company scientists and engineers performed more than eighty thousand analyses for copper, gold, and silver as well as more than seventy thousand rock density measurements. Once the work of exploratory drilling and metallurgical testing was done, they used computerized techniques to construct a geologic block model.

One thing became clear: after years of exploration effort Phelps Dodge had indeed located a major copper deposit—and one that compared favorably with ore bodies that already had yielded fortunes in copper and other minerals at Tyrone and Ajo. For that reason, even in its exploration phase, Candelaria captivated senior Phelps Dodge management in a way that Southern Peru Copper or the South African

An elevated view of the 1990s mine and concentrator complex at Candelaria. Safety and health practices at Phelps Dodge's Chile facility are the same as those in place at its mine and mill operations in the United States.

enterprises never really had. Executives were looking for any good news after the copper crisis of the 1980s, and that was why Candelaria was so important in the history of the corporation. It boosted morale first, and then the bottom line.

Phelps Dodge completed a final feasibility study during March 1991. Fifteen months later, the corporation decided to proceed with development and construction of its Candelaria open-pit mine. Not only had computer modeling figured into the corporation's decision, but computer-aided techniques designed the mine itself. Computer studies told Phelps Dodge where to mine first and even simulated the full landscape of production. Even after committing to the mine and concentrator complex, however, Phelps Dodge needed more than two years to complete financing arrangements that involved lenders from four countries.

Phelps Dodge crews commenced removal of the overburden on February 28, 1993, a process they completed in September 1994, only shortly before production of the first concentrate. Candelaria came into production about four months ahead of schedule and under budget at a cost of close to $570 million. By late 1994 the mine and concentrator (built by the construction giant Bechtel Corporation) operated twenty-four hours a day, seven days a week. No fewer than eight years had elapsed since Phelps Dodge first began systematic exploration of the area, and that time was reasonably short for such an accomplishment today.

Like other Phelps Dodge operations, the new open-pit mine used a computerized truck-dispatching system. From Candelaria's mill complex, trucks hauled 30 percent copper concentrate along the Pan American Highway to the Pacific Ocean. In January 1995 the first load went aboard ship at Punta Padrones, the new port facility Phelps Dodge had constructed on Caldera Bay. Providing maximum protection to the marine environment, the high-tech, totally enclosed port minimized both dust and the potential for accidental spills of concentrate, most of which was destined for Japan. "I really believe the port in Chile is probably the most incredible thing I've been responsible for building," observed Richard Rice, who oversaw construction worth at least $2.6 billion at Phelps Dodge. Chairman Yearley was only half joking when he referred to the impressive loading facility as his "Taj Ma-Port!"[17]

Compañía Contractual Minera Candelaria originally employed more than seven hundred people at the mine and mill site, the port facility, and its Santiago offices. Most of them were Chileans hired from Copiapó and surrounding communities; about 2 percent were expatriates. Phelps Dodge also constructed a number of ancillary community facilities at Copiapó, including the San Lorenzo school, roads, parks, and sports and recreational facilities. Apart from profits flowing to its American and Japanese partners, Compañía Contractual Minera Candelaria will likely inject more than $65 million each year into the Chilean economy through payrolls and the purchases of goods and services. In addition, the big mine and mill complex was predicted to generate billions more dollars in taxes for the Chilean government during the mine's life.

So rewarding was the Candelaria operation to all parties that on May 1, 1996, Phelps Dodge

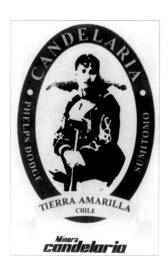

⟶ LOGO FOR THE
Candelaria mine and mill complex.

announced plans to double production at its concentrator. Its contractor finished the work in November 1997, more than eight months ahead of schedule and approximately 10 percent below the budgeted cost of $337 million. "We believe that the 18 months it took to complete this expansion is the fastest completion ever of a concentrator project of this magnitude," emphasized Yearley. In all, the expansion boosted Candelaria's annual copper production to about 380 million pounds and added approximately two hundred employees, bringing the total to nine hundred. From any perspective—whether technological or financial—Candelaria is an impressive operation. Moreover, because its development was relatively recent in Phelps Dodge history, most of the people whose skills and ingenuity (and even luck) made it possible were still alive in the late 1990s to talk about it.[18]

There is an element of luck in all exploration, emphasizes Ryan, who remains modest about his own accomplishments in Chile and South Africa. Yet he is quick to paraphrase a fellow South African, the golfer Gary Player, who observed that "the more I practice the luckier I become." And speaking of luck: during the copper crisis of the 1980s Ryan almost took a job in South Africa with his former employer, Anglo American; "I was one of the rats" who could see the Phelps Dodge ship sinking, he said, believing that the company was only a few months from Chapter XI bankruptcy (voluntary reorganization). In the end, Munroe asked him to help Durham with the corporate turnaround. As it turned out, future events proved lucky for both Phelps Dodge and Ryan.[19]

Evolution of Phelps Dodge International

APART FROM ITS various mining and manufacturing operations in North America, Phelps Dodge by the late 1990s had assembled a far-flung empire of businesses abroad. Because the corporation's various enterprises outside the United States take so many different forms, the structure and scope of its international activities perhaps appear confusing to anyone not already familiar with its numerous landscapes of production. For example, Candelaria and other overseas mining and milling properties are part of Phelps Dodge Mining Company, based in the parent corporation's home offices in Arizona. Headquartered in the same high-rise building is Phelps Dodge Industries, one component of which is Phelps Dodge International, based in Coral Gables, Florida. In addition, several overseas carbon black plants are operated by its Marietta, Georgia–based Columbian Chemicals Company, and magnet wire plants in Austria and Mexico are part of Fort Wayne–based Phelps Dodge Magnet Wire Company, both basic segments of Phelps Dodge Industries.

For many years it was Phelps Dodge International that provided the parent corporation its greatest degree of involvement in countries outside North America. The story of International—which exercised responsibility for managing profitable and cash-generating plants that fabricated building wire, power cables, and telephone cables in more than two dozen countries, primarily in Central and South America and the Far East—dates back to the aftermath of World War II. That was when, like most United States industries, Phelps Dodge Corporation discovered a bonanza in the postwar export trade.

With America's Marshall Plan helping to rebuild Europe, it was easy for American enterprises to sell their products abroad. But by the early 1950s both Europe and Japan had started to regain their economic strength and were becoming less dependent on companies based in the United States. Further, high-priced American goods became less competitive abroad. At one time about 30 percent of all the copper products Phelps Dodge produced were exported, but that figure soon fell to just 2 percent. It became clear that the only way for Phelps Dodge to make money abroad was to set up overseas plants jointly operated with local entrepreneurs, national governmental institutions, or competing international manufacturers.

After World War II there was little interest in Latin American wire and cable manufacture because United States businesses focused so much of their attention on the work of rebuilding Europe. However, one man at Phelps Dodge sensed an opportunity. That was Ted Michaelsen who in 1956 initiated work leading to what became Phelps Dodge Copper Products International Corporation the following year. Its first international manufacturing operation was the Pheldrak plant located in Cuba, the result of a joint venture among Phelps Dodge, the Cuban National Development Bank, and Draka, a Dutch company. In addition to the Cuba plant, which produced its first cable and wire in May 1957, Phelps Dodge in rapid succession launched plants in Venezuela, the Philippines, Puerto Rico, and

Colombia. Those were the five pioneers, and a Spanish heritage was probably their common denominator.

In Cuba as in other nations, Michaelsen's way was to promote a joint venture between Phelps Dodge and local investors who knew their home country well. Because of him, Phelps Dodge became one of the first American cable and wire companies to adopt that strategy. It was not foolproof, however. For a short time after Fidel Castro and his Marxist rebels came to power in 1959, the new leader of Cuba cited Pheldrak as a model for foreign investment, but shortly after Uncle Sam imposed an embargo on

PHELPS DODGE'S FIRST overseas fabricating plant was in Cuba and earned a net profit of $27,000 the year after it opened in 1956, of which the parent company's share was $3,000.

Cuba, the Castro government retaliated by expropriating American properties, including Phelps Dodge's in 1960. That was a real jolt to the parent corporation, which also lost a big cable manufacturing plant to Iranian revolutionaries during the 1970s.

Over the years, Phelps Dodge closed a plant in Lebanon because of that country's increasingly unstable political climate. Allende's Marxists expropriated Phelps Dodge International's COCESA plant in Chile in 1973. Insurgents in Central America once kidnapped the company's El Salvador office workers at gunpoint and held them as pawns to persuade the government to free rebel prisoners. Currency devaluations in some countries, though less dangerous than kidnappings, posed a big problem too.

At one time, staff members of international cable-and-wire fabricating operations were trained in Phelps Dodge plants in the United States, but perhaps because of their American connections, joint ventures were sometimes criticized by native entrepreneurs for paying local workers too much. But to astute plant managers it made little sense to pay a person merely two dollars a day to run a machine costing half a million dollars. The most successful managers adapted quickly to local cultures, attended weddings and funerals, and became personally acquainted with employees' families.

When Phelps Dodge International was incorporated on December 20, 1966, its mission was to provide management, marketing assistance, technical support, engineering, and purchasing services to its associate companies worldwide, primarily manufacturers of electrical and telecommunications cables. It also contributed to the engineering and installation of telephone lines in several countries. By early 1969, Phelps Dodge's equity interest ranged from 10 percent to 61 percent in seventeen wire and cable companies operating in fifteen different countries. A decade later it had invested in thirty wire and cable plants located in twenty-two countries around the world from Argentina to Zambia. Joint-venture relationships worked well in most countries. Some started small, with an average investment by Phelps Dodge of less than $2 million, and subsequently expanded; others Phelps Dodge essentially built from the ground up. The corporation's basic strategy proved most successful in developing nations, but in England and France far less so, and from them Phelps Dodge International later withdrew.

Michaelsen, essentially the father of Phelps Dodge International, was very much at home in a worldwide environment. A native of Copenhagen and educated in Denmark and England, he was twenty-two years old when he joined Phelps Dodge Copper Products Corporation in June 1940. Michaelsen's father was Phelps Dodge's sales agent for copper rod in Scandinavia, and Wylie Brown, the president of Copper Products at the time, may have thought he was doing him a big favor by employing his son as a price clerk earning $25 a week. Whatever his motivation, Brown got young Michaelsen, who was half-Jewish, out of Denmark and thus out of harm's way from Nazi invaders.

When the newcomer reported for work, Brown instructed his colleague William Dunbar, "This is my friend Mr. Michaelsen's son. You take him out there and show him the business. You come in and tell me every night how he is doing." On one occasion, Dunbar shared with Michaelsen a group of records from the company's export file. That night, in yet another example of Brown's petty tyranny, he berated Dunbar for revealing too much information to their new hire. Impressed by his understudy, Dunbar later advised Brown, "I think he would be ideal to set up an export company," something Phelps Dodge lacked at the time. That concept later matured into Phelps Dodge International, probably Michaelsen's greatest legacy.[20]

Ted Michaelsen became export sales manager for Phelps Dodge Copper Products after World War II, a promotion that gave him insight into developing markets in Latin America. With a small amount of investment money he wrangled from Howard T. Brinton, Copper Products president, he launched Phelps Dodge in Cuba. Michaelsen became vice-president of newly formed Phelps Dodge Copper Products International in 1956 and its president in 1962. Five years later, he became the executive vice-president of newly formed Phelps Dodge Industries, Inc., created in 1966 to direct and coordinate the company's manufacturing activities. Michaelsen succeeded Ed Dunlaevy as president in 1968, and chief executive officer a year later, positions he held for more than a decade. In 1980, he became one of three Phelps Dodge Corporation vice-chairmen, along with Richard T. Moolick and L. William Seidman.

Over time, Michaelsen cultivated a personal style that was unusually flamboyant for Phelps Dodge. Handsome and suave, he "was a lavish spender." His Park Avenue office was by far the most extravagant in the corporation, easily surpassing that of the chairman George Munroe, in part because Michaelsen collected so many fine souvenirs during his numerous trips abroad. He had probably the best car and chauffeur. Michaelsen's philosophy was "always go first class." Though not a details person, he was a good administrator and a marketing genius. "He could sell you anything."[21]

Colleagues often considered Michaelsen a renaissance man because of his multiple talents. Without doubt he was a charismatic figure who had a special gift for identifying up-and-coming management talent and serving as their patron as they climbed the corporate career ladder. "That may have been his real contribution to the company," emphasized G. Robert Durham, who along with his successor as chairman, Douglas Yearley, was one of "Ted's guys."[22]

At one time Michaelsen established two or three joint ventures a year all over the world. Nearly all the companies had Phelps Dodge in their names, though it was later eliminated except for the Philippines and Thailand to lessen identification with the United States during a rising tide of anti-Americanism in the 1960s. At the time many people in Central and South America regarded Castro as a folk hero. By 1976 the number of Phelps Dodge's international wire and cable companies had climbed to twenty-six, but net profits had dropped from a high of $17 million in 1974 to less than $6 million. The net profit to Phelps Dodge in 1976 was only $2.5 million.

As president of Phelps Dodge Industries, Michaelsen grew concerned when his international subsidiary seemed to founder. "International was basically Ted's baby, and it

TED MICHAELSEN *is
flanked by a smiling Roberto
Millar, first general manager of
the Cuba wire and cable plant,
and possibly Eugenio Castillo, the
government official who first
attracted Phelps Dodge to Cuba.
The plant was successful, perhaps
in part because Michaelsen
persuaded Cuban legislators to
enact a protective tariff that
effectively barred anyone from
competing in Cuba except local
companies such as Pheldrak,
which had no real competitors.
Previously, the country's wire and
cable had all been imported from
Europe and North America.*

bothered him that it wasn't doing better than it was," recalled Durham, who helped to guide International's turnaround in the late 1970s. Michaelsen at the urging of Munroe called Durham, then in early retirement, in the fall of 1977 to inquire whether he might consider returning to work. Durham protested that he was "not an international type. I'm an old Midwestern businessman."[23]

Michaelsen responded that Durham's aluminum work earlier had earned him the respect of Phelps Dodge, and he invited the former executive to serve as an outside consultant to advise management about what to do about Phelps Dodge International. Durham agreed to come to New York for only ninety days. He gathered the heads of various International operations behind closed doors at the Waldorf-Astoria Hotel, located across Park Avenue from Phelps Dodge headquarters, for two weeks to determine what steps to take. Durham sought to act as a catalyst that would result in a plan to save International. He personally believed it had "significant strengths."

Durham's turn-around plan so pleased Michaelsen that he asked Durham to implement it by becoming chairman of Phelps Dodge International. "Now that wasn't the deal," his consultant protested. Durham enjoyed his early retirement. Because he owned a home near Saint Louis and a longtime second home in Florida's Long Boat Key, he stated emphatically, "For certain I won't live in New York City." Michaelsen told him to relocate International's headquarters to any place that made sense.[24]

What could Durham say? At that point he believed his health problems were behind him (see chapter 15), so he agreed to serve as chairman of International. Scouting around, Durham personally selected offices on Ponce de Leon Boulevard in the Miami suburb of Coral Gables, where Phelps Dodge International remains headquartered today. Coral Gables had the virtue of being both an attractive community and conveniently located close to Miami International Airport. In September 1978, the time of relocation to Florida, fewer than thirty people in the new Coral Gables offices oversaw an international operation that employed about six thousand workers and managers.

Durham's rescue plan was very precise, and in time he and his International colleagues did turn the subsidiary around. Durham himself became an internationalist, and he traveled the world. "I got far more personal value than I contributed," he modestly claimed. Durham "was perfect for the situation we had," recalled Orlando G. Gonzalez, who served as president under him. Gonzalez himself, who fled a job in commercial banking in Castro's Cuba in 1961 with little more than the sixty pounds of clothes he was legally permitted to take out of the country (and a clandestine English-Spanish dictionary), worked his way up the ranks of International, eventually serving as its chairman. He was one of several talented, business-minded, Spanish-speaking managers jokingly known as the Cuban Mafia, all refugees from Castro's Cuba who worked long and hard to develop and strengthen wire and cable operations in Central and South America. They may have been the best thing Castro ever did for Phelps Dodge.[25]

It took some surgery to return International to financial health. In 1978 the number of companies it oversaw was pruned back to nineteen, and then to fourteen the next year. The only year to show a net loss for Phelps Dodge International Corporation was 1982, at $32 million. Gonzalez believed that Phelps Dodge would gladly have sold International then had it been able to find a buyer willing to pay $50 million. Phelps Dodge may have been a pioneer of the joint-venture concept, but Gonzalez now wonders whether anyone in the early 1980s would have cared to purchase partial ownership of several different companies of greatly varying financial strength. In any case, Phelps Dodge International survived and eventually thrived, so that by the late 1990s it was worth at least fifteen times $50 million.

Since its first joint ventures in Cuba and the Philippines in the mid-1950s, Phelps Dodge International's lineup of wire-and-cable manufacturing businesses has evolved in many different ways. One subsidiary was Phelps Dodge Puerto Rico Corporation. Not long after it celebrated its twentieth birthday in 1978, that property was transferred from Phelps Dodge International to Phelps Dodge Cable and Wire to capitalize on the marketing of Puerto Rico–made products in the United States by the sales staffs of Phelps Dodge Cable and Wire and Phelps Dodge Copper Products.

In early 1993, through its Venezuelan associate company Alambres y Cables Venezolanos, C.A. (ALCAVE), in which Phelps Dodge had maintained a majority interest since 1957, it acquired three additional Venezuelan companies: Industria de Conductores Eléctricos C.A. (ICONEL); Conductores y Aluminio C.A. (CONAL); and Aislamientos Plásticos C.A. Among ICONEL's specialty products was submarine cable used to supply power to electric motors on the many oil rigs that dot Lake Maracaibo. Collectively, the four properties, which are all located within sight of the busy Autopista del Centro highway linking Caracas and Valencia, constitute one of the largest manufacturers of electrical and telecommunication copper and aluminum wires and cables in Venezuela and the Andean region. Continued growth of phone cable sales in Latin America, as well as telephone line installations, contributed to the Venezuelan group's excellent performance in the mid-1990s.

During that time the Latin American market

for wires and cables boomed, much as the United States market had half a century earlier with rural electrification programs. Further, in anticipation of a common market for all of South America, Phelps Dodge in the 1990s rationalized production so that each property, instead of making a broad line of products to serve a home market, specialized in a few products marketed across the continent. It had to overcome some long-standing national prejudices. Would Colombians think that cable manufactured in Ecuador, for example, was as good as that coming from their own country or from Chile? The adoption of international standards of quality assured that it was.

To serve what until the late 1990s were rapidly growing Asian markets, Phelps Dodge International Corporation opened a new $40 million wire and cable plant in Thailand in 1992. A year earlier it acquired a 20 percent interest in Keystone Electric Wire and Cable Company, Ltd., the largest manufacturer of electrical wires and cables in Hong Kong. That investment offered Phelps Dodge access to the People's Republic of China, a country that was potentially one of its most significant customers anywhere in the world and where modern electric systems were vital to the nation's expanding infrastructure and standard of living. To serve that market better, in November 1997 it inaugurated Phelps Dodge Yantai Cable Company's $18 million manufacturing plant, a

A SEEMINGLY ENDLESS *variety of wire and cable products came from Phelps Dodge's former CABLEC plant in Quito, Ecuador.*

joint-venture facility in Shandong Province that had as participants various Chinese electrical power bureaus, Phelps Dodge Corporation, and the Keystone Corporation in Hong Kong. The manufactory was intended to produce approximately seven thousand metric tons of copper and aluminum cables annually, mostly for underground installations to replace traditional overhead power lines in China's high-density cities.

When Phelps Dodge International Corporation celebrated its fortieth birthday in November 1996 it showed no sign of growing old or set in its ways. Just one month after it inaugurated its modern plant in China, Phelps Dodge International in December 1997 agreed to acquire a 60 percent interest in a Brazilian copper and aluminum wire and cable business owned by ALCOA. As one of the country's largest wire and cable manufacturers, it would do business under the name Phelps Dodge & ALCOA Fios e Cabos Eléctricos, S.A., and allow Phelps Dodge to establish a presence in the rapidly growing Brazilian market. In fact, at the time of its fortieth birthday, Phelps Dodge International Corporation had participated in more than eighty different joint ventures in twenty-seven countries.

Phelps Dodge as a Modern Global Enterprise

ALL FACILITIES OF Phelps Dodge Industries, the parent corporation's manufacturing arm, had in common that they produced engineered products principally for the transportation, communication, and electrical sectors. The division's forty-two operating locations in twenty-six countries (in early 1999) benefited from synergies in technology, marketing, and international experience. Phelps Dodge Industries and its associate companies employed several thousand people worldwide. In the late 1990s its Columbian Chemicals Company, for example, derived more than 50 percent of its income from outside the United States.

There once was a time when Phelps Dodge companies were all fiercely independent. But by the 1990s the emerging pattern in the international arena was for whichever company established a toehold in a country first to help its Phelps Dodge sisters, whether in manufacturing or in mining. "The industries group has been a wonderful resource for the exploration group," observed its onetime counsel Dave Colton. A spare office overseas might often provide a base for initial operations, and International's knowledge of local culture, politics, and financial risk was an invaluable source of information. Observed Chairman Yearley when Phelps Dodge inaugurated its joint-venture manufacturing plant in China in 1997, "This is a historic day for the corporation." In part, that was because Phelps Dodge Yantai Cable's "venture will allow us to enhance our understanding of China's culture and business practices and will position us to capitalize on future mining and manufacturing opportunities."[26]

J. Steven Whisler, who became president and chief operating officer of Phelps Dodge Corporation late in 1997, recalled that persuading employees to move from Morenci to

Tyrone was once a difficult task, and most people were even more reluctant to relocate to Chile. Now employees come to him to ask when they can go to Chile or Brazil. "Our next challenge is to get them to feel the same way about Indonesia or China or Kazakhstan." The ongoing internationalization of Phelps Dodge, in fact, complemented an initiative the corporation launched in the mid-1990s to evaluate and modernize, where appropriate, its traditional corporate culture. Together, internationalization and culture change influenced the way Phelps Dodge Corporation conducted business as it entered the new millennium.[27]

Chapter Seventeen

CHARTING A COURSE *for the*
NEW MILLENNIUM

MAKING CONNECTIONS. THAT is the essence of the Phelps Dodge story: connections between copper ore in the ground and various high-performance conductors and magnet wire, between industrial innovations that permitted Phelps Dodge to manufacture its copper and carbon black products more efficiently than before, between swings in the price of copper and the corporation's responses over the years. Even the name Phelps Dodge speaks to the personal connections forged by the founding families.

Perhaps the single most important connection running through nearly 170 years of Phelps Dodge history is how people worked together to turn vision into enterprise. They took dreams, both financial and technological, and made something practical as well as profitable. That was the essence of what Mr. Phelps, Mr. Dodge, and Mr. James did in their day when they were merchant pioneers in the North Atlantic trade of the fledgling United States; and no less than the founding partnership in the 1830s in New York and Liverpool, modern Phelps Dodge Corporation continues to turn vision into enterprise in its everyday activities—and that includes looking ahead to prepare for its future.

⌇ "OUR MISSION: *To Be The World's Best Mining Company." A sign along Phelps Dodge's industrial railroad at an entrance to the Morenci complex.*

391

A main difference today is that the process of making connections between vision and enterprise is far more formal than it was 170 years ago. After Phelps Dodge management initiated a program in the mid-1990s to modernize its corporate culture, participants came to describe the complicated process as designing a "road map to the future." They did not plan to draw a literal map, but for a corporation largely defined by its several impressive landscapes and for which map making was second nature, the metaphor was perfect.

The step-by-step process aimed to take an abstract concept called "corporate culture"—that intangible yet palpable attitude that pervades an organization—and treat it in a concrete fashion that related directly to the future well-being of Phelps Dodge, in much the same way that matters of technology and finance were treated. Traditionally, to define the company meant listing those attributes that could be most easily measured and quantified, such as its annual production of copper or carbon black, its quarterly profits, the number of people it employed, and the number of countries in which it operated. Those are all valuable measures of how well Phelps Dodge makes connections between vision and enterprise and how it has grown over the years.

Back in 1909, shortly after Dr. James Douglas became the corporation's first president, Phelps, Dodge & Company marketed 185 million pounds of copper, tallied $49 million in assets, and paid out $5.4 million in dividends; in 1997 Phelps Dodge Corporation, headed by Douglas C. Yearley, marketed 1.6 billion pounds of copper, totaled nearly $5 billion in assets, and

paid out $123 million in dividends. In Dr. Douglas's time, unlike our own, the hand of government regulation seldom extended into everyday business, and environmental issues infrequently attracted serious attention from anyone.

Despite the obvious changes—especially all the readily measured ones—there has been considerable continuity at Phelps Dodge too, much of it occurring in the hard-to-define area of corporate culture. Although there were positive attributes worth celebrating, Chairman Yearley and his management colleagues were also correct to ask whether other attributes of corporate culture encouraged thinking and behaving in time-honored ways that actually impeded Phelps Dodge as it prepared itself for the new millennium. That certainly was a matter worth studying. So questions about corporate culture formally occupied center stage at Phelps Dodge in the late 1990s as never before in its history, and the positive results of that attention may well be remembered as one of Yearley's most important legacies.

No one could predict exactly how Phelps Dodge corporate culture might continue to evolve in the coming years, yet study of its origins and evolution does offer valuable clues. This much is certain: Phelps Dodge is anything but a "soulless" corporation, to recall a term that critics once used to bash big business in America. Its unique corporate culture forms its soul—and its heart. Speaking of which, whatever else might change, one important measure of continuity at Phelps Dodge is its enduring commitment to philanthropy, a commitment that originated with its founders.

THE MODEST AND *soft-spoken Cleveland Hoadley Dodge (1860–1926) succeeded James Douglas as chairman of Phelps Dodge Corporation.*

Cultural Continuity: Family and Philanthropy

CONSIDER FOR A moment the connections that exist between the founding families of Phelps Dodge, basketball, and philanthropic activity, the last of which is among the more quantifiable attributes of the modern corporate culture at Phelps Dodge. This story begins with William E. Dodge, Jr., who contributed significantly to establishing a training school for directors of local branches of the Young Men's Christian Association, secretaries, as they were originally called. Located in Springfield, Massachusetts, the YMCA's School for Christian Workers evolved into Springfield College. Cleveland Hoadley Dodge took up his father's interest in the school—whose motto was "A Sound Mind in a Sound Body"—and it was he who apparently suggested to teachers that someone should develop a wholesome winter activity for students to play indoors. He worried that patrons of the YMCA's recently built gymnasium in New York City grew tired of calisthenics as their primary recreation all winter long. The person who responded most effectively to Dodge's challenge was young James Naismith, the school's Canadian-born physical education teacher, who during the winter of 1891 invented a brisk indoor sport called basketball.[1]

Cleveland H. Dodge's son, Cleveland Earl Dodge, continued his family's many philanthropic activities even as he took an active role in the business enterprise his great-great-grandfather Anson Greene Phelps and his great-grandfather William Earl Dodge launched in 1834. Although

Cleveland E. Dodge retired as a vice-president of Phelps Dodge in 1961 at the age of seventy-three, he remained a director of the corporation until 1967. From then until his death at age ninety-four in November 1982, he remained on the board as an advisory director. His formal career at Phelps Dodge began shortly after graduation from Princeton University in 1909. Before he took a desk at the firm's headquarters at Cliff and John streets in New York, he first worked underground at the Copper Queen Consolidated Mining Company in Bisbee. Employed as a common miner, he assumed the name Charles because he thought Cleveland "seemed too highbrow." He also worked in Douglas to gain firsthand knowledge of smelter operations. Except for military service during World War I, Cleve Dodge spent the next seven decades associated in one way or another with the great enterprise that bore his surname.[2]

His son, Cleveland E. Dodge, Jr., also served on the corporation board of directors (from 1968 until 1994), but he chose not to work directly for Phelps Dodge. Thus the passing of Cleveland Earl Dodge effectively ended four generations of uninterrupted service in the management of Phelps Dodge by members of the Dodge family.

The changing corporate role of the Phelps Dodge families suggests one more way the company evolved during the 1980s and 1990s, the decades when it pruned away some businesses and acquired new ones. Yet despite the many changes detailed in earlier chapters, Phelps Dodge Corporation held firm to certain aspects of its past as it approached the new millennium. Its systematic altruism, as distinct from tens of millions of dollars disbursed each year for taxes,

payrolls, and purchases, had an important civic impact across the Southwest, as well as in many other locations where Phelps Dodge had its facilities; and in that way it remained true to the vision of its founders. When Anson G. Phelps died in 1853, he left more than half a million dollars to charity, an amount greater than any New Yorker had bequeathed for good works before that time. It might be worth noting, too, that when Phelps died, young Andrew Carnegie still managed a telegraph office in Pittsburgh and young John D. Rockefeller did chores on his father's farm near Cleveland, Ohio. After the

Civil War, the personal fortunes of those two men and others grew many times larger than those of the original Phelps Dodge partners, but Anson G. Phelps gave away his money before American philanthropy was in vogue. He was a pioneer of philanthropy.

The corporation perpetuating his name launched the Phelps Dodge scholarship program in 1949 in cooperation with Arizona State College (now University) in Tempe and the University of Arizona in Tucson. In years past it awarded twelve undergraduate scholarships and two graduate scholarships annually, and though that has changed now, it continues to maintain basic commitment to education. However, education is only one component of a remarkable story of modern philanthropy.

At the suggestion of its board of directors at the December 1953 meeting, Chairman Louis S. Cates authorized a contribution of $1 million in seed money to a not-for-profit corporation known as the Phelps Dodge Foundation to provide financial support for "religious, charitable, scientific, literary or educational purposes." Cates noted that Phelps Dodge had already been making substantial annual gifts for charitable purposes and that establishment of a permanent fund was one way to insure the continuation of such gifts in the future.[3]

The fund received an additional $1 million from Phelps Dodge Copper Products Corporation and $225,000 from Phelps Dodge Refining Corporation. The foundation made it possible for Phelps Dodge to maintain a program of philanthropy without spending money needed for ongoing operations. Over the next forty years the assets of the foundation grew in market value to approximately $17 million (in 1998) with help from additional contributions by the corporation. The money it gives away each year pays for several four-year National Merit Scholarships

⌁ GIFTS GALORE ⌁

A LIST OF JUST *one year's recipients of Phelps Dodge philanthropy would be a very lengthy one. Among many beneficiaries in its home city were the Phoenix Art Museum, Heard Museum, Phoenix Symphony, Arizona Opera Company, and Arizona Theatre Company; but Phelps Dodge money is also distributed in many far-flung locations. Gifts both large and small went to the Ajo Historical Society, Bisbee Council of the Arts and Humanities, Bayard Public Library, Grant County Softball Team, Silver City Gospel Mission, Myrna Loy Theater in Helena, Montana, South Arkansas Symphony, Hopkinsville Community College, Fort Wayne Museum of Art, Akron Art Museum, Easter Seals of Norwich, John F. Kennedy Center for Performing Arts, Metropolitan Museum of Art, O'Higgins Sportive Club in Chile, and dozens of other organizations.*

During the 1990s, for example, Phelps Dodge helped underwrite the cost of the Smithsonian Institution's Janet Annenberg Hooker Hall of Geology, Gems and Minerals, which opened in the National Museum of Natural History in Washington, D.C., on September 20, 1997. The dazzling display included large dioramas illustrating four of the nation's greatest mines, one of which, naturally, was Bisbee's Copper Queen. But that was not all. Among the showstoppers on permanent exhibit were the Hope Diamond and Queen Marie Antoinette's diamond earrings.

that permit children of employees to attend colleges or universities of their choice. It supports undergraduate and graduate-level programs for minority college students too. In all, from 1949 to 1997 Phelps Dodge provided scholarships to more than twenty-five hundred students attending dozens of different educational institutions across the United States.

The Phelps Dodge Foundation, Phelps Dodge Corporation, and its various business units gave some $3 million per year in the late 1990s to charitable causes, and that sum does not include contributions of time and money by employees to their local communities. Nor does it include donations of land that has recreational value or that conserves the Southwest's priceless natural, historical, or archaeological heritage. As one of the largest property owners in Arizona and New Mexico, Phelps Dodge has a long-standing tradition of donating such land to private and public organizations, such as the Nature Conservancy and the National Park Service. One subsidiary, the Pacific Western Land Company, donated approximately seventy acres of land in Grant County, New Mexico, in 1981 to the Nature Conservancy for a riparian preserve. That land, located along a floodplain about thirty miles northwest of Silver City, was included in the Gila Riparian Project dedicated to protection and study of the abundant plant and animal life that is part of the river environment.

Of course, neither its many philanthropic activities nor its annual infusion of salary and tax revenues into state and local economies would continue for long if Phelps Dodge did not continually look to its future. Even as it mines ore today, the corporation must search for additional sources of copper and other minerals that will assure it long-term success; and that commitment too, no less than the enduring legacy of its founding families and the ongoing commitment to philanthropy, defines its corporate culture.

Culture Change: Pioneering the Creation of Value

IT WAS IN 1995 that Phelps Dodge formally launched a program to examine and adjust where necessary its existing corporate culture so as to prepare itself to meet challenges posed by the twenty-first century. You may ask, what is the essence of corporate culture? "It's history, in a word," responded Thomas St. Clair, the corporation's former senior vice-president and chief financial officer. And that history, noted Yearley, included a strong work ethic. "'Conservative.' That's another word that would clearly characterize us."[4]

One example of how history defined corporate culture could be seen in the way Phelps Dodge traditionally managed its employees, for the concept of corporate culture is intimately related to leadership at all levels. Until the early 1980s, that hadn't changed significantly since the 1920s. "There was something of a military tradition out there," recalled the past chairman George Munroe of Phelps Dodge's western mining operations. At all its branches a rigid hierarchy existed, and observers sometimes compared company mining towns to military bases. Yet, whenever longtime employees chose to comment on everyday life in Phelps Dodge

INCLUDED WITH THE original United Verde mine property were imposing pre-Columbian ruins called Tuzigoot. In the late 1930s, Phelps Dodge donated the historic site in the Verde Valley to the local school district to give to the National Park Service for excavation and preservation. Tuzigoot, an Indian word that means "crooked water," became a national monument by proclamation of President Franklin D. Roosevelt in 1939. Today its ancient ruins continue to fascinate thousands of visitors a year. Much more recently, Phelps Dodge donated nearby Hatalacva, a site even bigger than Tuzigoot, to the Archaeological Conservancy.

communities, they were usually positive. They recalled low rents and how company maintenance crews regularly provided for the upkeep of their houses. A very special time was Christmas. At Ajo an enormous fir or pine tree arrived from Oregon by rail for electrical personnel from the New Cornelia Mine to install as the centerpiece of their vividly lighted display in the town's plaza.[5]

Even the odd juxtapositions of company-town life are often remembered fondly. Morenci, for instance, had a dance floor, and people there wore formal attire, as they did in New York or anywhere else. The retired president Warren Fenzi recalled as a young engineer being dressed for a special occasion in a white shirt but no tie. "You'd dance and go outside and have a fist fight, make up, and go back in and buy the guy a drink. Even Harry Lavender got in fights."[6]

Lavender, like all old-style branch managers, had great authority. That was encouraged by the physical separation that historically existed between Phelps Dodge headquarters in New York and its nearly autonomous branches in Arizona and New Mexico. "The western manager was a law unto himself," noted Yearley, and old-timers occasionally refer to the branch managers of former times as "little gods." It was with a mixture of awe and fear that they recalled Walter Lawson, who managed Phelps Dodge's Western Operations from 1955 to 1969 from his office in Douglas. He always urged his company chauffeur to drive fast. That may have been standard procedure on the lonely roads of Arizona (which once had no daytime speed limit), but in Colorado, when a state trooper pulled over Lawson's chauffeur, the officer remarked with amazement that he had never before been passed

while chasing another speeder! The story, even if some details are apocryphal, still illustrates the great intensity of Lawson's style. In the mines "the pecking order was absolute."[7]

A similar autonomy prevailed in the corporate offices in New York. "I was shocked," recalled Munroe of his first days at Phelps Dodge, "to find the segregation or separation" between manufacturing and mining. In years past that separateness had permitted executives like Wylie Brown to hone the traditional command-and-control style of management into an intimidating form of tyranny.[8]

Not until 1990–91, when Phelps Dodge again had a strong balance sheet, did anyone in management talk seriously about strategies for long-term growth or culture change. Until then the corporation was focused mainly on the need to stay alive. But during Management Review Week in mid-1995, a biennial gathering in Phoenix of executives from all parts of the corporation, discussion turned to employee surveys that identified a rather hard-edged corporate culture at Phelps Dodge. Some observers thought that was simply a lingering legacy of the 1983 strike, while others believed it was far older than that: mining companies in general had traditionally cultivated a tough and rugged persona. There was more to it than that, however. To succeed at Phelps Dodge an individual needed to find a niche and "do a hell of a good job" in that area of specialty, recalled L. William Seidman of his years at the company.[9]

It was time to alter the way Phelps Dodge managed people. Management launched the culture change initiative, and the man appointed to head the team to study what kinds of changes were required to modernize the corporate culture was Ronald Habegger, who took a year's leave from his position as a senior vice-president of Phelps Dodge Magnet Wire in Fort Wayne to work at corporate headquarters in Phoenix. Habegger's perception was that during the grim years of the 1980s, when the battle for survival animated Phelps Dodge, command-and-control management was a positive virtue. At that time, for example, employees in the mining company instinctively looked to Leonard R. Judd for direction and unhesitatingly asked, "What do you want us to do?" The command-and-control culture's most negative aspect was that over time it inhibited employees from trying anything new. If that mindset prevailed for any length of time, "at some point businesses stop developing new roads and start maintaining streets," explained Habegger.[10]

Added Yearley, "A corporation has to reinvent itself periodically or it is going to die." Reinvention was necessary even for terms like *culture change*, which was quickly superseded by a more colorful and more comprehensive slogan, "Pioneering the Creation of Value." The work of Pioneering the Creation of Value is so important to Phelps Dodge that it became a formal part of the corporation's vision statement adopted in 1998. The phrase speaks to an ongoing need to effect clearly identified changes in Phelps Dodge culture and sharpen the corporate vision as leaders look to the years ahead; it speaks to the need to preserve all the best elements of a time-honored way of doing business while cultivating continuous improvement in company personnel. One major challenge is to encourage a people-oriented leadership style among its executives.

That is harder than it sounds because the mining industry in general has traditionally valued actions above feelings.

A question asked repeatedly during follow-up discussions among executives was how various types of activity added value to Phelps Dodge and its employees, customers, shareholders, and communities. Finding more efficient ways to turn dirt and rock into valuable metals was one very straightforward answer. Another was the "Zero and Beyond" safety initiative intended to reduce accidents companywide to zero: "That means zero fatalities, zero injuries, and zero accidents," explained the corporation president and chief operating officer J. Steven Whisler. How an emphasis on safety added value to Phelps Dodge was self-evident.[11]

By the mid-1990s, too, Phelps Dodge needed to refashion its corporate culture to fit its newly expansive way of thinking, especially if it hoped to respond quickly to mining and manufacturing opportunities abroad. That meant eliminating certain obsolete hierarchical traditions and abandoning concerns among leaders about "turf." In addition, Phelps Dodge needed to foster a corporate culture that was attractive to young people, a modern generation of employees who are neither as patient as their elders nor as likely to recall long hours worked by employees during the copper crisis of the mid-1980s. "Our expectation of a normal workday 'is not normal,'" noted the senior vice-president and chief financial officer Ramiro G. Peru, speaking of one lingering result of the old survival mode of thinking. "This company's challenge is to find a balance between family and personal life and your worklife."[12]

At the same time it must develop a more democratic "listen and lead" culture. But any move away from the traditional command-and-control style of management presented quite a challenge to the old-line enterprise that Yearley once described, tongue-in-cheek to an audience at his alma mater, Cornell University, as "a mining company founded by two Presbyterians. We don't do warm and fuzzy." Further, "the company was really big on status," recalled Peru, who grew up in Morenci where the geography of power was a fact of everyday life.[13]

In one sense, changes in corporate culture take place all the time, sometimes in response to internal initiatives like the one that spurred Phelps Dodge in 1995, and sometimes in response to the changing world in which businesses must operate. Formal internal changes were often long in coming because Phelps Dodge was so old school in so many ways. Warren Fenzi recalled a time when all goals in mining related to costs and entailed a very detailed type of cost accounting. "But we didn't care about how the copper was sold." His onetime boss, Ernest Wittenau, the general superintendent at Morenci, "was quite a stickler for accounts. He would even shut down crews at the end of the month because they were using too much powder for that month."[14]

In the 1950s and 1960s there was no such thing as benchmarking, no corporatewide budget committees, and no strategic planning committees beyond the senior officers. To the corporation's former president and chief operating officer, Len Judd, Phelps Dodge when he rejoined it in 1977 was not professionally managed. Prior to 1979, the mining side of the corporation had no formal budgets because no one seemed to know how to budget for the inescapable uncertainties of mine production. A manager might question, "What if it rains? What if it snows? It will disrupt production." As long as individual mines made money, corporate executives seldom seemed to care about formalities of budgets. Phelps Dodge Corporation itself created a formal planning committee only in 1981.[15]

"I found it amazing," reflected Peru, that "growing up in Morenci you thought of Phelps Dodge as some big, sophisticated company with the latest and greatest of everything." Later he discovered "when it came to management practices on the financial side, they just weren't there." Peru started at Phelps Dodge in 1979, the same year that formal budgets were introduced in western mining operations. Today, "I think our cost accounting system and metals accounting system in the Mining Company are second to none. I think that has been the secret of our ability to manage those businesses."[16]

Many people credit L. William Seidman with launching the modernization of budgeting and long-range planning at Phelps Dodge, and indeed when Munroe hired him in 1977 as chief financial officer those were among his principal assignments. His reforms represented "a breath of fresh air," recalled G. Robert Durham, who came to Phelps Dodge the same month as Seidman. A former White House economics adviser, Seidman viewed himself as a provocateur within the corporation and an agent of change. Of course, many people originally resisted strategic planning, and some joked that he had better be careful when he visited mining properties out West, where the initiatives from New York headquarters were not always welcomed. Many believed that the corporation's mining operations lagged well behind Phelps Dodge Industries in implementing the modernization Seidman advocated.

Seidman, who had family roots in Grand Rapids, Michigan, where the accounting firm of Seidman and Seidman was prominent, was Munroe's roommate at Dartmouth and his classmate at Harvard Law School. He had played high school football in Grand Rapids with Gerald Ford, who after he succeeded Richard Nixon as president of the United States, called Seidman to serve as his economics adviser. Following Ford's loss to Jimmy Carter in the 1976 elections, Munroe invited Seidman to come to Phelps Dodge where he served as chief financial officer and one of three vice-chairmen after Munroe created that office in 1980. Seidman left in 1982 to become dean of the business school at Arizona State University but continued on the Phelps Dodge board until 1985, when he became chairman of the Federal Deposit Insurance Corporation.

Regardless of the driving force behind the changes, Phelps Dodge streamlined most noticeably after the copper crisis of 1983–84, when distinctions between East and West eased and finally disappeared following relocation of the corporation's headquarters to Phoenix. The long-standing tendency of individual branches to regard themselves as little fiefdoms disappeared too, observed Whisler. That was, he believed, partly because Phelps Dodge once had a certain comfortable largess or fatness it could no longer maintain.

In matters of safety, however, Phelps Dodge Corporation was always in the vanguard. "Safety is the leading indicator of operating excellence," observed Yearley, and emphasis on safety sends the signal that "we care." Over the years the corporation continued to set the bar of safe production ever higher, which in the late 1990s became nothing less than Zero and Beyond. At many facilities even a close call or near accident warrants serious investigation. At every plant numerous posters exhort employees to work safely, and many slogans originate with employees themselves.[17]

Charles R. Kuzell is generally recognized as the father of Phelps Dodge's modern safety program. As manager of the United Verde Branch in the 1930s, Kuzell instituted a formal code of safe practices. First, he had each job analyzed for its hazards, and then he listed specific safety precautions required to insure maximum worker protection. In the early 1950s a "Master Code for the Elimination of Accidents" further defined what safe production meant in practice.

Many significant improvements have occurred since then, and they include meticulous recordkeeping and a system of rewards for both the best and most improved safety records. Phelps Dodge Tyrone, for example, felt proud when it won the national Sentinels of Safety Award twice in a row during the mid-1990s. The award is a historic bronze statue of a mother and child that represents family members who want each miner to return home safely (though today a good many people who work in mines are women). The Sentinels of Safety Award has since 1925 recognized open-pit mines having the best

MEMBERS OF THE
*Safety Department at Ajo in
1954. More recently, in 1989
Phelps Dodge instituted the
Chairman's Safety Awards for
the best safety performance and
most improved safety performance
at its North American operations;
and it introduced a similar
international program in 1992.
That year the winner of the
latter award was Columbian
International Chemicals
Corporation.*

safety records in the United States; each employee receives a special lapel pin as part of the honor.

Whenever Phelps Dodge acquired a new subsidiary, one of the most visible aspects of its being "PD-ized" was heightened emphasis on safety in all operations. In 1990, for example, after Chino employees lowered the recordable accident rate by 72 percent over the previous year, they won the coveted James Douglas Memorial Safety Trophy, a Phelps Dodge award given annually since 1926.

Over the years, culture change in an informal sense embraced how Phelps Dodge treated its numerous Hispanic employees. Until the civil rights revolution of the 1960s, how they fared both on the job and in the surrounding mining communities mirrored the prevailing prejudices of the American Southwest. In Morenci, some restaurants once posted signs that read "No Mexicans Allowed," and the Morenci Social Club once was rigidly segregated by race and ethnicity. On one side was the theater, with segregated seating; the adjacent library and parlor were off limits completely to people of Hispanic ancestry. Downstairs was the bowling alley, and except for men and boys hired to set pins, Mexicans and Mexican Americans were not allowed in there either. A local Elks Club refused them membership until its numbers dropped so low that it changed its admission policy, but by then Hispanics were reluctant to join.

In the mines, too, the old culture meant lower wages for Hispanic workers who did the same jobs as Anglos, or their being barred entirely from certain jobs. For many years the mines of Bisbee did not permit Hispanic workers

underground. That was not the case in Morenci, which had a much larger Hispanic workforce; but just the same, there were jobs and places fully off limits to people of Mexican ancestry. The railway brotherhoods were extremely unyielding and did not even permit Hispanic track workers to touch locomotives.

There once were separate change rooms too. Simon Peru, a retired Morenci employee, recalled that after he joined Phelps Dodge in 1952 he mistakenly headed into the wrong change room. Seventeen years later, when Arthur Kinneberg returned to Morenci as general superintendent in 1969, he found the change room at the smelter still segregated. Having spent seven years in Chile, he disliked racial and ethnic distinctions. Thus he ordered the partition removed to create one large room, though some of his non-Hispanic supervisors objected to the change.

Most Hispanic employees tended not to blame Phelps Dodge for discrimination but rather cited fellow workers from Arkansas, Oklahoma, and Texas, who were often poorly educated and prejudiced against people of African or Mexican ancestry. "They felt superior to some of the minority people" was a common complaint, and Hispanic workers adjusted as best they could. Simon Peru's father-in-law, Pete Gomez, recalled attending Spanish school in Morenci as a child during summers in addition to his regular school. His instructors, all from Mexico, taught Mexican history and related subjects.

Times have certainly changed for the better for Gomez, Peru, and other Phelps Dodge workers of Hispanic heritage. As of the late

1990s, they composed approximately 40 percent of the corporation's some ninety-five hundred employees in the United States, and they have many more Spanish-speaking counterparts in Central and South America. Together they form Phelps Dodge's largest minority. Pete Gomez's grandson, Ramiro G. ("Ramey") Peru is now a top corporate executive, although (and offering an ironic commentary on generational differences) he did not grow up in Morenci speaking Spanish! He struggled to learn it later as an adult.[18]

Women too made many gains in what once was considered a male-oriented industry. Historically, they held few jobs outside the plant offices, except during World War II or during the big strike of 1983. Today, as is true in American society generally, women at Phelps Dodge can be found doing nontraditional work, for example, as research engineers, hydrogeologists, and haul truck drivers. Women today serve as supervisors in various operations and as company executives. In 1997 Susan M. Suver became Phelps Dodge

THE NEW CORNELIA BRANCH *in May 1954 and Ajo's swimming pool. Ajo once had formally designated neighborhoods called Mexican Town and Indian Town, an arrangement inherited from the Calumet and Arizona Mining Company, and there were separate days reserved for Anglo, Hispanic, and Indian residents to enjoy the local pool.*

Corporation's vice-president for organization effectiveness and communications. She worked her way through Arizona State University, where she majored in communications. Suver came to Phelps Dodge after spending ten years in the international travel industry. Among her tasks is to educate the public, especially the media, about how mining contributes to everyday life, and especially to the high standard of living many people take for granted. "I like the challenge here," she observes of her diverse responsibilities.[19]

Back in the nation's capital is Linda D. Findlay, who succeeded Charles A. Burns as vice-president for government relations. Her office is located almost within sight of the White House, and among her jobs is to educate members of Congress about issues of interest to Phelps Dodge and the nation's mining industry generally. Women executives like Findlay and Suver historically did not begin their business careers intending to work for a mining company. Findlay, before coming to Phelps Dodge in 1992, had served fifteen years as legislative assistant for environmental issues to New Mexico's Senator Pete V. Domenici. The corporation's second female vice-president (after Anita H. Laudone, who was elected vice-president and secretary in 1984) majored in English at Kenyon College in Ohio and worked for a time as director of admissions at a girls school in Virginia. After joining Phelps Dodge, Findlay recalls, "I learned a lot very quickly."[20]

Patricia Young, who is a shift supervisor at Morenci, is a second-generation Phelps Dodge employee. Her work experience includes driving haul trucks at both Morenci and Tyrone, where she was the first female supervisor of dispatchers. Today she oversees thirty shovel operators, truck drivers, and cat skinners who operate the big earthmovers. Of her work as a haul truck driver, when asked "Does it give you a sense of power to be up there that high?" she quickly responded, "Oh, yes," but "you get used to the distance." Of her current job she adds, "It's a good company to work for. I like what I'm doing."[21]

In total, about 15 percent of the employees of Phelps Dodge Corporation in the late 1990s were women. "In this organization there are a lot more women moving up than there were just five years ago when I started," noted Sue Suver in 1998.[22]

Together Employees and Management Succeed

THE PERSON WHO presided over Phelps Dodge during the 1990s was the chairman and chief executive officer Douglas C. Yearley, who during much of the time held the title of president too. He firmly believes that among his legacies will be the 1995 culture change initiative, which he hopes will encourage employees to work closely together instead of jockeying for position in a hierarchical structure based on command and control.

Perhaps because of his strong desire to nurture a modern listen-and-lead culture as the corporation entered the new millennium, Yearley was comfortable discussing ways that Phelps Dodge perhaps fell short in the past. "I'm a realist," he stated during one of our several interviews for *Vision and Enterprise*. "As you go back into our history there's going to be some dirty laundry. There has to be. The rules of the game sixty or seventy years ago were totally different. We were basically following the rules, and perhaps even remaining a little better than the rules, but still by today's perspective some things look pretty shabby." He added, "We're in dirty businesses, and we still carry the legacy of [the Bisbee deportation]." His candor was not only refreshing, it was representative of the openness he sought to foster at Phelps Dodge.[23]

To understand the ways that Yearley shaped and guided Phelps Dodge during the ups and downs of the 1990s, it is important to know what experiences molded his own career in business and his personal outlook. The future chairman grew up in Oak Park, Illinois, a Chicago suburb once known to cynics as Saints' Rest because it contained many churches and not a single saloon. The community is perhaps best known today as the place where Frank Lloyd Wright maintained his famous studio for many years, and Oak Park's main landmarks are twenty-five surviving Prairie-style buildings that Wright designed. Among the other notable native sons of Oak Park was Ernest Hemingway.

The Oak Park of Yearley's youth continued to be a leafy, middle-class suburb. His mother was a high school English teacher who was very musically oriented; and music naturally became part of her oldest son's life. Yearley played a clarinet in his high school orchestra—and he was also a member of its football team. Classical music remains one of the great loves of his life. He seldom travels by car without listening to one of the classics, notably opera. In 1998 he realized a long-held dream when he conducted the

AMERICAN METAL MARKET
SUPPLEMENT, FEBRUARY 17, 1993

Copper Club

MAN OF THE YEAR
Douglas C. Yearley
Chairman, President and CEO
Phelps Dodge Corp.

THE NEW YORK–BASED *Copper Club recognized Douglas C. Yearley as its 1993 Copper Man of the Year. The award, bestowed each of the preceding thirty years, honors individuals who have made outstanding contributions to the industry. Yearley has fostered cooperation among world copper producers and promoted greater use of copper globally. In 1989 he helped found the International Copper Association, which he served several years as chairman. The Copper Club honor also paid tribute to Yearley's role in helping to engineer the dramatic turnaround of Phelps Dodge during the 1980s.*

Phoenix Symphony in Rossini's stirring *William Tell Overture.*

Along the way Yearley developed a love for history too. He enjoys well-crafted studies like Stephen Ambrose's *Undaunted Courage,* a popular biography of Meriwether Lewis. In fact, when I first discussed the possibility of writing a modern history of Phelps Dodge Corporation, Yearley mentioned how much he enjoyed David McCullough's biography of Harry Truman. I wondered whether he saw parallels between troubles that beset Phelps Dodge during the early 1980s, when he was one of Munroe's lieutenants, and those that Truman confronted as president of the United States during the tumultuous years of the late 1940s and early 1950s.

Like Truman, Yearley despises arrogance and pomposity. Perhaps for that reason he was reluctant to brag about personal accomplishments, even for the historical record. From other sources I learned that he was a driving force behind the International Copper Association, which world producers formed in 1989 to promote the use of copper. He is also disinclined to take himself too seriously. When, for instance, a Phoenix business publication asked him in 1996 what profession he would pursue if he were not running Phelps Dodge, he shot back, "A beach bum." He was kidding, of course. "I have a sense of humor and I don't forget to smell the roses." He is also a family man who met his wife, Anne, when they were students, he attending Cornell.[24]

Yearley's father, a college-trained metallurgist employed as a sweeper in a foundry during the harsh years of the Great Depression, was very pleased when Cornell University admitted his son to its engineering program. "I never saw Cornell before I went," Yearley recalled. A few days after he arrived on campus, professors warned the entering freshmen that only one in three of them would graduate from the demanding program in engineering. "I was a free spirit" but determined to be among the one-third who made it.[25]

Shortly after receiving his bachelor's degree in metallurgical engineering in 1958, Yearley interviewed with the legendary curmudgeon, Admiral Hyman Rickover, for a position in the reactors branch of the United States Navy. After experiencing the infamous Rickover humiliation of applicants, he elected instead to take a job at Electric Boat, where he worked on nuclear submarines in 1958 and 1959. At the time that was America's first line of defense. Because of the nation's near panic after the Soviet Union launched its Sputnik satellite, life at Electric Boat was chaotic, recalled Yearley, and a day's work there often lasted sixteen hours because of constant pressure to get the ballistic missile submarines into the water.

Yearley, who in 1960 accepted an offer to work for Phelps Dodge Copper Products, joined its Bayway facility in New Jersey in product development. In the "Hawaiian Room" (so named by fellow workers because machinery used in Yearley's experiments caused the place to vibrate and sway like a hula dancer, according to one account, or because those experiments left a person nicely tanned, according to another), he sought to cast copper tubing continuously, which would mean potentially huge savings to Phelps Dodge. He came away from that assignment with seven patents, but "none of which was worth a damn." He simply couldn't get the casting process properly refined.

"I didn't like engineering. I wasn't very good at it," he confessed. "I had two failed projects." Looking around for other work, he pondered returning to school to earn a master's degree in business administration. Instead he went to Harvard's Program for Management Development in 1968. The idea of sending a junior employee to school for advanced training was so foreign to Phelps Dodge at the time that the company paid only for tuition—though Yearley soon discovered that everyone else in the program was on an expense account. At the age of thirty-three, he commuted by car for eighteen weeks between New Jersey and Massachusetts, returning home to spend each weekend with Anne and their four children, who then ranged in age from one to nine. "It was a defining event in my life. I found I loved business. I loved the complexities of finance and marketing."

Afterwards, Yearley worked another nine months at Bayway. One day as he sat in his starkly functional office, he was surprised to receive a phone call from Ted Michaelsen, president of Phelps Dodge Industries, whose own office was located across the Hudson River in a posh section of Manhattan and a world away from gritty, industrial Bayway. "Doug, I want you to be in my office tomorrow morning at nine o'clock."

"Yes, sir. Can I prepare myself for anything?"

"Your future," and with that cryptic bit of advice, Michaelsen hung up.

"I wasn't sure what I was going into."

Michaelsen, with Munroe's blessing, put Yearley on a career track that exposed him to numerous different situations, many of which, his patron warned, would involve having to make personal sacrifices. On one occasion, Michaelsen called and said, "You are now operations manager for Asia and Africa for the International company." For Yearley that meant about every two months in the late 1960s and early 1970s he made a three-week trek to consult wire and cable plant managers in Lebanon, Zambia, India, Thailand, and Japan.

Back in the United States he was next assigned to run the Los Angeles Tube Mill of Phelps Dodge Brass Company, and in that capacity he saw how easy it was for a corporation and its individual operations to become complacent. He recalled his first day on the job. Yearley arrived at the plant early Monday morning and was surprised to see rats scurrying all around after workers started the noisy machinery. The operation, which made water and condenser tubes, obviously had problems. His tactic was to ask the employees for ways to make conditions better. In short order "the rats disappeared and the place started to look good. We went from losing money to making money." Of the Los Angeles Tube Mill of Phelps Dodge Brass Company, which he later headed, he adds, "It was an independent command because I was three hours away from New York."

Next he became vice-president for marketing at Phelps Dodge Cable and Wire Company. "That was a tough transition because I had never done any selling at all." Not until 1980, when George Munroe made him president of Phelps Dodge Sales Company, did Yearley cross to the parent corporation from Phelps Dodge Industries, but all those early field assignments were invaluable for what followed next.

His many years in copper sales helped Yearley understand the corporation's core business well: "I got to know all the customers, and got to know how the copper was produced." In those days you'd take your customers to the mines, and that also provided Yearley an opportunity to become acquainted with mining employees. By the time he became chairman of Phelps Dodge Corporation, he already possessed an insider's detailed knowledge of its operations.

On March 3, 1989, the *New York Times* headlined its account of Yearley's succession, "Another Cost-Cutter to Head Phelps Dodge." One tale of his cost cutting that achieved near legendary status dates from the time Yearley summoned his top 120 managers from around the world to Phoenix *in the summer* (and which he still does for the Global Leadership Forum). People do not ordinarily consider Phoenix an appropriate place to hold business meetings during the summer. "As soon as they step outside," Yearley chuckled, "it's 110 degrees. They can't believe the heat." But the new chairman loved the deal. He locked up a block of rooms at a posh Phoenix resort at a bargain rate. Besides saving money, the thrift sent a strong signal that Phelps Dodge was not about to get complacent about spending. *Industry Week* in 1990 profiled Yearley as one of "America's Unsung Heroes" because of his role in putting torque in the turnaround at Phelps Dodge.[26]

Chauffeured Cadillacs are only a memory now, a casualty of the mid-1980s thrift. Similarly, Phelps Dodge owned no jet until 1999, when it gained one in an acquisition that greatly enlarged its international presence. Until then, Yearley regarded such amenities as incompatible with Phelps Dodge's goal of becoming the "low-cost producer in everything we do." He had observed in 1997, "Just in the last three or four years have I been able to relax about expenses," a comment he made before the late 1990s plummet of copper prices required renewed belt-tightening.

One of Yearley's goals for changing corporate culture at Phelps Dodge was to improve communications throughout the company, a work-in-progress I observed during a brief research trip to Phelps Dodge Magnet Wire in Fort Wayne, Indiana. The corporate "Town Hall Meeting" held in early August 1997 was in some ways an extension of "Donuts with Doug," an

open forum he hosts several times a year for employees at the Phoenix headquarters. After a thirty-minute presentation on the state of the company, Yearley (speaking from Phoenix) fielded ninety minutes of nonstop questions phoned in from places as distant as Connecticut and Chile. Employees all over the corporation's sprawling landscape were free to participate. Some facilities, such as Phelps Dodge Tyrone, were particularly involved in the process, others much less so. Obviously it takes time to change corporate culture, but it seems to start with this form of improved communications.

How did Yearley develop his obvious interest in collaborative management? "Mostly from common sense. I am comfortable as a consensus builder." He recalled postgraduate management education at Harvard Business School as an early form of sensitivity training. "It is important to emphasize the people side. We have a better trained and more sophisticated group than fifteen years ago. We had a bunch of people in cubby holes who knew how to sell copper or break rocks, etcetera, fifteen years ago. There was no cross fertilization and there was no external training; there was no benchmarking."

During my stay in Fort Wayne I also observed another aspect of corporate culture, a Phelps Dodge Code of Ethics seminar regularly conducted to familiarize employees with what constitutes proper conduct in situations ranging from correct financial reporting to sexual harassment. The vice-president and corporate secretary, Robert C. Swan, posed a series of hypothetical questions intended to encourage employees to think ethically by considering three basic issues: Is it legal? Is it ethical? Is it good business practice? In Fort Wayne questions of ethics and security loom large because several magnet wire competitors are located in close geographical proximity to one another. What if, for instance, a manager overheard a discussion of trade secrets by competitors aboard a commercial flight? That was one issue actually discussed at the Fort Wayne seminar, and among the many responses it elicited was one lighthearted suggestion to cup an in-flight magazine to your ear so as to hear better!

Something else highlighted in Fort Wayne was that the beautiful new One Technology Center, the headquarters building that Phelps Dodge Magnet Wire opened in 1994 adjacent to its big manufacturing plant, had very few doors on its offices. That was because the architect sought to encourage openness and cooperation among employees. There were, of course, locked and guarded doors in the research laboratory and pilot production facility where Magnet Wire developed, tested, and evaluated process improvements and new products.

On a related front: an initiative launched at Morenci in 1997 and affecting all of Phelps Dodge Mining Company's nonunion employees in the United States sought to create an all-salaried workforce. A kindred goal was for close-knit crews to make many day-to-day decisions and generally operate with a minimum of formal supervision. Since 1991, Phelps Dodge Morenci has used the acronym TEAMS (Together Employees and Morenci Succeed) to describe that philosophy.

Exactly where the idea originated is not certain. "I was using teams in Los Angeles" at the Phelps Dodge tube mill, recalled Yearley.

"Didn't call them that." Some people credit the El Paso refinery for formalizing the basic concept in the late 1980s. But at Morenci the teams concept was specifically oriented toward employee groups in the mine and its support facilities, and it offered another example of the kinds of creative initiatives bubbling to the surface in various parts of Phelps Dodge that contributed to culture change. Further, as Yearley observed in the *Business Week* "President's Forum" in 1993, "Narrow specialties are as obsolete on the work floor as in the executive offices. Of course, I'm not suggesting that we rid our companies entirely of specialists. They serve a valuable need. We must, however, cross-train much of our work force and expose workers to as many aspects of business as possible." In practice that meant giving greater emphasis to effective communication, driving responsibility down in the organization, and even teaching business classes to employees to help them understand the big picture of how Phelps Dodge worked.[27]

A watershed for managers at Morenci was the 1983 strike, when a rigid hierarchy and union work rules hurt the corporation badly and opened eyes regarding traditional command-and-control culture. At that time promotions, for example, came not from performance but from union-protected seniority. How did a manager get a job? "It was his turn," noted the onetime Morenci mine superintendent Michael Allen. So, how many workers did it take to fix a truck? Often, two. A union mechanic would say, "I took the bolts off, but I'm not taking the wires off," recalled Allen in 1992. How many workers did a mine foreman supervise? In 1986, about six. In 1992? A remarkable twenty-one. What was the absentee rate in mine operations? Eighteen percent back in 1986, but less than 1 percent by 1992. Measurements of efficiency are no less impressive today.[28]

To cynics, all such changes may seem like good old-fashioned paternalism. In some ways they probably were, and at a corporation as history-minded as Phelps Dodge that was not necessarily a bad thing. Phelps Dodge has no union workers at Hidalgo, and at Morenci it has been nonunionized since the 1983 strike. In various ways, the task of keeping nonunionized employees happy imposes many demands on company managers that were formerly addressed by union officials.

Contrary to popular belief, Phelps Dodge did not eliminate trade unions from its far-flung operations during the dark days of 1983 when it operated its facilities in defiance of strikers. The pattern was far more complicated than that. Phelps Dodge Magnet Wire in Fort Wayne and Hopkinsville remained unionized in the late 1990s, as did several carbon black plants of Columbian Chemicals Company. Within Phelps Dodge Mining Company, the Hidalgo smelter never organized, and all unions at Tyrone had decertified by 1994. In the mining sector the only unionized property left in the United States by the middle of 1999 was the Chino operations, where approximately half its workers were dues-paying members.

When Phelps Dodge purchased a two-thirds interest in Chino Mines from Kennecott in late 1986, seven unions were in place, at least one of them since the 1930s. First negotiations were held in 1989. After that, five locals decertified, beginning with the Machinists union in 1990.

Eventually all that remained were a local of the International Brotherhood of Electrical Workers with about seventy-five members and Local 890 of United Steelworkers with about five hundred members. That was the Juan Chacón local, not only one of New Mexico's oldest unions but one that sees itself as the proud heir to the "Salt of the Earth" legacy of the 1940s, a battle with New Jersey Zinc as dramatized in the feature film by that name. Phelps Dodge Corporation had absolutely nothing to do with that strike, but it acquired the historical burden along with the Chino property.

In the late 1990s Local 890 had a meeting hall in the town of Bayard where bold letters on its outside wall proclaimed "Union Yes!" and a series of attractive murals inside detailed its members' version of local labor history. Because unionism in New Mexico was intimately linked to civil rights for Hispanic workers, it included elements of a religious commitment, as when labor leaders exhort, "This is the union your father and grandfather fought for." That kind of loyalty was one reason workers at Phelps Dodge's only unionized copper property in the United States voted by a two-to-one margin in 1996, 1997, and again in 1998 to maintain representation by the United Steelworkers of America. In November 1998, local members of the Steelworkers and Electrical Workers unions ratified new four-year contracts by a margin of more than 90 percent to replace agreements that had expired some two years earlier.

Quite clearly, the ways employees and management interact at Phelps Dodge will vary considerably between the mining and industries groups and between union and nonunion facilities. The ongoing search for creative strategies to enhance interaction between management and employees is one way Yearley and his successors have helped to prepare Phelps Dodge for the challenges certain to lie ahead.

Responding to New Challenges

I N RECENT YEARS, Phelps Dodge declared a new commitment to protection of the environment. It would no longer be simply reactive, as in the old days, but fully active. The corporation promised publicly in 1992 not merely to comply with government regulations but to exemplify the best contemporary industry practice in all environmental matters. "I consider myself an environmentalist," opined Yearley in May 1993. "I have children and grandchildren and I am concerned about the water they drink, the air they breathe and the land around them. I also want them to find good jobs, to live in a country with a strong economy, and be able to enjoy a good standard of living."[29]

As he explained to the Colorado School of Mines in November 1994 in its Distinguished Lectureship Series, "I am absolutely in favor of environmental regulation. These changes are not always easy for industry, and the real downside is that some of the changes have not led to a better environment but simply more cost." Nonetheless, Phelps Dodge committed millions of dollars each year to ensure its good stewardship of the environment. Yearley stated that when his company closed a mine, for example, it had to be made environmentally benign. "And we'll do that."[30]

During the 1970s the main focus of environmentalists was the corporation's older smelters. Eventually Phelps Dodge closed all of those in Arizona and constructed or acquired modern ones in New Mexico. The two New Mexico smelters must capture more than 90 percent of the contained sulfur in the concentrate, and both easily achieve that. At Chino in 1995 the capture rate was 96 percent; Hidalgo's was 95 percent.

By the 1990s considerable environmental concern came to focus on the mines themselves. One group of environmental radicals publicly proclaimed its goal as "Mine Free by '93." What such opposition meant for the industry was vividly illustrated in Montana, where Phelps Dodge sought for several years, without success, to obtain permits it needed to extract precious metals near Lincoln. In forty thick volumes the joint-venture partners sought to address all relevant issues raised by state and federal officials, yet still there was no end in sight for the process. Set atop one another the volumes stood an impressive thirteen feet high.

In the end, the Seven-Up Pete Joint Venture produced mainly the mountain of paper and an outburst of environmental hysteria. Opponents included an array of environmental extremists, some carrying assault weapons and sidearms and wearing paramilitary khaki, who crossed the line between wild-eyed militants and responsible environmentalists. In the end, the plodding pace of the permit process and a severe slump in the price of gold helped convince Phelps Dodge that it could earn a better return on its dollars elsewhere. In September 1997 it announced sale of its 72.25 percent stake in the project to its partner, Canyon Resources, a Colorado-based mining company.

At the same time that Phelps Dodge pulled out of Montana, it scaled back exploration activities within the United States. Effective January 1, 1998, the corporation reduced annual spending on domestic exploration for new mineral deposits from $5 million to $1 million. In addition, it closed the majority of its exploration offices in the United States and decreased the number of exploration employees based there from forty to ten.

Exploration to expand mineral reserves at existing Phelps Dodge operations in the United States would continue, as would exploration abroad, under the aegis of Phelps Dodge Exploration Corporation. Formed in 1994 to streamline and focus the company's activities in some twenty-six countries, including Madagascar, where it discovered a potentially momentous nickel-cobalt deposit, Exploration is based in Phoenix and headed by A. L. ("John") Lawrence, a Phelps Dodge vice-president. Holding both bachelor's and master's degrees in geology from Cambridge University, he worked for De Beers and Anglo American before joining Phelps Dodge as a project manager in South Africa.

The cutback within the United States recognized that a great deal of exploration had already been conducted there and that the regulatory climate made starting new mining operations ever more difficult and expensive. International opportunities had become more abundant and had greater profit and growth potential. By redirecting its exploration resources toward those opportunities, Phelps Dodge

intended to ensure the ongoing success of its core mining business. As Patrick Ryan, onetime head of exploration, explained in late 1996, he was "extremely pessimistic" about new mines in the United States. He cited the Montana gold project, which may have cost the corporation as much as $60 million by the time of the 1997 sale to Canyon Resources and yielded nothing.[31]

Obtaining permits for new mines in the United States only grew increasingly difficult, and a single land exchange with the federal government for a minable tract of land could take years. With the rest of the world opening up to exploration, domestic companies refocused their angles of vision overseas. "That's where people are going to open mines," emphasized Ryan. "It is not a case of wanting to run away from one's environmental responsibilities," contrary to the claims of some critics. Phelps Dodge already had demonstrated responsibility in "engineering the best environmental protections for Candelaria." Further, mining companies benefited from a certain amount of geographic diversification, political unrest in one country being a case in point. Added Ryan, "To say we're only going to mine in the United States would be insane."

He noted too what a daunting challenge it is for exploration personnel to find the new copper Phelps Dodge needs just to replenish what it mines each year—some 180 million tons of material. And Ryan's former colleague Phil Matthews, manager of Geophysical Services for Phelps Dodge Exploration Corporation, underscores the seriousness of the pursuit when he says, "We're of course somewhat biased, but we really feel that the life's blood of this corporation is in our hands."[32]

In practical terms, what they're saying is that Phelps Dodge exploration teams must find an ore body roughly the size of Candelaria in Chile every six years just to have copper enough in the future to replace what it mines today. Further, to enlarge the mining enterprise would require even more frequent or larger mineral discoveries. Do the intimidating statistics cause the corporation's president to lose sleep? "Every night," quips Steve Whisler without betraying any overstatement or exaggeration.

Whisler is certainly a key player who will help determine where Phelps Dodge pursues exploration next. The onetime legal counsel to the exploration group as well as to its parent corporation, he served as president of Phelps Dodge Mining Company for much of the 1990s, until Timothy R. Snider, one of the early advocates of solution extraction–electrowinning at Tyrone, succeeded him in September 1998. Late the previous year, at age forty-three, Whisler was promoted to president and chief operating officer of Phelps Dodge Corporation.

Like three of his recent predecessors as president, Len Judd, Richard Moolick, and Warren Fenzi, Whisler is a son of the West: he was raised and educated in the shadow of the Colorado Rockies. Though the youngest president in Phelps Dodge history, he brings a potent combination of education and experience to his office. Just after he graduated from the University of Colorado in Boulder with a bachelor's degree in accounting, Whisler launched his business career in 1975 as a summer intern at Western Nuclear in Denver—and that internship may rank among the ill-fated

enterprise's most positive contributions to Phelps Dodge! The uranium company subsequently financed Whisler's way through the University of Denver College of Law, where he specialized in natural resources law, business planning, and taxes. Afterwards he added a master's degree in mineral economics at the famed Colorado School of Mines.

As an attorney Whisler continues in the tradition of Page and Munroe; as a mining man he walks in the footsteps of the Douglases, Cates, Fenzi, Moolick, and Judd; as a trained businessman he follows the course of Judd and Durham. Like Yearley he attended Harvard's program in advanced management. However, Whisler's own combination of legal, mining, and business education is unique for any president of Phelps Dodge. It might be added that he gained valuable additional experience via the school of hard knocks, thanks to his self-appointed mentor, Moolick: "I sent Whisler out to get his nose bloodied in the 1983 strike at Morenci. Here was an opportunity for a guy to see what the real world was like."[33]

Just two years earlier, in April 1981, when Whisler was still working for Western Nuclear, he received a call from Moolick at the corporation's headquarters. "We need you in New York tomorrow. I don't know what your job or salary will be." A few minutes later Moolick called back to ask, "Do you have a passport?" In that rather casual way, Whisler became legal counsel to the exploration group.[34]

Colleagues describe Whisler as both intense and relaxed. In most executives that would be a contradictory combination, but he

⌣ J. STEVEN WHISLER
*became president and chief
operating officer of Phelps Dodge
Corporation in 1997. He brought
an impressive combination of
college and university training
and hands-on experience to his
work. Whisler has served on the
board of directors of Burlington
Northern Santa Fe Corporation
and UNOCAL Corporation as
well as those of numerous civic
groups and organizations. He has
been a director of Phelps Dodge
Corporation since 1995.*

possesses an ability to focus all his energy on solving a problem—he is, in his own words, a perfectionist—and yet he somehow remains relaxed about any public assertion of his authority. After he became head of Phelps Dodge Mining Company in 1991, he spent a week at each facility doing manual jobs, including some that were grimy and physically demanding, just to show employees "he was for real" and that he had not forgotten his roots as a construction worker in Colorado. Whisler drives himself to work in a pickup, and to date his preferred leadership style has been to labor behind the scenes and give others a generous measure of credit for the positive results. He often compliments his management colleagues by referring to them as his "thoroughbreds."

Like many others at Phelps Dodge, Whisler as president and chief operating officer must struggle to juggle demands of work and a normal home life; he has a wife, Ardy, and two young children. One way he copes with the inevitable stresses is by maintaining his sense of humor, a

trait he has in common with his predecessors Yearley and Judd.

During his more than two decades at Phelps Dodge, Whisler gained important insights into the strengths and weaknesses of the corporation: "One of Phelps Dodge's greatest strengths is that it is 170 years old. One of Phelps Dodge's greatest weaknesses is that it is 170 years old." The challenge for present and future generations, he recognizes, will be to keep those two facts in proper perspective—much as employees did during the late 1990s culture change—and to use personal knowledge of the company's rich and complex history to best advantage as they make decisions that shape its future business landscapes. Whisler, who together with senior executives coined the slogan "Pioneering the Creation of Value" and helped make the concept central to the corporation's vision for itself, leaves no doubt that the important work of culture change begun by Yearley will continue as the enterprise charts its course for the new millennium.[35]

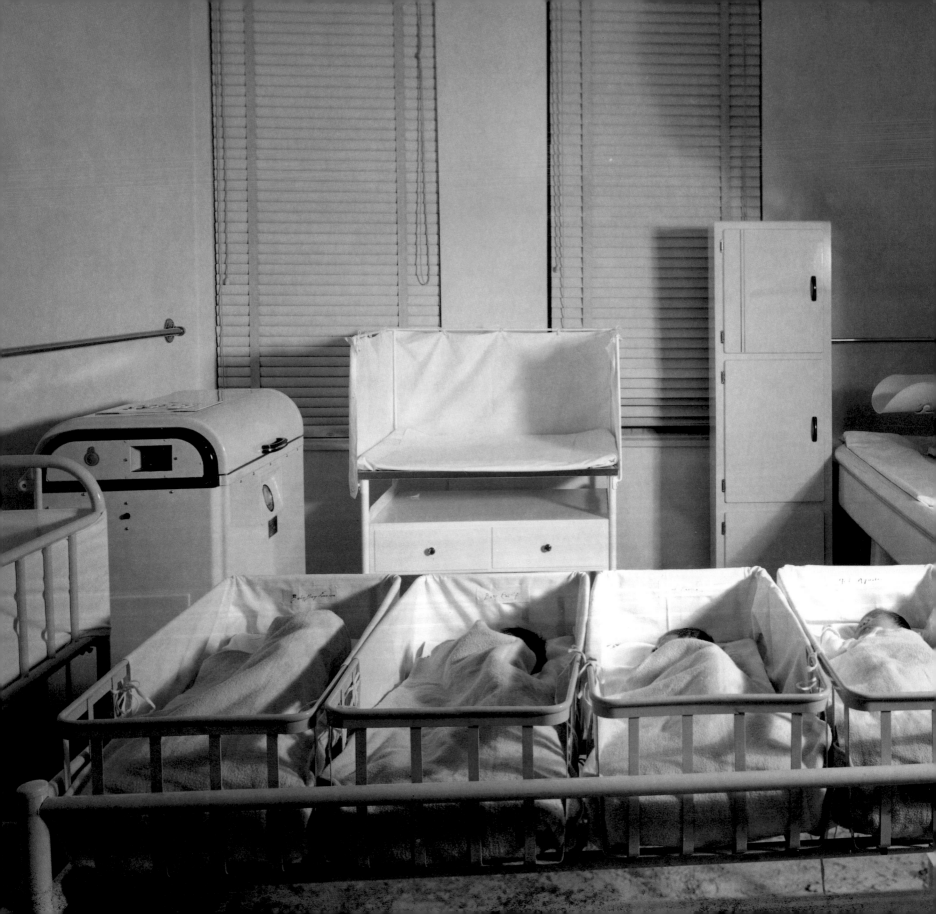

~ *Chapter Eighteen* ~

BACK *to the* FUTURE

I N THE POPULAR Hollywood movie *Back to the Future*, time travel takes place in a converted DeLorean automobile. The teardrop-shaped General Motors EV1 was hardly a time machine in the literal sense, but it provided an impressive link between past and future, between the unfulfilled promise of electric cars in the early twentieth century and the vehicles likely to become commonplace on the roads and highways of the twenty-first century.

Phelps Dodge leased an EV1, the first electric vehicle mass-produced in recent years in the United States, for employees to use. Individuals took turns commuting in it for several days and offering their evaluations. The corporation's green model, a muted color chosen specifically to emphasize the conservative nature of Phelps Dodge, was in the shop the day I took a brief test drive in mid-1997, so I enjoyed scooting around the streets of Phoenix in a racy candy-apple-red model. It was a sleek and quiet automobile, and it cornered like a low-slung sports car. It was no oversized golf cart, either. It could easily accelerate from zero to sixty miles per hour in under nine seconds.

~ BACK TO THE FUTURE: *this image of newborn babies in the nursery at Phelps Dodge's Douglas hospital reminds us that today is always born out of yesterday. That is true too for Phelps Dodge Corporation.*

Its inside climate was controlled by a heat pump and ozone-friendly refrigerant. The entire motor drive unit weighed 150 pounds, about 66 percent less than the typical gas-powered four-cylinder engine, it had only one moving part, and it never required an oil change. It used seven different onboard computers. All in all, General Motors filed twenty-three new patents in connection with development of the EV1. Its most noticeable drawback in the late 1990s was the capacity of its twenty-six storage batteries, which limited the EV1's range between charges to approximately seventy miles. If the sleek automobile, or some kind of successor model, ever becomes a popular consumer item, it bodes well for Phelps Dodge Corporation because of the copper its electric motors and support facilities will use.[1]

Even without the EV1, automobiles require more copper than ever before in history—about fifty pounds today versus only thirty pounds just a decade ago. Although by the 1990s, aluminum radiators replaced copper ones in about 75 percent of American-made automobiles, that was not the case in Japan and many other foreign countries, where a large majority of new cars still contained copper radiators. Those same automobiles used many dozens of small motors (needing countless miles of magnet wire) to roll down windows, adjust seats, activate antennae, lock doors, and do a myriad of other tasks. A new Lexus, for example, utilizes some 165 electric motors. Copper is also used extensively in automobile air conditioners and audio systems.

The same increase in copper usage was noticeable in residential and commercial construction. By the mid-1990s the substitution problem once posed by plastic pipes and aluminum building wires had diminished. In fact, by that time copper had regained much of the market once lost to aluminum. Copper consumption outside the United States expanded rapidly too, skyrocketing 1,200 percent between 1970 and 1992 in Asia, excluding China and Japan, and within those two nations it still grew by a respectable 224 percent. Japan and several rapidly industrializing nations of the Far East, the so-called economic tigers, accounted for fully one-third of global demand by the mid-1990s. That was the primary reason, when Asian economies faltered in late 1997, the world price of copper dropped from $1.23 a pound in June 1997 to an average of $0.70 a pound during the last quarter of 1998. The next year promised no improvement, the price falling in January 1999 to $0.65 a pound on the London Metal Exchange.

"We're dealing with the lowest annual average copper price, on an inflation-adjusted basis, of the century," observed Timothy R. Snider, president of Phelps Dodge Mining Company. The abrupt nose-dive caused the corporation to trim global copper production by 100,000 tons and its workforce at Chino by approximately three hundred. In two market-driven cutbacks, the first since 1985, it closed its last working properties underground at Ojos del Salado and at Cobre Mining Company in New Mexico.

But what about the challenge of fiber optics? Glass was never a viable competitor until thin filaments proved to have impressive conductivity. Glass substitution hurt copper in communication wiring, the very market in which it had seemed least exposed. Fiber optics

TIMOTHY R. SNIDER *was elected president of Phelps Dodge Mining Company and a senior vice-president of the corporation in September 1998 at age forty-eight. A graduate of Northern Arizona University (1979) with a degree in chemistry and geology, he began his climb to the top of the mining company as a laborer at the Copper Queen Branch in 1970. Snider championed the development of solution extraction–electrowinning operations at Tyrone and Morenci, where he later held the position of president, Phelps Dodge Morenci, Inc. He attended the Wharton Advanced Management Program in 1996.*

became important in transmission of data by telephone; however, electricity that powers the machines that generate data for fiber optics still relies heavily on copper, as do local connections—including those likely to be found inside a future generation of computers linked together by globe-girdling electronic networks.

From Molecule to Market

IN SEPTEMBER 1997 IBM announced that it had developed a way to make computer chips by using copper instead of aluminum, the metal long favored by manufacturers. The innovative product, aptly code named the Coppermine series by Intel, could ultimately trim the cost of chips by as much as 30 percent and speed calculations by perhaps as much as 40 percent. Further, the new technology will enable chips to operate with less electricity, making them especially useful for laptop computers and other battery-operated electronics products. IBM's breakthrough use of copper in chips will boost red metal's reputation but probably not its production because only a tiny amount is required to make each chip. If nothing else, the symbolism is important.

Who could predict what discoveries might emerge from Phelps Dodge's own research laboratories? Research has long been important to the corporation. As recently as the 1970s, for example, it staffed and equipped research laboratories at manufacturing facilities in Yonkers, New York; Elizabeth and South Brunswick, New Jersey; and Fort Wayne, Indiana, to study process improvements and new products. The Blue Box was the nickname for its

mammoth cube-shaped laboratory at Yonkers used in connection with the Electric Research Council's program on extra-high-voltage and cryogenic cable development.

At the end of the twentieth century, Phelps Dodge Corporation operated three state-of-the-art research laboratories (as distinct from quality control or assay offices): one in Fort Wayne next to the main offices of Phelps Dodge Magnet Wire; another in Marietta, Georgia, at the new headquarters building of Columbian Chemicals Company; and a third in Safford, Arizona, for Phelps Dodge Mining Company. The Fort Wayne laboratory, for instance, "is partly intended to be impressive from the standpoint of showing our customers our commitment," notes Daniel E. Floryan, Magnet Wire's vice-president for quality and technology, "but also it is a very functional tool. We think it is one of the best laboratories dedicated to magnet wire, certainly in this country, and probably in the world." Floryan enjoys walking customers through the facility just to "wow" them with its top-flight research and problem-solving capabilities. Although the mass media today tends to focus on flashy high-tech communications and computer breakthroughs, Floryan believes there is considerable room for innovation in basic manufacturing processes, such as making electric motors more energy efficient, which requires innovations in magnet wire coatings.[3]

Although each of the Phelps Dodge research facilities has a distinct mission and history, over the years all have discovered ever larger expanses of common ground. Research in particle science shows how Phelps Dodge is extending its reach

Historic laboratory interior *at the New Cornelia Copper Company in Ajo. Phelps Dodge today has several modern research facilities. Two of the most impressive are at Phelps Dodge Magnet Wire in Fort Wayne and Columbian Chemicals Company in Marietta, Georgia (see pp. 358–59). Research is vitally important in mining too, as the 1998 relocation of the laboratory of Phelps Dodge Mining Company to modern quarters in Safford attests.*

literally from molecule to market. What researchers in Georgia using Columbian's million-dollar field emission microscopes learn about basic properties of carbon black particles may well prove useful to scientists in Arizona who ponder afresh the fundamental aspects of producing copper and other minerals, including something so deceptively simple as the explosive force needed to blast copper ore from benches in an open-pit mine.

To nonspecialists the job of setting the explosive charges required to break ore into what will become successively smaller-sized pieces and particles may seem like uncompromisingly low-tech activity. Yet Phelps Dodge Mining Company researchers at the Safford laboratory have learned that using a satellite-based global positioning system allows workers to drill a close-to-perfect array of holes that, with precise preparation, can cause a blast to yield material of optimum size for maximum processing efficiency. This research can save energy both at the site of the blast and in subsequent processing of the ore. In short, innovation in "blast fragmentation" equals many dollars saved.

At a time when about half of all Phelps Dodge copper derives from the solution extraction–electrowinning process, its Safford research center examines ways to optimize existing SX/EW facilities as well as develop SX/EW processing units portable enough to make small or remote ore bodies profitable. That is one of several promising projects involving liquid-to-liquid extraction of metals that someday may make energy-hungry smelters obsolete. Another goal is to extract red metal in a cost-effective way from progressively leaner grades of

ore. "The real challenge for the future," observed Robert E. Johnson, a metallurgical engineer and scientist who oversaw approximately thirty-five researchers at the Process Technology Center in Safford, "is that ore bodies of the past are not long going to be there." In the United States exploration teams will probably not find copper deposits like those that made Butte and Morenci famous, but Johnson, now retired, remains confident that mining company researchers (and like-minded people throughout the industry) can give Phelps Dodge the technological tools it needs to assure its future success in mining.[4]

The list of research activities at all three facilities is a lengthy one, and the search for ways to improve existing processes and for new or better products is a serious quest. Nonetheless, all forms of innovation, no matter how esoteric, have their lighter moments. On one occasion, technicians in Fort Wayne bought a cigarette-wrapping machine in hopes of using it to make a special type of paper-insulated magnet wire. They failed to achieve fully the desired results, but just the same they rigged it up with lots of "bells and whistles" to create the illusion of a finely tuned mechanism, held an open house, and thus encouraged puzzled technicians from the competition to duplicate Magnet Wire's "success."

Like mine exploration personnel, with whom they have so much in common, the corporation's researchers sincerely believe they hold the prosperity of Phelps Dodge in their hands. For them this makes the challenge of Pioneering the Creation of Value especially meaningful: "We embrace it fully," says John O. Marsden, vice-president for technology, Phelps Dodge Mining

Company, and a graduate of the Royal School of Mines in London. One indication of long-term support for the contributions that laboratory workers make: during cutbacks at Phelps Dodge Mining Company in 1999 because of copper's low price, nothing was trimmed from the budget for research. Much will inevitably change at the corporation during the coming years, but Phelps Dodge's commitment to its own world-class research teams appears solid.[5]

The Certainty of Ever-changing Landscapes

PHELPS DODGE'S OWN landscapes of production, no less than the world markets for copper, have witnessed major changes in recent years. On November 16, 1996, demolition experts sent the Morenci smelter's two 600-foot-high stacks crashing down at precisely eight o'clock on that Saturday morning and forever changed the local skyline. Both had been impressive landmarks, the oldest one dating from 1941 when it ranked as the world's tallest industrial stack. Again, on February 12, 1997, just three months after Morenci's two smokestacks disappeared, a carefully placed series of explosives felled a long-unused smelter stack located just south of Clifton. For years, the tall sentinel greeted motorists as they dropped down into the San Francisco River valley. It was an imposing remnant of the Arizona Copper Company smelter built in 1913 and last operated by Phelps Dodge in 1938. Workmen had long since removed the buildings, but the solitary stack remained in place for several decades to jog the memories of old-timers who recalled the original landscape of production.

Nineteen miles south and across the mountains from Clifton and Morenci a different kind of landscape evolution was proposed for an open plain just north of the cotton-growing town of Safford. There never was a mine settlement on the site, but no one could say with certainty what the future might bring. Safford was where Phelps Dodge hoped to launch a $370 million copper-mining project by the early 2000s. For far longer than most people could recall, the Safford project had been a sleeping beauty for Phelps Dodge, always full of alluring promise as perhaps the most significant copper find in North America since World War II but never awaking to its full potential. Perhaps that will have changed by the time this book is published—but then again a 30 percent slump in the price of copper during 1997 and 1998 caused many plans to be scaled back or put on hold. Even so, "Safford is a great ore deposit," according to the retired senior vice-president Patrick J. Ryan, who first looked it over back in 1971.[6]

Phelps Dodge holdings in the Safford mining district, a rectangle in the foothills of the Gila Mountains that is roughly twelve miles long by three miles wide, encompass some twenty thousand acres of patented land and four significant bodies of ore. These include the Lone Star, a huge copper deposit buried beneath six hundred feet of overburden, together with the San Juan, Sanchez, and Dos Pobres ore bodies. The near surface oxide portion of the deposits is today "the most significant undeveloped copper resource in the United States," stated John Broderick, onetime project manager, in 1997.

THE LANDSCAPE OF the Safford project in 1974. In the early 1980s the facility consisted of two shaft sites ten thousand feet apart that Phelps Dodge planned to connect underground. The company attempted a block-caving operation by carving out a room the size of a football field. But the explosion and caving resulted in blocks the size of houses, something not desirable. Phelps Dodge suspended work during the corporate restructuring of the mid-1980s and allowed water to flood the underground passageways. It currently expects to mine the ore by open-pit methods.

Some people even predicted a new Morenci. More realistically, a new mine complex at Safford would probably employ about 250 people, and operating at full capacity, the Dos Pobres and San Juan deposits would provide about two hundred million pounds of copper each year for at least fourteen years.[7]

The corporation's interest in copper in the Safford area actually dates back to September 1957 when it optioned a group of claims located north of town. Phelps Dodge conducted exploratory drilling two months later, exercised its option to purchase the claim group in 1960, and commenced sinking an 1,875-foot shaft into a deeply buried sulfide ore body in November 1968. The Dos Pobres deposit proved a disappointment, and Phelps Dodge actually lost $85 million in a futile attempt to mine it by block-caving, then seemingly the most appropriate technology for a mine that deep. During corporate cutbacks of the mid-1980s it appeared doubtful that the Safford property had any future.

Yet over the years Phelps Dodge added three more copper deposits in the area, including Lone Star, which it gained as part of the Chino Mines Company wheeling and dealing with Kennecott in 1986. Lone Star came to Phelps Dodge along with a curious history: it was where Kennecott had planned to detonate an underground nuclear blast in the 1960s. The company had actually obtained a permit from the Department of Defense to do so.

It was all part of Project SLOOP, which now seems like science fantasy; but like the atomic-powered bomber and similar exotic projects that never progressed beyond the drawing boards, it dated from an era of popular fascination with all things nuclear—and that included Phelps Dodge's own infatuation with uranium and its Western Nuclear subsidiary. As part of Project SLOOP, the Atomic Energy Commission, Lawrence Radiation Laboratory, United States Bureau of Mines, and Kennecott conducted a two-year study of how to use a nuclear explosion to mine copper. The proposed blast twelve hundred feet beneath the surface would be equivalent to twenty thousand tons of TNT and would in theory pulverize an underground deposit containing eight million pounds of copper that Kennecott could then leach, presumably without problems from lingering radiation. The explosion never took place, and the copper is still there.

Federal and state laws today require that mining companies protect, relocate, or otherwise mitigate the effects of their operations on plants and wildlife. Thus around 10 percent of any money Phelps Dodge spends to develop its open-pit operation at Safford will go toward environmental protection. For example, a leach pad liner to prevent pregnant leach solution from touching undisturbed soil costs $35 million. Phelps Dodge, which for years had no formal environmental department, is now determined to remain well ahead of any environmental mandates or regulations.

On the opposite side of Arizona is Phelps Dodge's Copper Basin property, a medium-sized porphyry copper deposit located about midway between Skull Valley and Prescott. Despite its name, Copper Basin is actually situated nearly a mile above sea level. Phelps Dodge has held several claims in that area since the 1880s, when William E. Dodge became president of the

Copper Basin Mining Company, but it never became a big producer compared to Bisbee or Morenci. Over the years, Phelps Dodge drilled the area extensively to define a deposit of about seventy million tons of ore running approximately .5 percent copper. If the corporation should acquire additional land by purchase or exchange so it could conveniently locate its shops, concentrator, dumps, and tailing sites, the property might be a viable producer.

Meanwhile, a search for more copper around the corporation's active mining sites goes on. Back in 1988, Phelps Dodge officials had estimated that mining and milling operations would cease at Tyrone by the early 1990s and that nearby Chino had reserves sufficient to keep it operating about twenty more years, until 2008. As a result of outstanding efforts by exploration and mine staffs, in the late 1990s Chino still has twenty years of reserves and Tyrone at least ten.

At some properties, no one can be sure what the future holds. On November 29, 1995, for instance, Phelps Dodge had the tall smelter stack at Ajo demolished and remnants of the old smelter and concentrator complex removed from the landscape. The time-honored Tucson, Cornelia & Gila Bend Railroad seemed destined for the scrap heap as well, but early in 1997 everything changed. Phelps Dodge announced plans to invest $238 million to bring the moribund New Cornelia Branch back to life in 1999 after a fourteen-year hiatus. That meant obtaining a variety of permits and making major upgrades to the infrastructure, including building a state-of-the-art concentrator and refurbishing the railroad so that it could once again haul concentrates forty-four miles from Ajo to Gila Bend.

That was not happy news for retirees who imagined they had moved to a small, quiet town when they bought one of the nine hundred homes Phelps Dodge sold in Ajo in the early 1990s. (Many home buyers seemingly forgot that Phelps Dodge had advised them that it might later restart operations.) But hardly had anyone had time to voice a strenuous complaint before Asia's economic troubles cooled demand for copper and caused a price slump that forced Phelps Dodge to suspend the planned renovation work. Whenever it resumed depended on copper's recovery. As always, in Ajo, Safford, and every other would-be copper-producing community, the name of the game is the Price Is Right!

The mines of Bisbee could come back to life too with the right technology and a high enough price for copper. Development of its Cochise deposit would likely require a combination of open-pit mining and the solution extraction–electrowinning process that has proved such a blessing to Phelps Dodge since the mid-1980s. As for the United Verde property at Jerome, there seems little likelihood of revival. It is a topographically complex site, yet Phelps Dodge and Cominco Ltd. did launch a joint venture there in the 1990s to search for copper and zinc deposits. "You never say never in this business," emphasized J. Steven Whisler, president of Phelps Dodge Corporation.[8]

That was one reason no one dared to predict how the mining landscapes of Phelps Dodge might evolve in the future. There may be some big surprises along the way. It is not inconceivable, for example, that if company or industry researchers perfect a process that permitted Phelps Dodge to process copper ores of all types into refined cathodes at its individual mine properties, smelters might vanish entirely from the corporation's familiar landscape of production.

Mining's Life Insurance Policy

THE MANUFACTURING LANDSCAPE continues to evolve as well. In 1994, for instance, Phelps Dodge bought a magnet wire plant in El Paso, Texas, and spent $14 million on its expansion. Two years later Phelps Dodge Magnet Wire began construction of a $42 million plant in Monterrey, Nuevo León, Mexico. That facility, completed in 1998, was intended to serve the growing transborder maquiladora market and free up capacity in plants north of the border to meet growing demand for magnet wire within the United States. Columbian Chemicals Company continued to expand too; for instance, early in 1999 it acquired an 85 percent stake in a South Korean carbon black–manufacturing firm for $73.1 million. That permitted Phelps Dodge Industries to gain important market share in Asia, much as it did in South America with acquisition of the gigantic Copebras plant in Brazil a year earlier.

Already by the mid-1990s Phelps Dodge Industries accounted for about a third of the parent corporation's operating earnings, and in that way it provided a "life insurance policy for mining," emphasized Manuel Iraola, who became its president in 1995. He observed too that Phelps Dodge Industries was not an old-fashioned conglomerate but rather a collection of compatible process-oriented businesses that fabricate basic products used by a diverse group of manufacturers.[9]

The theory Iraola described in early 1997 (and which the Committee of Five had first enunciated in the mid-1980s) became reality during the second quarter of 1998, when copper prices that dropped from an average of $1.33 a pound in 1995 to $0.75 a pound in 1998 caused Phelps Dodge Corporation to cash in its "insurance policy." For the first time ever, Phelps Dodge Industries earned more than Phelps Dodge Mining Company ($46.1 million versus $34.9 million), or 67 percent of total operating income. That unprecedented pattern continued for the remainder of 1998. In all, when the corporation closed its books for the year, the Industries side accounted for operating income of just under $354 million, and Mining, slightly more than $110 million. Just a year earlier their relationship had been reversed, operating income from the latter in 1997 totaling more than twice that of the former ($459 million to $208 million).

One caveat: somewhat more than half the operating income for Phelps Dodge Industries during 1998 derived from the its first quarter sale of Accuride Corporation, its wheel and rim business. Accuride, with 70 percent of the market share in North America, dominated the overall medium-heavy wheel market and the sixteen-inch commercial light truck market, but Phelps Dodge concluded that it had better ways to invest its dollars. For a time, back in 1994, Hudson International, failing to perform well because of a downturn in federal defense spending and other reasons, appeared destined for sale; Phelps Dodge seemed ready to give it away. But Iraola, the new head of Phelps Dodge Industries, flew to South

MANUEL J. IRAOLA *was elected president of Phelps Dodge Industries, the diversified manufacturing business of Phelps Dodge Corporation, in 1995. He also serves as a senior vice-president of the corporation and a member of its board of directors. Born in Havana, Cuba, Iraola holds a B.S. degree in industrial engineering from the University of Puerto Rico, and an M.B.A. degree from Sacred Heart University in Fairfield, Connecticut. He attended Pennsylvania State University's Executive Management Program in 1991. Iraola came to his current position by way of Columbian Chemicals Company, where he was the senior vice-president and chief financial officer at the time Phelps Dodge acquired it in 1986, and later Phelps Dodge International Corporation, where he served as president.*

Carolina instead and, after studying the problem, installed a new management team headed by a young and energetic Glenn Decker. The facility, reborn as Phelps Dodge High Performance Conductors, once again headed down the often winding road to success.

Legacy: Triumph over Adversity

A NY ATTEMPT TO explore the many landscapes of Phelps Dodge presents a challenge because they are constantly changing. Could it be otherwise for an active mining and manufacturing enterprise? I write these words early in 1999 knowing full well that before *Vision and Enterprise* appears in print a year from now, familiar components of Phelps Dodge Corporation may well be gone and new ones added. The same will be true for personnel.

Among the more intriguing of its modern landscapes, and a prime example of why it is impossible to anticipate what changes lie ahead, is one that Phelps Dodge proposed back in the late 1980s for the area near Clarkdale, Arizona, where tailings, or finely ground rock, from the old United Verde concentrator had been deposited in a bend in the river until the early 1950s. Phelps Dodge proposed to cap the tailings over and landscape the area to create an attractive golf course and real estate development called Verde Valley Ranch. There was precedent for this because in the 1920s, when William Andrews Clark owned the mine, his company built an extensive employee recreation facility nearby that included a clubhouse and a nine-hole golf course that used cotton hulls for its greens. He also built

a two-story house that Phelps Dodge anticipated using as a museum. Alas, though the process of creating the Verde Valley Ranch began in 1987, applications for the necessary permits were still caught in a political and environmental thicket more than a decade later.

Some things, however, remain remarkably consistent through the years, and for that reason *Vision and Enterprise* concludes with a late 1990s example of Phelps Dodge's ability to triumph once again over adversity. The incident is reminiscent of the day back in 1832 when the recently built Phelps & Peck warehouse collapsed in New York; from that tragedy emerged a new business partnership two years later, Phelps, Dodge & Company.

The more recent misfortune occurred at Morenci just two days before Christmas 1997. Shortly after the morning shift change at seven, the Phelps Dodge Morenci president, H. M. ("Red") Conger, was on his way to work when he received word that a power line had gone down. That did not concern him greatly until he learned why: after steady service for more than ten years, 180 feet of the conveyor system that spanned Highway 191 had collapsed suddenly and without warning, taking out the power line. One moment the conveyor carried crushed ore as usual from the pit to stockpiles for the concentrators; then a joint failed, and the next moment it was a massive pile of twisted steel blocking the main highway. A vital artery in the production of copper—Morenci produced more then a billion pounds of copper in 1997—was severed.

Emergency measures had to be taken without delay. First, Morenci personnel made certain that no one was trapped beneath the wreckage. Then Conger, employing the teams approach, formed eight or nine groups to deal with different aspects of the recovery process. Soon flaggers rerouted traffic through the mine; the State of Arizona gave Phelps Dodge permission to extend a haul road along the top of the existing highway. Large trucks worked around the clock to supply the concentrators with ore that formerly traveled by conveyor.

On Christmas day one team gingerly picked apart the hulking remains. It was dangerous and "very nerve wracking" work, recalled Conger; the metal pieces were bent and twisted to form something like a gigantic spring, and removing one piece could release potentially lethal energy. Meanwhile, a fabricator in Salt Lake City built new conveyor sections, including a 100-foot truss that it carefully trucked along interstate highways much of the way to Morenci.

Conger, who had worked there only since October 1, drew upon his own training as an engineer at the Colorado School of Mines and upon earlier experience he had gained when an important conveyor at the Chino Mines Company went down after a motor caught fire. That misfortune also happened on December 23! The remarkable coincidence caused Steve Whisler to quip to Jim Madson, then vice-president and general manager of Phelps Dodge Mining Company, "Don't ever send this guy to another property that has a conveyor." Conger recalled that, with a little humor and a lot of teamwork by hundreds of employees and retirees who toiled through the holidays, they had the vital conveyor system back in action in exactly thirty days. Through it all the Morenci mine never shut down.

What happened at Morenci in late 1997 and early 1998—like the mid-1990s storm-caused roof collapse at the sprawling Phelps Dodge High Performance Conductors plant in Inman, South Carolina, and additional disasters at Phelps Dodge facilities on other occasions—illustrates that success in business includes employees pulling together to help a company triumph over adversity. By that measure, Phelps Dodge Corporation should be able to anticipate a good future.

The long-term outlook is bright for other reasons too. At Morenci, for example, where Phelps Dodge and its predecessors have mined mountains of copper for well over a century, may lie the world's largest chalcocite blanket ore body. Even now, "we haven't defined the edges of this thing," said Conger of its "fantastic opportunities" for copper production as he looked ahead to the twenty-first century. No wonder he added—with words commonly used by many other Phelps Dodge employees— "I really like this business."[10]

WHEN DOUGLAS C. YEARLEY retires at the beginning of the new millennium, the legacy of his long tenure at Phelps Dodge Corporation, including twelve years as chairman, can be measured in many different ways. Foremost among his contributions is that he, a metallurgist by training, engineered a new corporate "alloy" that was tougher and more durable than the sum of its parts. In other words, since 1989 when he succeeded G. Robert Durham, Yearley and his lieutenants have consistently labored to give Phelps Dodge the strength and resilience it required to respond successfully to future challenges.

CRUSHED ORE DROPS *onto a stockpile from the in-pit conveyor, as photographed in Morenci on a snowy day in January 1997. Eleven months later the portion of the system spanning the highway just out of sight to the left suddenly collapsed.*

Yearley once termed that process "managing for the downside—even on the way up." In late 1998 he observed to his industry colleagues: "When times are good, nothing earns money like a mine. I submit to you that this is one of our industry's biggest problems. Those of us in the mining business are like surfers—we can't control or even predict the commodity price wave. The best we can do is stay on our boards and ahead of the curl. But in the euphoria of riding a wave of $1.35 copper, it's easy to forget to look around at the tides of supply and demand, political forces and global economics, and not see the breaking wave until it has washed over you."[11]

During the 1990s, Yearley promoted low-cost production of copper, the corporation's core business, as well as prudent diversification into noncopper areas intended to provide downside protection during unavoidable slumps in the price of copper. In many ways large and small he and his lieutenants sought to strengthen the balance sheet to increase the value of Phelps Dodge to its shareholders and keep it a blue-chip enterprise. They pushed forward the process of globalization that Durham earlier fostered, achieving success in the Candelaria mine in Chile.

Remarked Gary Loving, vice-president for South American operations, "I think the vision at the top is one that understands how to stay strong in this business." That conviction is affirmed by many financial observers outside the corporation: "PD management has done a superb job of running the company over the last ten years," noted Morgan Stanley Dean Witter in late November 1998.[12]

Yearley and his leadership team have supported positive changes in Phelps Dodge's corporate culture as part of their larger goal of seeking "operating excellence." They have worked to foster better communication throughout the corporation as part of an openness they believe will lead to greater employee participation in management at all levels, which in turn will produce positive results for everyone. They continue Phelps Dodge's long-standing commitment to philanthropy in student scholarships and a variety of community betterment programs.

The vision that Yearley and other employees of Phelps Dodge translated into profitability during the good years of the early and mid-1990s greatly strengthened the nearly 170-year-old corporation and positioned it well to triumph over adversity. That vision also encouraged individuals to discover for themselves that "I really like this business." And "this business," it must be added, included production of about as much carbon black as copper, or approximately 1.8 billion pounds each in 1998.

In fact, the corporate-restructuring program that Yearley and his colleagues set in motion back in the mid-1980s makes the name Phelps Dodge increasingly synonymous with both copper and carbon black as the new millennium approaches. That remarkably synergistic combination is likely to reward investors and employees alike, along with all others who benefit from Phelps Dodge's formal commitment to Pioneering the Creation of Value. Anson Phelps certainly would have understood and approved what that pledge means; and as a pioneer himself in the world of multinational business, he would surely be proud of the enterprise that carries his name and vision into the twenty-first century.

BACK IN THE early 1880s when Phelps Dodge's Ansonia Clock Company plant burned to the ground in Brooklyn, it used the symbol of the Phoenix in its products catalog to inform customers of its rebirth. That remains an apt symbol for a corporation with a proud history of bounding back from adversity.

New Clock Factories, Brooklyn.

~~ *Chapter Nineteen* ~~

UPDATE 1999:
THE PACE OF CHANGE

VISION AND ENTERPRISE ends its story of Phelps Dodge Corporation early in the summer of 1999, when the book manuscript began the process leading to its publication the following year. However, like a fast-flowing river, the pace of change at Phelps Dodge did not moderate during that interval. In fact, it picked up considerable speed, and like a river full to its banks, it cut some significant new channels and bypassed old ones.

The following, in brief, are several key events that impacted Phelps Dodge Corporation in mid-to-late 1999. All of these add to the history told on the previous pages.

A major milestone for Phelps Dodge during the past twenty-five years was construction of the Hidalgo smelter and the town of Playas in New Mexico as part of a successful corporate commitment to clean air. The massive smelter, along with the one in Chino, supplied a steady flow of copper anodes to the refinery in El Paso. For a variety of reasons, the familiar landscape of production changed in 1999, as it had so often in years past. Phelps Dodge Mining Company took steps during the first week of September 1999 to shut down (at least temporarily) its Hidalgo smelter because of

~~ THE FACES OF CHANGE. *No longer can mining operations be thought of as a "man's business." A collective portrait of women at work at Phelps Dodge Morenci in 1999: Sara Skoff, SX/EW metallurgist (upper left); Connie Waddell, mine engineer (upper center); Patricia Young, supervisor change agent (lower right); and Rachel McCarthy, haul truck driver (lower left).*

433

rock-bottom copper prices and a noticeable slackening of concentrate production in the Southwest. When it closed the smelter, it mothballed the settlement of Playas.

Further, Phelps Dodge Mining Company placed on standby the smaller of two concentrators at its Morenci, Arizona, mining complex. The giant mine would continue to yield approximately 800 million pounds of copper a year. In fact, by the year 2001 the company expected to use solution extraction–electrowinning technology for all copper from Morenci, up from the present 50 percent. The $220 million conversion will shave seven to nine cents a pound off production costs, an important saving in times of low copper prices—and a blessing when prices climb again. The anticipated change-over to all SX/EW production at Morenci further contributed to the shutdown of the Hidalgo smelter. Outside Arizona and New Mexico, Phelps Dodge Mining Company sold its Witkop mine and related fluorspar-producing facilities in South Africa.

Phelps Dodge Industries took steps to combine Phelps Dodge Magnet Wire Company and Phelps Dodge International Corporation (which included Phelps Dodge High Performance Conductors) into a single wire and cable manufacturing division. Phelps Dodge further announced that it would cease all wire and cable manufacturing in Ecuador and close a small rod mill and magnet wire facility in Venezuela. It closed its magnet wire plant in Hopkinsville, Kentucky, as well as two small High Performance Conductors facilities in Montville and Fairfield, New Jersey.

The late 1990s downturn in world copper

JULIA MAZANEK ON *the job in Phoenix in 1999. Probably the longest-serving employee in the history of Phelps Dodge Corporation, she recalls many of the landmark events that occurred during the fifty years since she started work at the Douglas, Arizona, offices in 1947.*

prices created several major headaches for American producers and long-term structural changes in the entire industry, not just at various Phelps Dodge operations. For example, back in 1996 at the top of the market boom, Broken Hill Proprietary, the Australian mineral giant, paid $2.4 billion for large copper mines in Nevada and Arizona, but in mid-1999 it closed those properties, which included the modern San Manuel smelter in Arizona. On the other hand, as had often been the case in the past, the months of prolonged adversity presented Phelps Dodge with some unusual growth opportunities.

At the 1999 Global Leadership Forum (the successor to Management Review Week) in Phoenix in July, top executives publicly noted that they were exploring various ways to reinvent the corporation. That meant altering the historic pattern of a major restructuring at Phelps Dodge every fifty years to a fifteen-year cycle. What form reinvention might take became clear only a few weeks later when Phelps Dodge announced a takeover offer for both Cyprus Amax Minerals Company and ASARCO Incorporated. The bold move compared favorably

CYPRUS AMAX MINERALS COMPANY
1998 ANNUAL REPORT

Our Employees:
committed to their Quest for quality...

⟿ The FRONT COVER *of the last annual report issued by Cyprus Amax Minerals Company, which became a wholly owned subsidiary of Phelps Dodge Corporation late in 1999.*

Phelps Dodge to Acquire Cyprus Amax

Deal Valued at $1.79 Billion; Hostile Struggle to Buy Asarco Set to Continue

---

WALL STREET JOURNAL, OCT. 1. 1999

with how Phelps Dodge acquired Calumet and Arizona Mining Company and United Verde Copper Company in response to the depression of the 1930s.

Two industry giants, ASARCO and Cyprus Amax, announced plans on July 15, 1999, for a $1.86 billion merger as Asarco Cyprus Inc. Had they been successful they would have created the world's largest publicly traded copper firm. But when the stock market reacted with apathy to the proposed merger, Phelps Dodge saw the chance to create a copper colossus with itself in the driver's seat.

When the two companies rejected its plan for a three-way combination, Phelps Dodge countered with a public offer for both copper companies that provided their shareholders a 40 percent premium. On September 24, the same day that Phelps Dodge received the needed anti-trust clearance from the United States Department of Justice, ASARCO's largest shareholder, Grupo Mexico S.A., greatly complicated the merger scenario by launching a tender offer for

ASARCO. The following week Phelps Dodge Corporation announced that it had signed a $1.8 billion cash-and-stock buyout agreement for Cyprus Amax, effective immediately. It declined to trump Grupo Mexico in a bidding war for ASARCO.

This ended the six-week takeover drama, which Chairman Douglas Yearley aptly characterized as an "emotional roller-coaster" and as intellectually challenging as "three-dimensional chess" for his management team. When it gained Cyprus Amax, Phelps Dodge Corporation became the largest publicly traded copper company, accounting for approximately 12 percent of the world's total annual output of the red metal. Among all producers, it ranked behind only CODELCO, owned by the government of Chile. Further, when it completes its integration of Cyprus Amax operations by the end of 2001, Phelps Dodge's new landscape of production should yield savings of at least $100 million a year.[1]

Cyprus Amax, based in Englewood, Colorado, had 4,000 employees at the time Phelps Dodge acquired it in 1999. It ranked as the world's largest producer of molybdenum, as well as a major player in copper. Its copper properties included the El Abra mine in Chile (51 percent ownership), Cerro Verde mine in Peru, and the Bagdad, Sierrita, and Miami open-pit mines in Arizona. Among its several processing facilities were a copper refinery and rod mill in Miami, Arizona, and a rod mill in Chicago, Illinois.

The abbreviated family tree for Phelps Dodge's newest member starts with the formation of the American Metal Company (Limited) in 1887. Seventy years later the firm merged with the Climax Molybdenum Company, organized in 1918, to form what became AMAX Inc. American investors organized Cyprus Mines in 1912 to explore for metals on the island of Cyprus. The two firms combined as Cyprus Amax Minerals Company in 1992.

Of course, it gave Chairman Yearley great personal satisfaction to know that after the acquisition was completed, the Phelps Dodge name continued. He quickly noted, however, that while the time-honored name was emotionally important to him, paying homage to the corporation's historical legacy "can't ever get in the way of good business judgment." Yearley observed that while "the Cyprus people are sad to see their name retired, . . . these things happen. I'm delighted that we've been able to survive this one." With the latest growth of Phelps Dodge, he said, "it is going to be harder and harder for someone to acquire us and take our name away."[2]

Exactly how the Phelps Dodge landscape of production will change during the next year or two is a story that remains to be told. As for the headquarters location, well before its acquisition of Cyprus Amax, Phelps Dodge announced that when its current lease expired, it planned to relocate its corporate offices from a familiar address at 2600 North Central Avenue to Square One, a twenty-story copper-trimmed building proposed for downtown Phoenix. The move could take place as early as November 2001.

Finally, the two executives who did most to open the dramatic new chapter of Phelps Dodge

⁓ PORTRAIT OF *David L. Pulatie, who became a member of the Senior Management Team in March 1999. After a 34-year career with Motorola Inc., he joined Phelps Dodge Corporation as senior vice president, human resources. Pulatie holds a bachelor of science degree in psychology and personnel management from Arizona State College and a master of arts degree in education from Northern Arizona University.*

history that started with acquisition of Cyprus Amax were Douglas C. Yearley and J. Steven Whisler. In June 1999 the Phelps Dodge Board of Directors announced that it intended to name Whisler to succeed Yearley as chairman and chief executive officer of the corporation. Whisler became chief executive officer on January 1, 2000, and chairman four months later at Phelps Dodge's annual meeting held on May 3, 2000, when Yearley retired after more than forty years of service. For both men, the acquisition of Cyprus Amax will stand as a major milestone in any account of their leadership at Phelps Dodge Corporation.

It will be up to Whisler, a man some company insiders describe as "intense and brilliant" to make Phelps Dodge's acquisition of Cyprus Amax work. The challenge, he notes, is "to continue to grow and prosper in a world that is increasingly global and competitive. No one, whether Phelps Dodge, ASARCO or Cyprus or a mom and pop shop, can afford to be small and inefficient."[3]

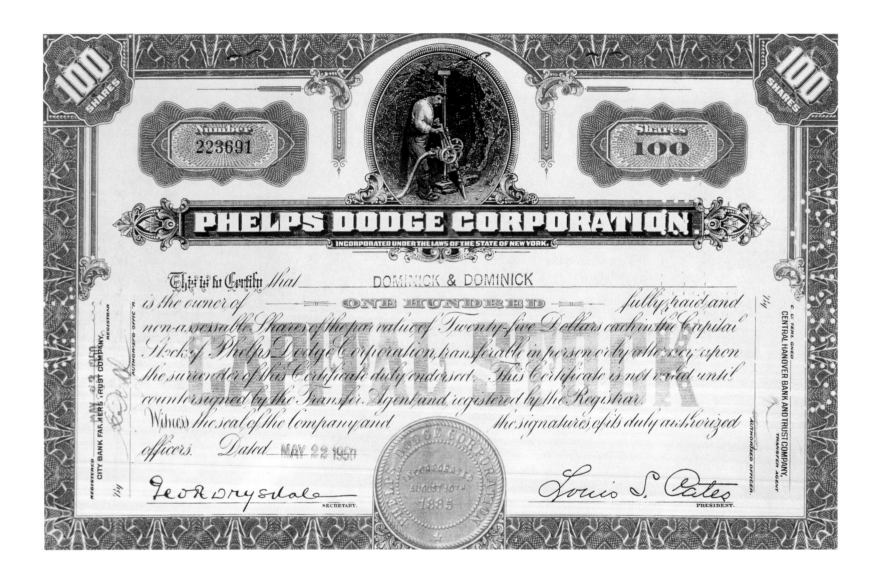

NOTES

All interviews, unless otherwise noted, were conducted by the author.

Preface

1. Interview with Douglas C. Yearley, Scottsdale, Ariz., February 17, 1997.
2. *Pay Dirt*, December 1993, p. 4A.
3. As stated in *Purchasing*, March 8, 1990, p. 70.
4. As quoted in *Enterprise*, March–May, 1992, p. 14.
5. Interview with Pedro Gomez, Safford, Ariz., January 5, 1997.

Chapter 1
The View from Dr. Douglas's Desk

1. The story of first encounters in 1881 is derived from a typescript reminiscence by Arthur Churchill in Phelps Dodge Corporation archives, Phoenix; and James Douglas, *Notes on the Development of Phelps, Dodge and Co.'s Copper and Railroad Interests* (n.p.: January 1906), n. pag.
2. Details of Douglas's early years are found in H. H. Langton, *James Douglas, a Memoir* (Toronto: University of Toronto Press, 1940).
3. Douglas, *Notes on the Development of Phelps, Dodge and Co.'s Copper and Railroad Interests*; Phelps Dodge

Papers, Huntington Library, San Marino, Calif. At the time I used them in 1998, those documents had not yet been organized.
4. Interview with Richard W. Rice, Phoenix, June 1, 1998.

Chapter 2
Family Ties, or the Religion of Business

1. Phyllis B. Dodge, *Tales of the Phelps-Dodge Family: A Chronicle of Five Generations* (New York: New-York Historical Society, 1987), 44.
2. Dodge, *Tales of the Phelps-Dodge Family*, 64.
3. Richard Lowitt, *A Merchant Prince of the Nineteenth Century: William E. Dodge* (New York: Columbia University Press, 1954), 68, 78–79.
4. Richard H. Phelps, *A History of Newgate of Connecticut* (Albany, N.Y.: J. Munsell, 1860), 15–17.
5. *Ansonia Clock Company Catalogue* in Phelps Dodge Corporation archives, Phoenix.
6. As quoted in Lowitt, *A Merchant Prince of the Nineteenth Century*, 164, 339–40.

CHAPTER 3

Forging the Copper Kingdom

1. *Letters Written by Henry Lesinsky to his Son* (New York: n.p., 1924).

2. *Letters Written by Henry Lesinsky to his Son*, 51.

3. Arthur Train, Jr., and John L. Lawler, "Men and Mines: The Story of Phelps Dodge," manuscript (ca. 1945), 134, Phelps Dodge Corporation archives, Phoenix.

4. Train and Lawler, "Men and Mines: The Story of Phelps Dodge," 138.

CHAPTER 4

A Landscape Designed for Production

1. *Pay Dirt*, November 1992, p. 12ff.; *Tucson Star*, December 22, 1968; *Tucson Daily Citizen*, January 30, 1976.

2. James Colquhoun, *The History of the Clifton-Morenci District* (London: John Murray, 1924), 14–15.

3. James Douglas, *Notes on the Development of Phelps, Dodge and Co.'s Copper and Railroad Interests* (n.p.: January 1906), n. pag.

4. Arthur Train, Jr., and John L. Lawler, "Men and Mines: The Story of Phelps Dodge," manuscript (ca. 1945), 211, Phelps Dodge Corporation archives, Phoenix; H. H. Langton, *James Douglas, a Memoir* (Toronto: University of Toronto Press, 1940), 95 (quotation).

5. Robert Glass Cleland, *A History of Phelps Dodge, 1834–1950* (New York: Alfred A. Knopf, 1952), 154.

CHAPTER 5

Copper's Urban Outposts

1. Albert W. Atwood, "Absentee Capitalism," *Saturday Evening Post*, March 24, 1923, p. 54.

2. Albert W. Atwood, "Where Have the Miners Gone?" *Saturday Evening Post*, March 10, 1923, p. 74.

3. As quoted in Tom Vaughan, "Everyday Life in a Copper Camp," in *Bisbee: Urban Outpost on the Frontier*, ed. Carlos A. Schwantes (Tucson: University of Arizona Press, 1992), 57.

4. P. G. Beckett to A. W. Liddell, June 21, 1933, Phelps Dodge Mercantile records, Phoenix.

5. Albert W. Atwood, "Absentee Capitalism," *Saturday Evening Post*, March 24, 1923, p. 54.

6. A. W. Liddell to P. G. Beckett, July 25, 1933, Phelps Dodge Mercantile records.

7. Louis S. Cates to P. G. Beckett, November 20, 1931; P. G. Beckett to Louis S. Cates, December 9, 1931, Phelps Dodge Mercantile records.

8. Don Blanding as quoted in John Hays Hammond, *The Autobiography of John Hays Hammond*, 2 vols. (New York: Farrar and Rinehart, 1935), II, 89.

CHAPTER 6

Landscapes of Labor

1. As quoted in Liping Zhu, "Claiming the Bloodiest Shaft: The 1913 Tragedy of the Stag Cañon Mine, Dawson, New Mexico," *Journal of the West* 35 (October 1996): 59. A valuable source.

2. As quoted in James R. Kearney, "Phelps Dodge Corporation: A Twentieth-Century Business Odyssey," manuscript (ca. 1996), 66, Phelps Dodge Corporation archives, Phoenix. See also Toby Smith, *Coal Town: The Life and Times of Dawson, New Mexico* (Santa Fe: Ancient City Press, 1993), 51.

3. *Phelps Dodge Annual Report* (1913), 53.

4. Report of coroner's jury published in *Arizona Mining Journal*, March 1, 1923, p. 44.

5. *Arizona Mining Journal*, February 15, 1923, p. 18.

6. *Sixth Annual Report of the Secretary of Labor*, 1918 (Washington, D.C.: Government Printing Office, 1918), 13.

7. As quoted in George G. Suggs, Jr., *Colorado's War on Militant Unionism: James H. Peabody and the Western Federation of Miners* (Detroit: Wayne State University Press, 1972), 23.

8. "Preamble of the Industrial Workers of the World,"

reprinted in *Songs of the Workers to Fan the Flames of Discontent*, 34th ed. (Chicago: Industrial Workers of the World, 1973).

9. For further details see James R. Kluger, *The Clifton-Morenci Strike: Labor Difficulty in Arizona, 1915–1916* (Tucson: University of Arizona Press, 1970).

10. "Christians at War," composed by John F. Kendrick and sung to the tune of "Onward Christian Soldiers," first appeared in the ninth edition (1913) of the *IWW Songs*, the official songbook of the union. It was reprinted in Bisbee-area newspapers shortly before the Wobblies' deportation from town.

11. *American Heritage*, December 1996, p. 28.

12. Barbara Tuchman, *The Zimmermann Telegram* (New York: Dell, 1965 reprint of 1958 edition), 136.

13. Walter Douglas to Nicholas Murray Butler, July 6, 1917, Phelps Dodge Papers, Huntington Library, San Marino, Calif.

14. See the federal investigation of the Everett massacre as presented in William J. Williams, "Bloody Sunday Revisited," *Pacific Northwest Quarterly* 71 (April 1980): 50–62.

15. Felix Frankfurter in "Report on the Bisbee Deportations," President's Mediation Commission, November 6, 1917, p. 6, copy in Phelps Dodge Corporation archives.

Chapter 7

The Modern Corporation from Mine to Market

1. *Engineering and Mining Journal-Press*, January 20, 1923, 127–28.

2. James Colquhoun, *The Passing of the Arizona Copper Company* (n.p.: 1921), 1.

3. *Phelps Dodge Annual Report* (1923), 2, and (1924), 2.

4. *Phelps Dodge Annual Report* (1928), 2.

5. *Engineering and Mining Journal*, May 4, 1929, p. 738.

6. *Engineering and Mining Journal*, August 9, 1930, p. 129. The full story of the Douglas-to-Cates succession will probably never be known, but these details come from my interviews with Cleveland E. Dodge,

Jr., Pownal, Vt., August 12, 1998, and William Rea, Pittsburgh, Pa., April 7, 1998. Both men served on the Phelps Dodge board of directors, as did their fathers.

7. "Presbyterian Copper," *Fortune*, July 1932, p. 104.

8. Recounted in interview with George Bailey, El Paso, Tex., November 25, 1996.

9. *Engineering and Mining Journal*, December 31, 1921, p. 1041.

10. *Mining Journal*, February 28, 1930, p. 12.

11. *Engineering and Mining Journal*, May 18, 1929, p. 787.

12. Roy M. Bates and Kenneth Keller, "Dudlo: Wire Wizards," *Old Fort News*, Winter 1970.

13. Phelps Dodge Papers, Huntington Library, San Marino, Calif.

14. As quoted in James R. Kearney, "Phelps Dodge Corporation: A Twentieth-Century Business Odyssey," manuscript (ca. 1996), 92, Phelps Dodge Corporation archives, Phoenix.

15. Interview with William Dunbar, Bay Head, N.J., July 25, 1996.

16. Interview with Warren Fenzi, Santa Barbara, Calif., April 15, 1997.

17. Interview with Naomi Kitchel, Scottsdale, Ariz., November 21, 1996.

Chapter 8

Meeting the Challenge of the Great Depression

1. *Phelps Dodge Annual Report* (1930), 7.

2. *Phelps Dodge Annual Report* (1931), 1, and (1932), 7.

3. As quoted in James R. Kearney, "Phelps Dodge Corporation: A Twentieth-Century Business Odyssey," manuscript (ca. 1996), 97, Phelps Dodge Corporation archives, Phoenix.

4. "Fuller Dymaxion Bathroom file," in Phelps Dodge Corporation archives.

5. *Douglas* (Ariz.) *Dispatch*, September 6, 1931.

CHAPTER 9

Evolving Enterprise in War and Peace

1. William E. Leuchtenburg, *Franklin D. Roosevelt and the New Deal, 1932–1940* (New York: Harper and Row, 1963), 150–52.
2. Interview with Denison Kitchel, Scottsdale, Ariz., November 21, 1996. The full text of *Phelps Dodge Corporation v. National Labor Relations Board* is reported in Archibald Cox and Derek Curtis Bok, *Cases and Materials on Labor Law*, 6th ed. (Brooklyn: Foundation Press, 1965), 261–69.
3. Interview with Julia Mazanek, Phoenix, Ariz., October 1, 1998.
4. *Mining and Contracting Review*, March 15, 1942, p. 15.
5. William A. Evans, *Two Generations in the Southwest* (Phoenix: Sims Printing, 1971), 101–109; interviews with Cynthia Chandley, Phoenix, March 19, 1997, and June 1, 1998; interview with Kenneth Bennett, Phoenix, March 18, 1997; Schwantes's personal flyover of Phelps Dodge water system accompanied by Jerry Haggard and Barry Holt, July 1, 1997; *Phelps Dodge Today*, December 31, 1984, pp. 1–3.
6. Interview with William Dunbar by James Kearney, Bay Head, N.J., May 18, 1995, as quoted in James Kearney's "Phelps Dodge Corporation: A Twentieth-Century Odyssey," manuscript (ca. 1996), Phelps Dodge Corporation archives, Phoenix; interview with William Dunbar, Bay Head, N.J., July 25, 1996.
7. As quoted in Phyllis B. Dodge, *Tales of the Phelps-Dodge Family: A Chronicle of Five Generations* (New York: New-York Historical Society, 1987), 203.
8. *Phelps Dodge Annual Report* (1945), 5, 9.

CHAPTER 10

In Search of Easy Street

1. Interview with Walter Ainsworth, Fort Wayne, Ind., July 31, 1997; *Fort Wayne Journal-Gazette*, August 24, 1966; *Fort Wayne News-Sentinel*, August 24, 1996.

2. *Forbes*, August 1, 1958, p. 10.
3. *Forbes*, August 1, 1958, p. 10.
4. *New Mexican* (Santa Fe), September 8, 1998.
5. *Forbes*, December 1, 1964, p. 42.
6. *Phelps Dodge Annual Report* (1957), 10; *Forbes*, April 1, 1958, pp. 10–11.
7. *Forbes*, July 1, 1962, p. 20.
8. *Iron Age*, November 12, 1966, p. 122.
9. *Forbes*, November 1, 1968, p. 68.
10. *Forbes*, November 1, 1968, p. 8; *Phelps Dodge Annual Report* (1963), 12–13, and (1964), 12, and (1965), 12.
11. *Forbes*, June 1, 1962, p. 24.
12. *Iron Age*, June 1, 1972, p. 67.
13. Interviews of George B. Munroe by James Kearney, Phoenix, Ariz., February 28 and March 1, 1994, as quoted in James Kearney's "Phelps Dodge Corporation: A Twentieth-Century Odyssey," manuscript (ca. 1996), Phelps Dodge Corporation archives, Phoenix, Ariz.
14. See *Forbes*, October 15, 1966, p. 48.

CHAPTER 11

A Decade of Uncertainty

1. James Burke, *Connections* (Boston: Little, Brown, 1978), 166–69.
2. Duane A. Smith, *Mining America: The Industry and the Environment, 1800–1980* (Lawrence: University Press of Kansas, 1987), 136–48.
3. *Engineering and Mining Journal*, September 29, 1928, p. 493.
4. Interview with Richard T. Moolick, Glenwood Springs, Colo., October 1, 1997.
5. Interview with Richard W. Rice, Phoenix, Ariz., June 1, 1998.
6. Effie Watkins, "The Smelter," in *Prospector*, March 1949, p. 11, copy in Phelps Dodge Corporation archives, Phoenix.
7. Interview with L. William Seidman, Washington, D.C., May 1, 1998.

8. *Engineering and Mining Journal*, December 3, 1921, p. 913.

9. *Phelps Dodge Today*, June 1981, p. 3.

10. Interviews with George B. Munroe by James Kearney, Phoenix, February 28 and March 1, 1994, quoted in James Kearney's "Phelps Dodge Corporation: A Twentieth-Century Business Odyssey," manuscript (ca. 1996), Phelps Dodge Corporation archives; interview with George B. Munroe, Albuquerque, N.Mex., July 24, 1996.

11. *Avenue*, October 1978, p. 58.

12. *Forbes*, April 1, 1967, p. 68.

13. Telephone interview with Charles A. Burns, October 1, 1998.

14. Interview with L. William Seidman, Washington, D.C., May 1, 1998; *Phelps Dodge Annual Report* (1971), 15.

15. Interview with Jack Ladd, Bisbee, Ariz., February 20, 1997.

16. *Forbes*, May 1, 1972, p. 34.

17. Telephone interview with Leonard Judd, July 6, 1998.

18. Interview with George B. Munroe, Albuquerque, July 24, 1996.

CHAPTER 12

Red Metal Blues

1. *Forbes*, December 21, 1981, p. 51.

2. *Phelps Dodge Today*, March 1983, p. 3. This publication was distributed to employees of Western Operations from 1979 to the end of 1984.

3. *New York Times*, April 8, 1982.

4. Interview with George Munroe, New York City, May 4, 1998.

5. Interview with George Munroe, New York City, May 4, 1998.

6. *Business Week*, July 21, 1980, p. 66; interview with L. William Seidman, Washington, D.C., May 1, 1998.

7. *Business Week*, April 12, 1982, p. 5, and July 26, 1982, p. 59.

8. *Phelps Dodge Today*, June 1982, p. 7.

9. *Phelps Dodge Today*, September 1982, p. 3.

10. *Phelps Dodge Today*, June 1980, p. 3.

11. *Phelps Dodge Today*, March 1982, p. 5.

12. *Forbes*, August 1, 1958, p. 10.

13. Interview with George B. Munroe, May 4, 1998.

14. Remarks of George B. Munroe to 1984 Annual Meeting of Stockholders as reported in *Phelps Dodge Today*, June 1984, p. 8; *New York Times*, April 27, 1984; *Phelps Dodge Today*, March 1984, pp. 4–5.

15. *Business Week*, July 26, 1982, p. 59.

16. Interview with Richard T. Moolick, Glenwood Springs, Colo., October 1, 1997.

17. *Business Week*, July 26, 1982, p. 60.

18. Interview with Douglas C. Yearley, Scottsdale, Ariz., February 17, 1997; *Business Week*, July 26, 1982, p. 60.

19. *Business Week*, July 26, 1982, p. 60.

20. *Phelps Dodge Today*, March 1982, pp. 5, 9.

21. *Phelps Dodge Today*, June 1982, p. 3.

CHAPTER 13

Hammered on the Anvil of Adversity

1. Televised comments by Bruce Babbitt when he sent the National Guard to Morenci.

2. Interview with Douglas C. Yearley, Scottsdale, Ariz., February 17, 1997.

3. Interview with George Munroe, New York City, August 20, 1998.

4. Interview with Richard T. Moolick, Glenwood Springs, Colo., October 1, 1997.

5. *Business Week*, July 26, 1982, p. 58.

6. *Business Week*, May 20, 1967, p. 156. The summary of events from the 1940s to the 1967 strike derives from James Kearney's manuscript history of Phelps Dodge.

7. *Iron Age*, December 21, 1967, p. 35.

8. *Phelps Dodge Today*, September 1980, p. 3, and December 1980, p. 3.

9. *Phelps Dodge Today*, March 1980, p. 4; *El Paso Times*, July 1, 1984.

10. *Phelps Dodge Today*, June 1983, p. 3.

11. Interview with Jack Ladd, Bisbee, Ariz., February 20, 1997.

12. Interview with Jack Ladd, Bisbee, February 20, 1997.

13. Interview with Richard T. Moolick, Glenwood Springs, October 1, 1997.

14. *Fortune*, August 22, 1983, p. 107.

15. *Arizona Republic* (Phoenix), October 3, 1983.

16. *Fortune*, August 22, 1983, pp. 106–108.

17. *Fortune*, August 22, 1983, pp. 106–108; *Phelps Dodge Today*, September 1980, p. 3; interview with Matthew P. Scanlon, Phoenix, Ariz., March 22, 1986.

18. Interview with Richard T. Moolick, Glenwood Springs, October 1, 1997.

19. Interview with J. Steven Whisler, Phoenix, January 3, 1997.

20. *Phelps Dodge Workers Strike Bulletin 1983*, July 6, 1983.

21. *Phelps Dodge Workers Strike Bulletin 1983*, July 18, 1983.

22. Interview with Leonard Judd, Paradise Valley, Ariz., March 20, 1996.

23. Interview with George B. Munroe, Albuquerque, N.Mex., July 24, 1996; *Fortune*, August 22, 1983, p. 107.

24. "Procedures for Crossing a Picket Line," handout in Phelps Dodge Morenci archives, Morenci.

25. Written recollections of Matthew P. Scanlon to Carlos A. Schwantes, September 3, 1998.

26. *Fortune*, August 22, 1983, p. 107.

27. Interview with Simon Peru, Safford, Ariz., January 5, 1997.

28. As quoted on KPHO news, August 9, 1997; *Daily Labor Report*, August 10, 1983; interview with Matthew P. Scanlon, Phoenix, March 22, 1996.

29. Written recollections of Matthew P. Scanlon to Schwantes, September 3, 1998.

30. Written recollections of Matthew P. Scanlon to Schwantes, September 3, 1998.

31. Interview with Matthew P. Scanlon, Phoenix, March 22, 1996.

32. Interview with Arthur Kinneberg, Phoenix, March 16, 1998.

33. Interview with Matthew P. Scanlon, Phoenix, March 22, 1996; interview with Richard T. Moolick, Glenwood Springs, October 1, 1997.

34. Interview with Matthew P. Scanlon, Phoenix, March 22, 1996; interview with Richard T. Moolick, Glenwood Springs, Colo., October 1, 1997.

35. Interview with Matthew P. Scanlon, Phoenix, March 18, 1998.

36. Interview with Richard T. Moolick, Glenwood Springs, October 1, 1997.

37. Interview with Richard T. Moolick, Glenwood Springs, October 1, 1997.

38. Interview with Matthew P. Scanlon, Phoenix, March 18, 1998.

39. *New York Times*, July 30, 1984.

40. Interview with James Madson, Phoenix, January 2, 1997.

41. *New York Times*, May 7, July 30, 1984.

42. Interview with Richard T. Moolick, Glenwood Springs, October 1, 1997.

43. Interview with J. Steven Whisler, Phoenix, July 13, 1998; interview with L. William Seidman, Washington, D.C., May 1, 1998. The retired vice-president and general manager Arthur Himebaugh made a similar assessment of Moolick's importance in an interview, Phoenix, August 27, 1998: "Right or wrong," he made decisions.

Chapter 14

"Fix It, Sell It, or Shoot It"

1. Interview with G. Robert Durham, Phoenix, Ariz., November 13, 1997.

2. As quoted in *Wall Street Journal*, November 24, 1989.

3. Interview with Richard W. Rice, Phoenix, June 1, 1998.

4. Interview with G. Robert Durham, Phoenix, November 13, 1997; interview with Douglas C. Yearley, Phoenix, October 1, 1998.

5. D. C. Yearley, "Confidential Review of World Copper Industry, August 1984," pp. 5–6, typescript copy in Phelps Dodge Corporation archives, Phoenix.

6. Interviews with Douglas C. Yearley, Scottsdale, Ariz., February 17, 1997, and Phoenix, October 1, 1998.

7. *Wall Street Journal*, July 28, 1983; memo from George Munroe to Carlos A. Schwantes, August 19, 1998.

8. Interview with George Munroe, New York City, May 4, 1998; memo from George Munroe to Schwantes, August 19, 1998.

9. Interview with George Munroe, New York City, May 4, 1998; memo from George Munroe to Schwantes, August 19, 1998.

10. Interview with G. Robert Durham, Phoenix, November 13, 1997; interview with George Munroe, New York City, May 4, 1998.

11. Interview with George Munroe, New York City, May 4, 1998.

12. Interview with Ramiro G. Peru, Phoenix, October 1, 1998; interview with Leonard R. Judd, Scottsdale, Ariz., October 25, 1998.

13. Interview with G. Robert Durham, Phoenix, November 13, 1997; interview with Leonard R. Judd, Scottsdale, October 26, 1998; memo from George Munroe to Schwantes, August 19, 1998.

14. Interview with G. Robert Durham, Phoenix, November 13, 1997; interview with Douglas C. Yearley, Scottsdale, February 17, 1997; interview with Leonard R. Judd, Scottsdale, October 26, 1998.

15. Interview with Leonard R. Judd, Scottsdale, October 25, 1998.

16. Interview with Leonard R. Judd, Scottsdale, October 26, 1998.

17. *Wall Street Journal*, July 28, 1983.

18. Interview with G. Robert Durham, Phoenix, November 13, 1997; *Pay Dirt*, December 1984, p. 12A.

19. "Highly Confidential to Mr. Munroe," September 26, 1984, photocopy in Schwantes's possession; interview with George Munroe, New York City, May 4, 1998.

20. Undated notes from Richard T. Moolick to Carlos A. Schwantes (sent late 1998).

21. Interview with Arthur Kinneberg, Phoenix, March 16, 1998.

22. Interview with Arthur Kinneberg, Phoenix, March 16, 1998.

23. Speech by G. Robert Durham to the American Mining Congress, October 8, 1986, p. 3, copy in Phelps Dodge Corporation archives.

24. Interview with Patrick J. Ryan, Scottsdale, November 21, 1996.

25. Interview with G. Robert Durham, Phoenix, November 13, 1997.

26. Interview with G. Robert Durham, Phoenix, November 13, 1997.

27. Interview with William C. Tubman, Phoenix, July 1, 1997; interview with G. Robert Durham, Phoenix, November 13, 1997.

28. Interview with George Munroe, New York City, May 4, 1998.

CHAPTER 15
Turnaround Time

1. *Fortune*, October 28, 1985, p. 42; interviews of Ramiro G. Peru, Phoenix, Ariz., October 1 and 2, 1998; interview with Leonard R. Judd, Scottsdale, Ariz., October 26, 1998.

2. Interviews with Ramiro G. Peru, Phoenix, October 1 and 2, 1998; interview with Leonard R. Judd, Scottsdale, October 26, 1998.

3. *Fortune*, May 15, 1995; *Brewery Gulch Gazette* (Bisbee, Ariz.), March 17, 1993.

4. Interview with Richard W. Rice, Phoenix, Ariz., June 1, 1998.

5. Interview with Leonard Judd, Paradise Valley, Ariz., March 20, 1996; interview with G. Robert Durham, Phoenix, November 13, 1997.

6. *Pay Dirt*, September 30, 1986; interview with Leonard Judd, Paradise Valley, Ariz., March 20, 1996.

7. Transcript of Phelps Dodge Corporation 1985 Annual Meeting.

8. Interview with Douglas C. Yearley, Scottsdale, Ariz., February 17, 1997; *Financial World*, December 23, 1986, p. 101.

9. *Brewery Gulch Gazette*, March 27, 1993.

10. Memo from Jack Walsh to Robert C. Swan, August 24, 1998, in Swan's possession.

11. Interview with Douglas C. Yearley, Scottsdale, February 17, 1997; *Morenci Copper Review*, May 1989, p. 2; Richard Pendleton, Jr., to Richard Moolick, February 25, 1998, in Moolick's possession; *American Metal Market*, Supplement, February 17, 1993, p. 4A.

12. Interview with Leonard R. Judd, Scottsdale, October 26, 1998.

13. *Industry Week*, December 3, 1990, p. 17.

14. *American Metal Market*, Supplement, February 17, 1993, p. 4A.

CHAPTER 16

From Copper Kingdom to Global Enterprise

1. Interview with G. Robert Durham, Phoenix, Ariz., November 13, 1997; interview with Leonard Judd, Paradise Valley, Ariz., March 20, 1996.

2. Interview with G. Robert Durham, Phoenix, November 13, 1997.

3. *Wall Street Journal*, November 24, 1989; interview with Douglas C. Yearley, Scottsdale, Ariz., February 17, 1997.

4. Interview with George B. Munroe, Albuquerque, N.Mex., July 24, 1996.

5. Speech of Douglas C. Yearley to American Mining Hall of Fame, December 2, 1995, typescript copy in Phelps Dodge Corporation archives, Phoenix; interview with Patrick J. Ryan, Scottsdale, March 19, 1998; interview with J. Steven Whisler, Phoenix, January 3, 1997.

6. Interview with Patrick J. Ryan, Scottsdale, November 21, 1996.

7. *S.A. Mining and Engineering Journal*, August 1978, p. 24.

8. Interview with Robert Jenkins, Tucson, Ariz., June 2, 1998.

9. Interview with David Colton, Phoenix, June 1, 1998;

interview with A. L. Lawrence, Phoenix, March 20, 1998.

10. Interview with David Colton, Phoenix, June 1, 1998.

11. Undated Johannesburg, South Africa, newspaper clipping supplied by Patrick J. Ryan; "The History of Phelps Dodge in Southern Africa," paper prepared by M. J. Evans, A. L. Lawrence, and P. J. Ryan, copy in Phelps Dodge Corporation archives.

12. Interview with G. Robert Durham, Phoenix, November 13, 1997; interview with David Colton, Phoenix, June 1, 1998.

13. *Forbes*, December 1, 1964, p. 42.

14. Richard T. Moolick to Carlos A. Schwantes, June 26, 1998.

15. Interview with John ("Phil") Matthews, Tucson, Ariz., June 2, 1998; interview with Robert Jenkins, Tucson, June 2, 1998; interview with José Luis Gorrini, Santiago, Chile, May 26, 1988; *Wall Street Journal*, July 23, 1983; interview with Leonard R. Judd, Scottsdale, October 26, 1998.

16. Interview with Patrick J. Ryan, Scottsdale, March 19, 1998.

17. Interview with Richard W. Rice, Phoenix, June 1, 1998; interview with Douglas C. Yearley, Phoenix, July 13, 1998; P. J. Ryan and J. L. Madson, "Candelaria—Low Cost Copper-Gold Producer in Chile," *Mining Engineering*, August 1996, pp. 35–41, and September 1996, pp. 44–51.

18. Phelps Dodge news release, November 3, 1997.

19. Interview with Patrick J. Ryan, Scottsdale, November 21, 1996.

20. Interview with William Dunbar, Bay Head, N.J., July 25, 1996; phone interview with William Dunbar, August 3, 1998.

21. Interview with Douglas C. Yearley, Scottsdale, February 17, 1997.

22. Interview with G. Robert Durham, Phoenix, November 13, 1997.

23. Interview with G. Robert Durham, Phoenix, November 13, 1997.

24. Interview with G. Robert Durham, Phoenix, November 13, 1997. The "Plan for Restructuring Phelps Dodge International Corporation" is dated June 1, 1978, copy in Schwantes's possession.

25. Interview with G. Robert Durham, Phoenix, November 13, 1997; interview with Orlando Gonzalez, Coral Gables, Fla., October 13, 1997.

26. Interview with David Colton, Phoenix, November 10, 1997; Yearley as quoted in Phelps Dodge news release, November 14, 1997.

27. Interview with J. Steven Whisler, Phoenix, January 3, 1997.

CHAPTER 17

Charting a Course for the New Millennium

1. Phyllis B. Dodge, *Tales of the Phelps-Dodge Family: A Chronicle of Five Generations* (New York: New-York Historical Society, 1987), 231–32.

2. "Reminiscences of Cleveland E. Dodge," typescript manuscript provided by Clee and Phyllis Dodge, Pownal, Vt.

3. As quoted in G. Robert Durham, *Phelps Dodge Corporation: "Proud of Its Past, Prepared for the Future"* (New York: Newcomen Society of the United States, 1989), 23.

4. Interview with Thomas St. Clair, Phoenix, March 18, 1997; interview with Douglas C. Yearley, Scottsdale, Ariz., February 17, 1997.

5. Interview with George Munroe, New York City, May 4, 1998.

6. Interview with Warren Fenzi, Santa Barbara, Calif., April 15, 1997.

7. Interview with Douglas C. Yearley, Scottsdale, February 17, 1997.

8. Interview with Douglas C. Yearley, Scottsdale, February 17, 1997; interview with George B. Munroe, Albuquerque, N.Mex., July 24, 1996.

9. Interview with L. William Seidman, Washington, D.C., May 1, 1998.

10. Interview with Ronald Habegger, Phoenix, March 21,

1996; interview with Ramiro G. Peru, Phoenix, October 1, 1998.

11. Interview with Douglas C. Yearley, Scottsdale, February 17, 1997; J. Steven Whisler as quoted in *Workscapes*, February 1998, p. 3.

12. Interview with Ramiro G. Peru, Phoenix, November 26, 1996.

13. Speech by Douglas C. Yearley at Cornell University, October 12, 1995, copy in Phelps Dodge Corporation archives, Phoenix; interview with Ramiro G. Peru, Phoenix, November 26, 1996.

14. Interview with Warren Fenzi, Santa Barbara, April 15, 1997.

15. *Business Week*, July 26, 1982, pp. 58–60; interview with Ramiro G. Peru, Phoenix, October 2, 1998.

16. Interview with Ramiro G. Peru, Phoenix, October 2, 1998.

17. Interview with Douglas C. Yearley, Scottsdale, February 17, 1997.

18. Interview with Simon and Velia Peru, Safford, Ariz., January 5, 1997.

19. Interview with Susan Suver, Phoenix, March 18, 1998.

20. Interview with Linda Findlay, Washington, D.C., April 30, 1998.

21. Interview with Patricia Young, Morenci, Ariz., March 21, 1997.

22. Interview with Susan Suver, Phoenix, March 18, 1998.

23. Interview with Douglas C. Yearley, Scottsdale, February 17, 1997.

24. *Phoenix Business Journal*, April 5, 1996, p. 28; interview with Douglas C. Yearley, Scottsdale, February 17, 1997.

25. Interview with Douglas C. Yearley, Scottsdale, February 17, 1997.

26. *New York Times*, March 3, 1989; *Enterprise*, March–May, 1992, p. 11; interview with Douglas C. Yearley, Scottsdale, February 17, 1997; *Industry Week*, December 3, 1990, pp. 17–18.

27. Interview with Douglas C. Yearley, Scottsdale, February 17, 1997.

28. *Enterprise*, March–May, 1992, p. 19.

29. *Chairman's Quarterly Letter*, May 1993.

30. Interview with Douglas C. Yearley, Scottsdale, February 17, 1997.

31. Interview with Patrick J. Ryan, Scottsdale, November 21, 1996.

32. Interview with John ("Phil") Matthews, Tucson, Ariz., June 2, 1998.

33. Interview with Richard T. Moolick, Glenwood Springs, Colo., October 1, 1997.

34. Interview with J. Steven Whisler, Phoenix, January 3, 1997.

35. Interview with J. Steven Whisler, Phoenix, July 13, 1998.

CHAPTER 18

Back to the Future

1. "EV1 Electric," undated brochure prepared by General Motors.

2. Phelps Dodge Corporation news release, October 21, 1998.

3. Telephone interview with Daniel E. Floryan, February 10, 1999.

4. Interview with Robert E. Johnson, Morenci, Ariz., June 11, 1998.

5. Telephone interview with John Marsden, February 10, 1999.

6. Interview with Patrick J. Ryan, Scottsdale, Ariz., November 21, 1996.

7. *Tucson Monthly*, December 1997, pp. 34–39.

8. Interview with J. Steven Whisler, Phoenix, Ariz., January 3, 1997.

9. Interview with Manuel J. Iraola, Phoenix, February 17, 1997.

10. Interview with H. M. ("Red") Conger, Morenci, June 11, 1998.

11. Remarks by Douglas C. Yearley to CRU Benchmarking Base Metals Operations Conference, Tucson, Ariz., November 5, 1998.

12. Telephone interview with Gary A. Loving in Santiago, Chile, July 13, 1998.

CHAPTER 19

Update 1999: The Pace of Change

1. Telephone interview with Douglas C. Yearley, October 22, 1999.

2. Telephone interview with Douglas C. Yearley, October 22, 1999.

3. *Arizona Daily Star* (Tucson), September 26, 1999.

GLOSSARY

acid plant. A *smelter* facility that recovers sulfur dioxide from discharged gases and manufactures sulfuric acid from it.

adit. A nearly horizontal entrance to an underground mine.

anode. a: A slab of fire-refined copper weighing about 750 pounds that has been formed in a mold. A typical one at Phelps Dodge is about 99.5 percent pure. From *smelters* at Chino (and Hidalgo when it was in operation) it goes to El Paso for *electrolytic refining.* b: In the *SX/EW* process it is the positive electrical pole to which direct current is applied.

assay. Chemical evaluation of an *ore* sample to determine the value of a mineral deposit.

ball mill. A large steel cylinder partially filled with steel balls that cascade as the mill rotates to grind *ore.*

blister copper. A form of copper that develops a blistered surface after casting due to gases generated during solidification. Because it is only 96 to 98 percent pure, it is further refined at the *smelter* and cast into *anodes* that go to the *refinery* for additional purification.

block-caving. A process whereby large blocks of ore are undercut and allowed to fall and fracture into smaller pieces that can be loaded through chutes into mine cars and taken to the surface. It is the lowest-cost underground mining method.

blow in. To put a *smelter* furnace into operation.

carbonate ore. A carbonate of copper that contains carbon dioxide and is typically found in the uppermost portions of *ore* bodies.

cathode. Copper 99.99 percent pure that results from the *SX/EW* process or from *electrolytic refining* of *anodes.*

cement copper. A sludge of wet, powdery copper particles, containing 50 percent to 80 percent copper, that is sent to a *smelter* for further treatment.

claim. A tract of land with defined boundaries that includes mineral rights extending downward from the surface.

concentrate. Copper-bearing material produced by the *flotation* process that contains 15 percent to 30 percent copper plus various quantities of sulfur, iron, and other impurities; it is sent to a *smelter* for further treatment.

concentrator. A plant where barren material is rejected as *tailings*, thus concentrating the copper in the *ore* before it goes to a *smelter* for further treatment.

converter. In a *smelter*, this is a large brick-lined

cylindrical vessel in which molten copper *matte* from a *reverberatory* or flash *furnace* has impurities such as iron and sulfur removed by blowing air through the liquid materials. As the converter is tilted, the *slag* and molten metal are poured off separately.

drift. A horizontal passageway in an underground mine; it typically follows a vein of ore.

electrolytic refinery. A facility in which fire-refined copper *anodes* are immersed in an acid solution containing *cathode* starter sheets. An electric current passed between the *anode* and *cathode* causes copper to transfer from the *anodes* and be redeposited as 99.99 percent pure copper on the *cathodes*.

flotation. A method of combining finely ground *ore* with water and chemical reagents to create a frothy mixture that separates metallic particles from other minerals; the metallic particles are then collected and dried and the resulting *concentrate* is sent to the *smelter* for fire-refining.

flux. A substance that promotes melting in a *smelter*; it lowers the melting point of *gangue* thus helping to create a more liquid *slag* from which the desired metal can escape and drop to the bottom of the furnace.

furnace. Melts the copper *concentrates* from which *slag* is drawn off and molten copper-bearing *matte* tapped for further processing.

gangue. Nonvaluable minerals associated with *ore*; that portion of *ore* rejected as *tailings* in the *flotation* process.

glory hole. A large caved-in entrance to an exhausted underground mine.

grubstake. An advance of money, food, and supplies to a prospector in return for a share of any mineral finds.

headframe. A structure erected over a shaft from which to lift or lower a cage or platform to a desired level in a mine.

hydrometallurgy. The treatment of *ores*, *concentrates*, and other metal-bearing materials by a wet process, such as *SX/EW*. Distinct from the pyrometallurgy of a *smelter*.

in-pit crusher. Machines used to crush *ore* in an open-pit mine preparatory to sending it to a *concentrator* for further treatment.

leach dump. A pile of rock material consisting of low-grade copper *ore*, either *oxides* or *sulfides*, through which acid-rich waters are percolated to dissolve and subsequently precipitate the copper. Copper precipitation is the process whereby copper in solution is deposited on iron. The precipitate when largely drained of water is known as *cement copper*, and it is then sent to the *smelter* for further treatment.

leaching. A process used to remove soluble minerals by percolating solutions through low-grade *oxide* and *sulfide ores*.

lode. A continuous mineral-bearing deposit or vein.

matte. Mixture of sulfur, iron, and copper tapped from the primary *furnace* in a *smelter*.

mill. A building in which *ore* is crushed and ground to extricate valuable minerals.

mining district. An area of land described for legal purposes and containing valuable minerals in payable amounts.

open-pit mining. Surface mining method in which overlying rock, or *overburden*, is removed to expose the *ore* body, which is then drilled, blasted, and loaded into trucks, or formerly railroad cars, for haulage from the pit.

ore. Rock containing minerals of sufficient concentration, quantity, and value to be mined at a profit. The definition changes as technology improves, today's ore being yesterday's valueless pile of rock.

overburden. Rock material of little or no value that overlies an *ore* deposit and must be removed before *ore* mining can begin.

oxide ore. Ore containing copper minerals that have been altered by oxidation or weathering process; they contain oxygen and are usually high in copper content.

pinch out. The narrowing of a vein or deposit.

porphyry. A term originally applied to an igneous rock of highly variable composition and structure, containing a combination of fine-grained minerals with some larger-grained mineral constituents. Since porphyries occasionally contain finely disseminated grains of copper *sulfides*, especially chalcopyrite, in small quantities, they constitute the *ore* body for low-grade copper deposits. In many places "porphyry" has come to be used to describe any large-tonnage, low-grade, highly altered copper deposit.

prospects. Mineral workings of unproved value.

raise. A mine shaft driven from below upward.

refinery. A facility in which pure *electrolytic* copper is made by electrolysis and the precious and rare metals are recovered. Two of importance in the history of Phelps Dodge were located at Laurel Hill (in New York City) and El Paso.

reverberatory furnace. A furnace with a ceiling that radiates heat back onto the surface of the material being treated.

shaft. A vertical or nearly vertical opening into the earth for mining.

skip. A container used to lift *ore* through a *shaft* and to the surface above a mine.

slag. The waste product of a *smelter*.

smelter. A metallurgical complex in which material is melted in order to separate impurities from pure metal.

stope. Underground opening from which *ore* is extracted.

sulfide ore. *Ore* composed of copper, sulfur, and usually iron along with various other minerals.

SX/EW. Abbreviation for the *hydrometallurgical* process known as solution extraction–electrowinning that produces *cathodes* of 99.9 percent pure copper.

tailings. Finely ground rock materials left after milling is complete; distinct from the old waste dumps, today's low-grade ore stockpiles, that contain rock not currently of sufficient value to warrant milling.

tailings ponds. Built-up or diked ponds in which *tailings* slurry water is impounded to settle and evaporate. When filled, another level is developed above the former pond, creating a terraced effect.

workings. A general term indicating any mining development.

(Adapted from *Arizona Highways*, October 1975, p. 37; and Philip Varney, *Arizona Ghost Towns and Mining Camps*, Phoenix: Arizona Highways, 1994, p. 81.)

SOURCES AND SUGGESTIONS FOR FURTHER READING

DURING MY EXPLORATION of Phelps Dodge's lengthy and complex history I received information and other forms of help from numerous individuals. On the pages that follow, I wish to acknowledge their contributions, without which this book would have been impossible to research and write. If I missed a contributor, I apologize for my unintentional oversight.

The list includes librarians and archivists across the United States—including the kindly and efficient staffs of the American Antiquarian Society, Worcester, Massachusetts; Ansonia Public Library, Connecticut; Arizona Historical Society, Tucson; University of Arizona Library, Tucson; Arizona State University Library, Tempe; University of Idaho Library, Moscow; Huntington Library, San Marino, California; and New York Public Library—as well as numerous individuals who shared their insights and personal information about Phelps Dodge during interviews.

In addition to searching the records stored in Phelps Dodge's corporate archives in Phoenix, I pursued other forms of historical information by traveling to the far-flung operations of Phelps Dodge Mining Company located in Ajo, Bisbee, Jerome, Morenci, and Safford, Arizona; Tyrone, Hurley, Santa Rita, and Playas, New Mexico; El Paso, Texas; Norwich, Connecticut; and Candelaria in Chile; Phelps Dodge Magnet Wire plants in Fort Wayne, Indiana; Hopkinsville, Kentucky; and El Paso, Texas; Phelps Dodge High Performance Conductors plants in Trenton, Georgia; Inman, South Carolina; and Elizabeth, New Jersey; Phelps Dodge International headquarters in Coral Gables, Florida, and plants in Venezuela, Ecuador, and Chile; and the Columbian Chemicals Company headquarters in Atlanta, Georgia, and the Moundsville, West Virginia, plant.

During mid-1997 I visited the Accuride Corporation plant in Henderson, Kentucky, which at that time was still part of Phelps Dodge Industries. I'll not soon forget the periodic plant-shaking "whump" of a massive press that stamped out wheels for trucks and buses; during one of the interviews it caused me to think of the hulking dinosaurs in the movie *Jurassic Park*. It was only one of many memorable sounds and sights I recorded on tape or film. All together I amassed hundreds of hours of interviews and thousands of photographs that served as research notes later to jog my memory. There are, of course, many more offices and plants composing Phelps Dodge, but to complete this book in timely fashion meant visiting only these representative landscapes. Each tour of mining or manufacturing facilities enabled me to understand the corporation that much better.

Because it would require a full chapter to be more specific, permit me simply to list in alphabetical order (and without reference to job titles or locations) the names of all who contributed in a major way to this project. Most are current or retired employees, but some people listed here never worked for the corporation, and a few actively opposed it on the picket line during labor disputes. If I have omitted anyone, please forgive. At Phelps Dodge there is a tradition of addressing colleagues formally in correspondence, but here I will use familiar first names because each of these individuals became part of an ever-widening circle of friends.

My special thanks go to Lynne Adams, Andrés Aguirre, Walter Ainsworth, Patricio Albán, Armando Alsina, Juan Carlos and Carmen Gloria Altimiras, George Anderson, Fernando Araneda, Isaac and Patricia Aránguiz, Ernie Arriola, Ross Bacho, George Bailey, Dennis Bartlett, Jack Bell, Lester Bell, Ken Bennett, Jim Berresse, Sharon Bice, Judy Blackwell, Jim Bodette, Clayton Boyd, Suzy Boyd, Gabriel Bracho, Bill Brack, John Brack, James Brock, John Broderick, Dave Brooks, Mike Brophy, Charlie Brown, Dale Brunk, Joe Burgess, Jim Bush, Edwardo Bustamante, Ginny Calderón, Reed Carlock, Emilio Cerra, Cindy Chandley, Sonia Cifuentes; Patricio Ciuchi, Judy Clifton, Betty Cloudt, Ted Cogut, Nell Collins, Dave Colton, Bill Conger, Red and Candace Conger, Jack Coulter, Scott Crozier, Roger Dancause, Kathleen Davis, John de Armas, Glenn Decker, Lucy Diana, Dennis Dinges, Nathan Doctor, Clee and Phyllis Dodge, Bill and Libby Dunbar, Bull Durham, Lou Elkins, Larry Endsley, Paty Enriquez, Julian Espinosa, Marcello Feldstein, Warren and Eleanor Fenzi, Amelia Fernandes, Mark Fields, Linda Findlay, Dan Floryan, Daisy Gallardo, Federico and Carolina Gana, Liz Gary, Rob Gibbs, Barb Gibson, Lynne Gilmore, Greg Gluchowski, Jr., Pedro Gomez, Orlando Gonzalez, José Luis Gorrini, Gerald Grandstaff, Kenny Greene, Ron Habegger, Jerry Haggard, John Hague, Joe Hanley, Tom Hethmon, Steve Higgins, Art Himebaugh, Gina Hirt, Bob Hodge, Stan Holmes, Barry Holt, John Hozenthaler, Manolo Iraola, Jamie Ivey, Bob Jenkins, Bob Johnson, Debbie Jordan, Len Judd, Gary Juno, Hyman Kelley, Kirk Kemmish, Bruce and Linda Kennedy, Dale Kerr, Art Kinneberg, Dave Kinneberg, Kevin Kinsall, Barbie Kissell, Denison and Naomi Kitchel, Hank Konerko, Jack Ladd, Ralph Ladner, Jack Lasher, John Lawrence, Bill Little, Jack Little, Gary Loving, and Orlando Lujan.

The list continues: Larry and Juliet McDonald, Don McLoughlin, Jim Madson, Ken Malloque, Stu Marcus, John Marsden, Darwin Marshall, Bob Martin, Jack Masten, John Matera, Phil Matthews, Julia Mazanek, George Meseha, Art Miele, Duke Milovich, Bob Mock, Dick and Esther Moolick, Bob Moore, Cheryl Moore, Christián Morán, Manuel Moreno, Jerry Morris, Joe Mortimer, George Munroe, David Myers, David Naccarati, Fred Narayan, Theresa Nichol, Leslie Nielsen, Tony Oh, Jim O'Neil, Bob Onslott, Lola Pacheco, Luis Paradas, Dick Pendleton, Jr., Omar Perdomo, Ramey Peru, Simon and Velia Peru, Andy Peterson, Richard Peterson, Chuck Phelps, Fred Phillipi, Don Powell, Rasty Powell, Dennis Preisler, Leo Pruett, Carlos and Maria Isabel Quiroz, Lisa Rapps, Monica Rapps, Bill Rea, Richard Remaks, Bob Rennie, Harold Reynolds, Dick Rhoades, David Rice, Dick Rice, Bruce Richardson, Stan Rideout, Jorge Riquelme, Patricio Rodriguez, Manny Rodriguez-Fiol, Felix Romero, Don Rorick, Alistair Ross, Frank Ruedas, Hector Ruedas, Steve Ruth, Pat Ryan, Paul Rykard, Tom St. Clair, Mathias Sandoval, Floranne Sartorius, Pat Scanlon, Brad Schultz, William Seidman, Duane Sexton, Thurman Shannon, Dave Sheets, Sara Skoff, Tim Snider, Stevie Sorensen, Sam Sorich, Jr., Bill Spellman, Mark Spencer, Linda Stacy, Kim Sterling, Terry Stimmel, Carl Straw, Larry Strunk, Soren Suver, Sue Suver, Bob and Dixie Swan, Rodney Taylor, Dave Thornton, David Till, Ramon Trujillo, Alberto Tuberoso, Bill Tubman, Dietmar Voss, Jack Walsh, Doug Weaver, Steve Whisler, David Wiegman, A. D. Wilcox, Cris Williams, Mark Yarbro, Doug and Anne Yearley, Patty Young, John Zamar.

In addition, I want to thank the University of Arizona

Press; Christine Marín, Arizona State University Libraries; John Stanley, *Arizona Republic*, Phoenix; Jon Schwantes, *Indianapolis Star-News*; Carol Zabilski, manuscript editor; Trina Stahl, freelance book designer; Erik Nordberg, Michigan Technological University; Nancy Burkett, American Antiquarian Society; Eleanor Swent, Bancroft Library; Richard Graeme, Golden Queen Mining Company; Mason Coggin, Arizona Department of Mines and Mineral Resources; Carrie Gustavson, Bisbee Mining and Historical Museum; Kermit C. Parsons, Cornell University; Dave Kinneberg, retired Kennecott executive; Peter Neill and Charlie Sachs, South Street Seaport Museum; Gary Dillard and Cindy Hayostek, *Pay Dirt* magazine; Mimi Bowling, New York Public Library; Joe DeCamillo and Sharon Ann Jordan of Spear, Leeds and Kellogg; and Tom King, Standex International Corporation; Bud Webb of Hebbard and Webb; David Myrick; Howard I. Scott, Jr., United Steelworkers of America; Wayne Ranney, Yavapai College; Kirk Davis, CS Cattle Company; Pendleton Gaines of Fennemore Craig; and David V. Smalley of Debevoise and Plimpton.

Vision and Enterprise benefited greatly from the suggestions offered by three peer reviewers selected by the University of Arizona Press: Richard Francaviglia of the University of Texas at Arlington, Bob Spude of the National Park Service, and one anonymous scholar. I will take full responsibility for picking and choosing from among their comments.

Other Sources

FOUR SUBSTANTIAL HISTORIES of Phelps Dodge Corporation have been written to date. Those include the unpublished manuscript called "Men and Mines: The Story of Phelps Dodge," by Arthur Train, Jr., and John L. Lawler, who took the story to 1945. Building on the foundation provided by their largely anecdotal account, Robert Glass Cleland wrote *A History of Phelps Dodge, 1834–1950* (New York: Alfred A. Knopf, 1952). In the

mid-1990s, James R. Kearney completed six chapters of a manuscript history, "Phelps Dodge Corporation: A Twentieth-Century Business Odyssey," with an emphasis on economics. In addition, the commemoration of its first hundred years in the Southwest yielded a lively newspaper account by William Epler and Gary Dillard, *Phelps Dodge, a Copper Centennial, 1881–1981* (Bisbee: Copper Queen Publishing, 1981). Also invaluable for the early years is Phyllis B. Dodge's *Tales of the Phelps-Dodge Family: A Chronicle of Five Generations* (New York: New-York Historical Society, 1987); William E. Dodge, *Old New York* (New York: Dodd, Mead, 1880); and Richard Lowitt, *A Merchant Prince of the Nineteenth Century: William E. Dodge* (New York: Columbia University Press, 1954).

I have used all those sources, including some of their original notes and interviews, to write *Vision and Enterprise: Exploring the History of Phelps Dodge Corporation*. I originally wanted to emphasize what has happened at Phelps Dodge since publication of Cleland's history in 1952, but I soon found myself revisiting vital details of the corporation's story since 1834 to make sense of post-1952 events. It also made sense to provide a new one-volume history of the corporation, a book far more heavily illustrated than any of its predecessors.

Earlier accounts together with archives compiled by Phelps Dodge Corporation and other institutions, including the New York Public Library and the Huntington Library, proved invaluable—but so too did personal travel outside Phoenix to study firsthand the many landscapes Phelps Dodge created. Landscape study provided a form of insight that books and other documents alone couldn't. As much as we might wish it, we cannot travel back in time; but by visiting historic landscapes like the one preserved at New York's South Street Seaport area, where Phelps Dodge originated, an observer can experience a place and in an odd sort of way make emotional connections with the past. Such landscapes should be viewed as complements to books and articles. They should be read in conjunction with the printed word. To those who care to study them, the defining landscapes of Phelps Dodge Corporation are themselves chapters of an open book, revealing key aspects of its proud heritage.

Suggestions for Further Reading

THE FOLLOWING BIBLIOGRAPHICAL essay seeks to guide readers to my major published sources and suggest various opportunities for further study. Some sources, such as interviewees, are mentioned only in the notes or in the acknowledgments above.

Among the best books that provide general understanding of the world of business in which Phelps Dodge Corporation operated are Keith L. Bryant, Jr., and Henry C. Dethloff, *A History of American Business* (Englewood Cliffs, N.J.: Prentice Hall, 1983); Thomas C. Cochran and William Miller, *The Age of Enterprise: A Social History of Industrial America*, rev. ed. (New York: Harper and Row, 1961); Thomas C. Cochran, *Frontiers of Change: Early Industrialism in America* (New York: Oxford University Press, 1981); Edward Chase Kirkland, *Industry Comes of Age: Business, Labor and Public Policy, 1860–1897* (Chicago: Quadrangle, 1967 reprint of 1961 edition); Alan Trachtenberg, *The Incorporation of America: Culture and Society in the Gilded Age* (New York: Hill and Wang, 1982); Olivier Zunz, *Making America Corporate, 1870–1920* (Chicago: University of Chicago Press, 1990).

Books that aid general technical understanding of Phelps Dodge's various enterprises include David P. Billington, *The Innovators: The Engineering Pioneers Who Made America Modern* (New York: John Wiley and Sons, 1996); Daniel J. Boorstin, *The Republic of Technology: Reflections on Our Future Community* (New York: Harper and Row, 1978); Ruth Schwartz Cowan, *A Social History of American Technology* (New York: Oxford University Press, 1997); Joseph Gies and Frances Gies, *The Ingenious Yankees: The Men, Ideas, and Machines That Transformed a Nation, 1776–1876* (New York: Thomas Y. Crowell, 1976); David Freeman Hawke, *Nuts and Bolts of the Past: A History of American Technology, 1776–1860* (New York:

Harper and Row, 1988); Brooke Hindle and Steven Lubar, *Engines of Change: The American Industrial Revolution, 1790–1860* (Washington, D.C.: Smithsonian Institution Press, 1986); David A. Hounshell, *From the American System to Mass Production, 1800–1932* (Baltimore: Johns Hopkins University Press, 1984); and Walter Licht, *Industrializing America: The Nineteenth Century* (Baltimore: Johns Hopkins University Press, 1995).

The list of useful studies of technological development continues with Robert B. Gordon and Patrick M. Malone, *The Texture of Industry: An Archaeological View of the Industrialization of North America* (New York: Oxford University Press, 1994); Leo Marx, *The Machine in the Garden: Technology and the Pastoral Ideal in America* (New York: Oxford University Press, 1964); David F. Noble, *American by Design: Science, Technology, and the Rise of Corporate Capitalism* (New York: Oxford University Press, 1980 reprint of 1977 edition); Carroll Pursell, *The Machine in America: A Social History of Technology* (Baltimore: Johns Hopkins University Press, 1995).

On the history of electrification, so much a part of the Phelps Dodge story, are Neil Baldwin, *Edison: Inventing the Century* (New York: Hyperion, 1995); John Winthrop Hammond, *Men and Volts: The Story of General Electric* (Philadelphia: J. B. Lippincott, 1941); Thomas P. Hughes, *Networks of Power: Electrification of Western Society, 1880–1930* (Baltimore: Johns Hopkins University Press, 1983); Carolyn Marvin, *When Old Technologies Were New: Thinking about Electric Communication in the Late Nineteenth Century* (New York: Oxford University Press, 1988); David E. Nye, *Electrifying America: Social Meanings of a New Technology* (Cambridge, Mass.: MIT Press, 1990).

Among the many books on mining history, the following were especially helpful: David Lavender, *The Story of Cyprus Mines Corporation* (San Marino, Calif.: Huntington Library, 1962); Isaac F. Marcosson, *Anaconda* (New York: Dodd, Mead, 1957); Robert H. Ramsey, *Men and Mines of Newmont: A Fifty-Year History* (New York:

Octagon Books, 1973); Raye C. Ringholz, *Uranium Frenzy: Boom and Bust on the Colorado Plateau* (New York: W. W. Norton, 1989); Duane A. Smith, *Mining America: The Industry and the Environment, 1800–1980* (Lawrence: University Press of Kansas, 1987); Clark C. Spence, *British Investments and the American Mining Frontier, 1860–1901* (Ithaca: Cornell University Press, 1958); and Clark C. Spence, *Mining Engineers and the American West: The Lace-Boot Brigade, 1849–1933* (New Haven: Yale University Press, 1970).

Specific studies of the copper industry include Charles K. Hyde, *Copper for America: The United States Copper Industry from Colonial Times to the 1990s* (Tucson: University of Arizona Press, 1998); Christopher J. Huggard, "Environmental and Economic Change in the Twentieth-Century West: The History of the Copper Industry in New Mexico," Ph.D. dissertation (University of New Mexico, 1994); Ira B. Joralemon, *Copper: The Encompassing Story of Mankind's First Metal* (Berkeley: Howell-North Books, 1973); Theodore H. Moran, *Multinational Corporations and the Politics of Dependence: Copper in Chile* (Princeton: Princeton University Press, 1974); Thomas R. Navin, *Copper Mining and Management* (Tucson: University of Arizona Press, 1978); Breandán Ó hUallacháin and Richard A. Matthews, "Restructuring of Primary Industries: Technology, Labor, and Corporate Strategy and Control in the Arizona Copper Industry," *Economic Geography* 72 (April 1996): 196–215; A. B. Parsons, *The Porphyry Coppers* (New York: American Institute of Mining and Metallurgical Engineers, 1933); Ronald Prain, *Copper: The Anatomy of an Industry* (London: Mining Journal Books, 1975).

Among the in-house publications are two that Phelps Dodge Morenci sponsored in the 1990s, *Copper Today* and *Morenci Copper Review*, both of which provide valuable historical perspectives. Also of value is "The Columbian Chemicals Company Story," an unpublished manuscript dating from 1997. *Workscapes* published in the late 1990s contained news from throughout the corporation, as did the *Chairman's Quarterly Letter*.

Finally, annual reports and quarterly reports are a rich source of information, as is *Phelps Dodge Today*.

Offering insights by two of the corporation's executives during the 1980s are G. Robert Durham, *Phelps Dodge Corporation: "Proud of Its Past, Prepared for the Future"* (New York: Newcomen Society of the United States, 1989); and L. William Seidman, *Full Faith and Credit: The Great S &L Debacle and Other Washington Sagas* (New York: Times Books, 1993).

On the portion of the American West where Phelps Dodge has played such a big role, see J. Ross Browne, *Adventures in the Apache Country: A Tour through Arizona and Sonora, with Notes on the Silver Regions of Nevada* (New York, 1871); John S. D. Eisenhower, *Intervention! The United States and the Mexican Revolution, 1913–1917* (New York: W. W. Norton, 1993); Michael P. Malone and Richard W. Etulain, *The American West: A Twentieth-Century History* (Lincoln: University of Nebraska Press, 1989); D. W. Meinig, *Southwest: Three Peoples in Geographical Change, 1600–1970* (New York: Oxford University Press, 1971); Thomas E. Sheridan, *Arizona, a History* (Tucson: University of Arizona Press, 1995); Duane A. Smith, *Rocky Mountain Mining Camps: The Urban Frontier* (Lincoln: University of Nebraska Press, 1974 reprint of 1967 edition); Peter Wiley and Robert Gottlieb, *Empires in the Sun: The Rise of the New American West* (Tucson: University of Arizona Press, 1985 reprint of 1982 edition).

Among studies of Phelps Dodge communities are Arlene Allen, "A Brief History of the Burro Mountain Copper Company and the Town of Tyrone, New Mexico," M.A. thesis (Western New Mexico University, 1990); Lynn R. Bailey, *Bisbee: Queen of the Copper Camps* (Tucson: Westernlore Press, 1983); William C. Conger, "History of the Clifton-Morenci District," in *History of Mining in Arizona*, ed. Michael J. Canty and Michael N. Greeley (Tucson: Mining Club of the Southwest Foundation, 1987); Margaret Crawford, *Building the Workingman's Paradise: The Design of American Company Towns* (London: Verso, 1995); Glenn S. Dumke,

"Douglas, Border Town," *Pacific Historical Review* 17 (August 1948): 283–98; James M. Patton, *History of Clifton* (Clifton, Ariz.: Greenlee County Chamber of Commerce, 1977); Carlos A. Schwantes, ed., *Bisbee: Urban Outpost on the Frontier* (Tucson: University of Arizona Press, 1992); Mark C. Vinson, "Vanished Clifton-Morenci: An Architect's Perspective," *Journal of Arizona History* 33 (Summer 1992): 183–206; Toby Smith, *Coal Town: The Life and Times of Dawson, New Mexico* (Santa Fe: Ancient City Press, 1993); Don Dedera, *In Search of Jesús García* (Payson, Ariz.: Prickly Pear Press), 1989, a popular study focusing on Nacozari.

On some of the railroads of the Southwest that relate to Phelps Dodge, see James Douglas, *Notes on the Development of Phelps, Dodge and Co.'s Copper and Railroad Interests* (n.p., January 1906); Don L. Hofsommer, *The Southern Pacific, 1901–1985* (College Station: Texas A and M University Press, 1986); David F. Myrick, *Railroads of Arizona*, Vol. 1 (San Diego, Calif.: Howell-North, 1975), Vol. 2 (San Diego: Howell-North, 1980); and Vol. 3 (Glendale, Calif.: Trans-Anglo Books, 1984).

Among the books dealing the history of labor and safety issues are Mark Aldrich, *Safety First: Technology, Labor, and Business in the Building of American Work Safety, 1870–1939* (Baltimore: Johns Hopkins University Press, 1997); Ronald C. Brown, *Hard-Rock Miners: The Intermountain West, 1860–1920* (College Station: Texas A and M University Press, 1979); James W. Byrkit, *Forging the Copper Collar: Arizona's Labor-Management War, 1901–1921* (Tucson: University of Arizona Press, 1982); Melvyn Dubofsky, *We Shall Be All: A History of the Industrial Workers of the World* (New York: Quadrangle, 1969); James C. Foster, ed., *American Labor in the Southwest: The First One Hundred Years* (Tucson: University of Arizona Press, 1982); George H. Hildebrand and Garth L. Magnum, *Capital and Labor in American Copper, 1845–1990: Linkages between Product and Labor Markets* (Cambridge, Mass.: Harvard University Press, 1992); Robert Kern, ed., *Labor in New Mexico: Unions, Strikes, and Social History since 1881* (Albuquerque:

University of New Mexico Press, 1983); Barbara Kingsolver, *Holding the Line: Women in the Great Arizona Mine Strike of 1983* (Ithaca, N.Y.: ILR Press, 1989); James R. Kluger, *The Clifton-Morenci Strike: Labor Difficulty in Arizona, 1915–1916* (Tucson: University of Arizona Press, 1970); Vernon H. Jensen, *Heritage of Conflict: Labor Relations in the Nonferrous Metals Industry up to 1930* (New York: Greenwood Press, 1968 reprint of 1950 edition); Jonathan D. Rosenblum, *Copper Crucible: How the Arizona Miners Strike of 1983 Recast Labor-Management Relations in America*, 2d ed. (Ithaca: Cornell University Press, 1998); Mark Wyman, *Hard Rock Epic: Western Miners and the Industrial Revolution, 1860–1910* (Berkeley: University of California Press, 1979).

Published studies dealing with some personalities that figured in Phelps Dodge history include Walter R. Bimson, *Louis D. Ricketts (1859–1940)* (Princeton: Princeton University Press, 1949); Isabel Shattuck Fathauer, with Lynn R. Bailey, *Lemuel C. Shattuck: "A Little Mining, a Little Banking, and a Little Beer"* (Tucson: Westernlore Press, 1991); *Warren E. Fenzi: Junior Engineer to President, Director of Phelps Dodge, 1937 to 1983*, an interview conducted by Eleanor Swent in 1995 (Berkeley: Bancroft Library Regional Oral History Office, 1996); H. H. Langton, *James Douglas, a Memoir* (Toronto: University of Toronto Press, 1940); Richard Oliver, *Bertram Grosvenor Goodhue* (New York and Cambridge, Mass.: Architectural History Foundation and MIT Press, 1983); T. A. Rickard, *Interviews with Mining Engineers* (San Francisco: Mining and Scientific Press, 1922); and C. L. Sonnichsen, *Colonel Greene and the Copper Skyrocket* (Tucson: University of Arizona Press, 1974).

Finally, on using landscapes to study history, see Michael P. Conzen, ed., *The Making of the American Landscape* (London: HarperCollins, 1990); Richard V. Francaviglia, *Hard Places: Reading the Landscape of America's Historic Mining Districts* (Iowa City: University of Iowa Press, 1991); and Christopher J. Huggard, "Reading the Landscape: Phelps Dodge's Tyrone, New Mexico, in Time and Space," *Journal of the West* 35 (October 1996): 29–39.

ILLUSTRATION CREDITS

INDEX

About the Author

CARLOS A. SCHWANTES is Professor of History at the University of Idaho and Director of the Institute for Pacific Northwest Studies. In addition to teaching classes on the history of the Pacific Northwest and the twentieth-century West, he is the author or editor of twelve books, including *The Pacific Northwest: An Interpretive History; In Mountain Shadows: A History of Idaho; Long Day's Journey: The Steamboat and Stagecoach Era in the Northern West;* and *Railroad Signatures across the Pacific Northwest*, which received the Railway and Locomotive Historical Society's George W. and Constance M. Hilton Book Award as an outstanding work of lasting value to the interpretation of North America's railroading history for 1992–1994. Awards committee chair William L.

Withuhn, curator of transportation at the Smithsonian Institution, described *Railroad Signatures* as a "stellar contribution to both scholarly and popular audiences: well-written and lavishly produced, the book tells the story of railroading's indelible imprint, its 'signature,' on everyday life in America."

Schwantes received his doctorate in American history from the University of Michigan in 1976 and has been a faculty member at the University of Idaho since 1984. He is an avid photographer and had a collection of his images published in 1996 as *So Incredibly Idaho: Seven Landscapes That Define the Gem State*. Schwantes has served on the editorial advisory boards of five history journals.